SPECIFIC ION EFFECTS

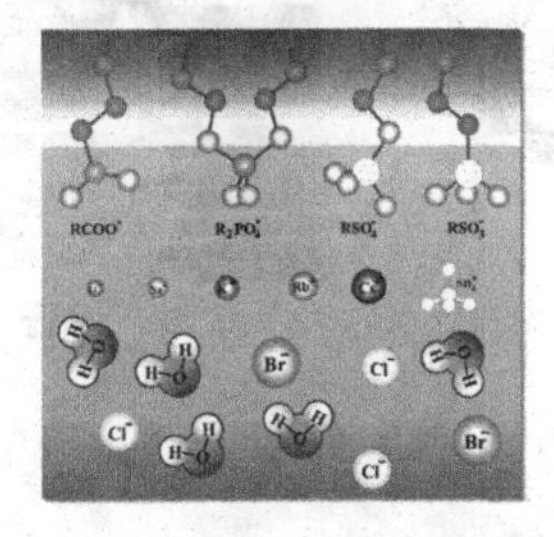

SPECIFIC ION EFFECTS

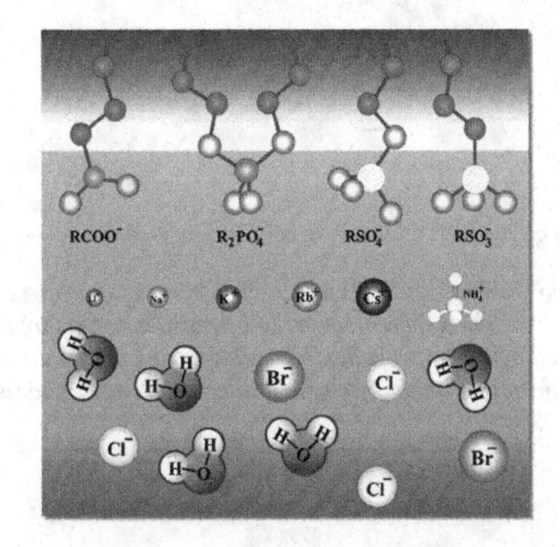

edited by

Werner Kunz

University of Regensburg, Germany

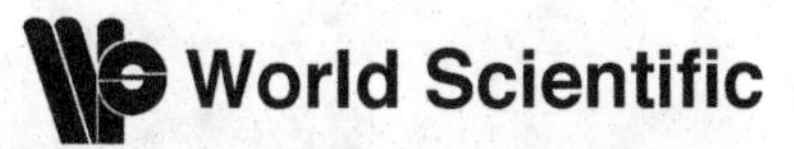

NEW JERSEY • LONDON • SINGAPORE • BEIJING • SHANGHAI • HONG KONG • TAIPEI • CHENNAI

Published by

World Scientific Publishing Co. Pte. Ltd.

5 Toh Tuck Link, Singapore 596224

USA office: 27 Warren Street, Suite 401-402, Hackensack, NJ 07601

UK office: 57 Shelton Street, Covent Garden, London WC2H 9HE

British Library Cataloguing-in-Publication Data
A catalogue record for this book is available from the British Library.

Cover image from Dr. Roland Neueder.

SPECIFIC ION EFFECTS

ISBN-13 978-981-4271-57-8
ISBN-10 981-4271-57-8

Typeset by Stallion Press
Email: enquiries@stallionpress.com

Printed in Singapore.

Contents

Foreword

At a symposium celebrating the 100th anniversary of the Nobel Prizes in Chemistry, Aaron Klug, Nobel Laureate of 1982, made the remark that it was well known to biologists that the Debye–Hückel theory, and by implication its extensions, was limited to slightly contaminated water. This remark on the relevance of physical chemistry to biology is widely held. Biologists and biochemists, long familiar with specific ion effects, have quite reasonably been bemused by the failure of physical chemistry to deal with effects they know are universal. Another biochemist participant remarked that mechanisms of enzyme action were all understood. There was no need for further work.

This was news to physical chemists.

Klugg's view is fair enough, at one level. After all Hofmeister was a pharmacologist, and the biologists could claim him as their own. Hofmeister's work on specific ion effects goes back to the 1870s. It predates Arrhenius and van 't Hoff. It predates pH and buffers. It was put to one side by colloid scientists and physical chemists and only came back to centre stage again a decade or so ago.

Theorists have been uneasily well aware that Hofmeister effects are as important in the scheme of things as Mendel's work was to genetics. And that they were not accommodated by the classical theories of colloid and physical chemistry. But they could not explain the results.

At another level the Kluggian view is absurdly limited. It admits no role for physical chemistry in biology, or for that matter in chemical engineering. Chemistry is specific, and that is that. This view is pre-Mendeleev.

But biologists and biochemists do use the tools and concepts of physical chemistry. These have their roots in the 'slightly contaminated' water of Debye–Hückel theory and its extensions to include 'ion size' and double layers. This is so for concepts like pH, buffers, pKa's, ion binding, interactions of ions with surfaces, membrane and zeta potentials, colloidal interactions that are deeply embedded in our collective psyche. The foundations of these have only been questioned over the last decade. The theories used to interpret these things all involve Debye–Hückel theory and its decorations. Indeed the dictum of the IUPAC commission on best practice in pH measurement — that relies on the validity of extended Debye–Hückel theory and the equivalent Poisson Boltzman distribution for charged surfaces — has the unhelpful advice that it is better to restrict pH measurements to pH somewhat less than 0.1 molar. The interpretation of standard measurements generally depends on classical theory. So if the theory is flawed, so is the meaning of the measurement. Various unknown ion-specific mechanisms, words like hydrophobic, hydrophilic, 'hydration', 'secondary hydration', Stern layers, site binding models, 'water structure', have perforce substituted for 'explanation' of 'non-DLVO' forces and disguise our ignorance.

We have had huge frustration with the muddle and mishmash of unexplained phenomena and the lack of systemisation associated with specific ion effects.

The availability of powerful experimental techniques and advances in computation and computer simulation over the last few decades have changed matters. Add to the situation the fact that Klugg's real dirty water contains dissolved atmospheric gas and other solutes that strongly affect everything else, like hydrophobic interactions in a way that is itself ion specific. Then we have new beginning.

Some real progress has been made. New insights have emerged. There are opposing views that are coming together. These problems of physical chemistry we can hope are now well on the way to solution and a new synthesis.

From the truly vast confusing literature, this book has distilled a timely, fair, clear and useful statement and summary on the central problem of ion specificity, and of progress in understanding from the viewpoint of physical chemists proper. It is also appropriate as Pavel Jungwirth's laboratory is a stone's throw from Hofmeister's in Prague and that of Werner Kunz in Regensburg is a few hours' drive away.

Barry W. Ninham
Research School of Physics and Engineering
Australian National University
Canberra, Australia
June 2009

Preface

So many specific ion effects are described in literature in thousands of papers since the pioneering studies by Franz Hofmeister and his group in Prague, some 130 years ago. Theoretical work has been done over a century now, yet still the puzzle persists. What is the origin of the interactions of simple ions with water, other ions and solutes, and with interfaces, beyond electrostatics? Is it only the size and charge density or do dispersion forces play a major role? In what cases are the charge distribution and the geometry of ions important? Is the size dominating or is it the size plus polarisability?

Can we ever predict specific ion effects quantitatively and beyond empirical rules? Is it like long-term weather forecast or can we find some general principles behind it?

The fact that the so-called Hofmeister series exist is a sign that some generalisation should be possible. But there are several Hofmeister series, direct and reversed ones, or even bell-shaped, and they may change with temperature and ion concentration. All this seems to be hopelessly complicated, and part of the problem comes from apparently diverging experimental results.

As an example, we can consider the surface properties of a simple electrolyte solution as follows: the experimentally measured surface tension can be calculated, according to Gibbs' equation, as an integral over the ion profiles perpendicular to the surface from zero to infinite distance from the surface. However, some experiments mainly probe the first atomic and molecular layers near the surface. It was one of the major findings of the last several years that *locally* (close to an interface) accumulation of ions can occur, despite the fact that *globally* (integrated over a large distance) depletion is detected. Therefore the old perception that small ions are always pushed away from the interface turned out to be wrong. Ions can be attracted to the surface in the first layers, although the surface tension of the total solution is enhanced, compared to pure water.

So, one of the challenges now is to relieve the putative contradictions coming from different experiments that were formerly believed to probe the same properties. All the more, numerous sophisticated experiments were done over the last ten years, raising more questions than giving answers.

As far as interpretations, models, and theories are concerned, there has been a major step forward over the last ten years or so. It seems that this old topic becomes again fashionable among theoreticians. Computers are so much more powerful than they were twenty years ago, and significant new ideas came up, especially concerning the probable role of ion polarisabilities. We can be optimistic that in another ten years, the puzzle of ion specificity will ultimately be resolved. What we can say today is that there will not be only one parameter for one type of ion that describes its effect in all circumstances. There will be a map of different and subtly mixed interactions responsible for specific ion effects. Depending on the quantity and system discussed, these interactions are mixed in different ways. However, there will appear some generality, for example, for ion–protein interactions. It will be possible to predict salting-in and salting-out effects independently of denaturation effects, by just knowing the nature of the amino acid residues that are exposed to water and ions. It will be possible to predict the negative surface potential of water in the presence of ions, the surface active behaviour of acids and, perhaps, the secret of gas bubble–bubble coalescence in water as a function of salt composition.

All these will have a major impact on our understanding of processes and structures in living systems, as well as in technological applications. The prediction of phase equilibria in the presence of different salts can save a lot of money.

So should we wait another ten years before editing this book? I believe that the progress in the last few years is so significant that we can just start to draw a raw picture of specific ions effects. Of course, the different chapters will also present several contradictory points of views. Therefore, in the last chapter I will try to make a summary out of all this information showing what is commonly accepted today and what is still in debate, and I will also try to sketch what will probably be in the future.

In the introductory chapter, it seems worthy to make an attempt of a general overview. It is not intended to be done extensively, for many good reviews exist. It is rather the intention to show some exemplary experimental results and theoretical approaches that illustrate the broad range of ion effects, and to give an impression of the main axes of thinking about them.

Werner Kunz
Regensburg
July 2009

About Werner Kunz

Werner Kunz was born in 1960 in Krummennaab/Bavaria, Germany. He obtained his Ph.D. degree in Chemistry in 1988 at the University of Regensburg, Germany, working on *Vapour Pressure Measurements and Statistical-Mechanical Theories for the Determination of Thermodynamic and Structural Properties of Electrolytes in Acetonitrile and Methanol*, in the group of Prof. Dr. Josef Barthel.

He spent four years as a postdoctoral fellow with Prof. Pierre Turq at the Université Pierre et Marie Curie in Paris, France and in the Laboratoire Léon Brillouin, CEA, Saclay, France. In 1992 he got his french 'habilitation' working on the *determination of the structure and dynamics of solutions, especially with the help of neutron scattering experiments.* In the same year he was assistant professor at the Université de Technologie de Compiègne (UTC), France and worked as a research fellow at the CEA, Saclay, France.

In 1993 he was appointed professor at the UTC, France. He worked there until 1997, studying especially liquids, solutions, microemulsions and emulsions with industrial interest. In 1995 he created a new series of post-graduate courses (D.E.S.S.) on complex liquids and colloids.

In 1997 he moved to the Universität Regensburg as full professor. He has been a programme coordinator for the European Master of Science in colloidal and formulation chemistry since 2004.

He received grants and fellowships from the State of Bavaria, the Studienstiftung des Deutschen Volkes, the Fonds der chemischen Industrie, the NATO, the European Community, and Elf Aquitaine.

Since 2000 he has been a corresponding member of the European Academy of Sciences and Arts (Paris, London).

Acknowledgements

Twenty-five years ago, my Ph.D. supervisor Professor Josef Barthel initiated me into the field of ions and electrolyte solutions. I am deeply indebted to him as well as to my other mentors Pierre Turq and Barry W. Ninham. I greatly appreciate the constant help, encouragement and critical support of these three outstanding personalities.

I express my gratitude to all contributors of this book. They took the time to make a synthesis of their most recent ideas and achievements. I enjoyed the meetings and discussions with all of them.

For my own chapters, I acknowledge most valuable discussions with numerous colleagues. Among them I have to thank especially Kim D. Collins. Further thanks to Hans Lyklema, Didier Touraud and all my students and postdocs, who made so many significant contributions to my present knowledge about specific ion effects.

I should also mention meetings and symposia on this subject in Regensburg in 2004, in Prague in 2007 (thanks to Pavel Jungwirth) and in Munich in 2008 (thanks to Dominik Horinek and Roland Netz).

Finally, I would like to give my special thanks to the German 'Arbeitsgemeinschaft Industrieller Forschungvereinigungen Otto von Guericke e.V. (AiF)'. This German Federation of Industrial Research Associations significantly supported our research on specific ion effects over the past years.

Werner Kunz
Regensburg
July 2009

Acknowledgements

Twenty-five years ago my Ph.D. supervisor Professor [illegible] introduced me to the [illegible] ions and electrons [illegible] understand [illegible] as well as to [illegible] Tang and Henry W. [illegible]. I greatly appreciate the constant help [illegible] and a cordial atmosphere [illegible].

[illegible] this book. The [illegible] of them [illegible] and [illegible]. This is [illegible] with all of them.

For my own [illegible] knowledge [illegible] most valuable discussions with numerous colleagues. Among them I have to thank especially [illegible] Cecil [illegible]. Further thanks to Hans [illegible], Dr. [illegible] and [illegible] students and postdocs who made [illegible] significant contributions [illegible] my knowledge [illegible] effects.

[illegible] in 2004 [illegible] Paul [illegible] in Munich in 2008 [illegible].

Finally, I would like to [illegible] German [illegible] (AiF) [illegible] the German Federation of Industrial Research Associations [illegible] support [illegible] effects [illegible] the past [illegible].

Werner Kunz
Regensburg

List of Contributors

Dr. Luc Belloni
CEA Saclay
Direction des Sciences de la Matière
Service de Chimie Moléculaire, LIONS
91191 Gif-sur-Yvette Cedex
France
luc.belloni@cea.fr

Prof. Dr. Evaristo C. Biscaia Jr.
Programa de Engenharia Química
COPPE
Universidade Federal do Rio de Janeiro
21941-914 Rio de Janeiro, RJ
Brazil
evaristo@peq.coppe.ufrj.br

Dr. Mathias Boström
Division of Theory and Modeling
Department of Physics, Chemistry and Biology
Linköping University
SE-581 83 Linköping
Sweden
mabos@ifm.liu.se

Dr. John W. Brady
Department of Food Science
Cornell University
Ithaca, NY
USA
jwb7@cornell.edu

Dr. Ioulia Chikina
CEA Saclay
Direction des Sciences de la Matière
Service de Chimie Moléculaire, LIONS
91191 Gif-sur-Yvette Cedex
France
julia.chikina@cea.fr

Dr. Vincent S. J. Craig
Department of Applied Mathematics
Research School of Physics
Australian National University
Canberra, ACT 0200
Australia
vince.craig@anu.edu.au

Dr. Jean Daillant
CEA, IRAMIS, LIONS, CEA Saclay
F-91191 Gif-sur-Yvette Cedex
France
jean.daillant@cea.fr

Dr. Joachim Dzubiella
Physics Department T37
Technical University Munich
85748 Garching
Germany
jdzubiel@ph.tum.de

Maria Fyta
Physics Department T37
Technical University Munich
85748 Garching
Germany
maria.fyta@ph.tum.de

Luc Girard
ICSM UMR 5257
CEA Marcoule BP17171
30207 Bagnols sur Cèze Cedex
France
luc.girard@cea.fr

Christoph Held
Laboratory of Thermodynamics
Department of Biochemical and Chemical Engineering
Technische Universität Dortmund
Emil-Figge-Str. 70
44227 Dortmund
Germany
Christoph.Held@bci.tu-dortmund.de

Dr. Christine L. Henry
Department of Applied Mathematics
Research School of Physics
Australian National University
Canberra, ACT 0200
Australia
christine.henry@anu.edu.au

Jan Heyda
Institute of Organic Chemistry and Biochemistry
Czech Academy of Sciences
Flemingovo nam. 2
16610 Prague
Czech Republic
jan.heyda@uochb.cas.cz

Dr. Dominik Horinek
Physics Department T37
Technical University Munich
85748 Garching
Germany
dominik.horinek@ph.tum.de

Prof. Dr. Pavel Jungwirth
Institute of Organic Chemistry and Biochemistry
Czech Academy of Sciences
Flemingovo nam. 2
16610 Prague
Czech Republic
pavel.jungwirth@uochb.cas.cz

Immanuel Kalcher
Physics Department T37
Technical University Munich
85748 Garching
Germany
Immanuelkalcher@tum.de

Dr. Patrick Koelsch
Institute of Toxicology and Genetics
Karlsruhe Institute of Technology (KIT)
Hermann-von-Helmholtz-Platz 1
D-76344 Eggenstein-Leopoldshafen
Germany
patrick.koelsch@itg.fzk.de

Prof. Dr. Werner Kunz
Institute of Physical and Theoretical Chemistry
University of Regensburg
93040 Regensburg
Germany
Werner.Kunz@chemie.uni-regensburg.de

Prof. Dr. Epameinondas Leontidis
Department of Chemistry
University of Cyprus
P.O. Box 20537
1678 Nicosia
Cyprus
psleon@ucy.ac.cy

Dr. Eduardo R. A. Lima
Escola de Química
Universidade Federal do Rio de Janeiro
Cidade Universitária
CEP 21949-900 Rio de Janeiro, RJ
Brazil
lima.eduardo@gmail.com

Dr. Mikael Lund
Department of Theoretical Chemistry
University of Lund
P.O.B. 124
SE-22100 Lund
Sweden
mikael.lund@teokem.lu.se

Dr. Philip E. Mason
Department of Food Science
Cornell University
Ithaca, NY
USA
pem27@cornell.edu

Prof. Dr. Hubert Motschmann
Institute of Physical and Theoretical Chemistry
University of Regensburg
93040 Regensburg
Germany
Hubert.Motschmann@chemie.uni-regensburg.de

Dr. George W. Neilson
H. H.Wills Physics Laboratory
University of Bristol
Tyndall Avenue
BS8 1TL
UK
George.W.Neilson@bristol.ac.uk

Prof. Dr. Roland R. Netz
Physics Department T37
Technical University Munich
85748 Garching
Germany
netz@ph.tum.de

Dr. Roland Neueder
Institute of Physical and Theoretical Chemistry
University of Regensburg
93040 Regensburg
Germany
Roland.Neueder@chemie.uni-regensburg.de

Prof. Dr. Barry W. Ninham
Department of Applied Mathematics
Research School of Physical Sciences and Engineering
Australian National University
Canberra, ACT 0200
Australia
barry.ninham@anu.edu.au

Dr. Dmitri Novikov
HASYLAB at DESY
Notkestr. 85
D-22607 Hamburg
Germany
dmitri.novikov@desy.de

Prof. Dr. Gabriele Sadowski
Laboratory of Thermodynamics
Department of Biochemical and Chemical Engineering
Technische Universität Dortmund
Emil-Figge-Str. 70
44227 Dortmund
Germany
gabriele.sadowski@bci.tu-dortmund.de

Nadine Schwierz
Physics Department T37
Technical University Munich
85748 Garching
Germany
nadine.schwierz@ph.tum.de

Dr. Olivier Spalla
CEA, IRAMIS, LIONS, CEA Saclay
F-91191 Gif-sur-Yvette Cedex
France
olivier.spalla@cea.fr

Prof. Dr. Frederico W. Tavares
Escola de Química
Universidade Federal do Rio de Janeiro
Cidade Universitária
CEP 21949-900 Rio de Janeiro, RJ
Brazil
tavares@h2o.eq.ufrj.br

Prof. Dr. Gordon J. T. Tiddy
School of Chemical Engineering & Analytical Science
University of Manchester
PO Box 88
Manchester, M60 1QD
UK
gordon.tiddy@manchester.ac.uk

Dr. Padmanabhan Viswanath
CEA, IRAMIS, LIONS, CEA Saclay
F-91191 Gif-sur-Yvette Cedex
France
viswap@gmail.com

Part A

EXAMPLES, ION PROPERTIES AND CONCEPTS

Chapter 1

An Attempt of a General Overview

Werner Kunz* and Roland Neueder*

In the present chapter, selected examples of specific ion effects are discussed. We start with experimental evidences in simple systems and end with ion effects at complex biological surfaces. Some general rules are extracted from these examples. The chapter also contains a short overview of current theories and simulations aiming at a more fundamental understanding of ion specificity. This chapter is intended to be a general introduction to the field and a basis for the following chapters that deal with more specific aspects.

1. About Hofmeister Series, Kosmotropes, Chaotropes, and Related Concepts

During the 1880s and 1890s, Franz Hofmeister, Professor of Pharmacology at the University of Prague, and some of his co-workers published a series of seven papers in the *Archiv fuer experimentelle Pathologie und Pharmakologie*, which were all written in German. The title of the whole series was 'Zur Lehre von der Wirkung der Salze', which in English means 'About the science of the effect of salts'. The first paper carried the subtitle 'The behaviour of the proteins in the blood serum in the presence of salts'.[1] Interestingly, Hofmeister was not an author of this paper. It was his student S. Lewith, who presented these first and preliminary results on salt effects. Hofmeister was the author of the second paper with the subtitle 'About regularities in the protein precipitating effects of salts and the relation of these effects with the physiological behaviour of salts'.[2] Here Hofmeister completed Lewith's experiments and drew the first of his fundamental conclusions that constituted a breakthrough in electrolyte chemistry, much as did the work of van 't Hoff, Arrhenius and Pfeffer.

*Institute of Physical and Theoretical Chemistry, University of Regensburg, 93040 Regensburg, Germany.

The third contribution was also written by Hofmeister himself. Its subtitle was 'About the water withdrawing effect of the salts'.[3] This paper is perhaps the most interesting one, not only because of the remarkable experiments, but also due to the ingenious conclusions that Hofmeister drew from them. The fourth contribution was from R. v. Limbeck and had the subtitle 'About the diuretic effect of salts'.[4] This paper is mostly of pharmacological interest. The fifth contribution was again written by Franz Hofmeister. Its subtitle was 'Investigations about the swelling process'.[5] This paper is mostly a description of the careful experimental setup and the way to interpret swelling. It does not deal with the behaviour of different salts. Hofmeister was also the author of the sixth contribution with the subtitle 'The contribution of dissolved components to swelling processes'.[6] This paper describes the different swelling effects of a large series of salts, of alcohol and of saccharose. Finally, E. Münzer wrote the seventh contribution with the subtitle 'The general effect of salts'.[7] This paper is mostly a pharmacological study.

What was so revolutionary in the papers by Hofmeister and co-workers? First, these were the first extended systematic studies on specific ion effects beyond the effect of different charges. Second, Hofmeister considered quite different systems to derive the specific effects: precipitation of proteins, of colloidal ferric oxide, and of sodium oleate, collagen (isinglass), etc. Third, he was the first to draw some general conclusions about specific ion effects that allowed him to order salts according to their 'water withdrawing capability':

> ... the colloid precipitating effect of a salt is dependent on its water absorbing capability, one can expect that this activity against the various colloid substances should be constant in relation to other salts. For example, if sodium chloride proves to have a stronger precipitating effect on globulin than potassium chloride, the same behaviour should be valid for the isinglass, the ferric oxide etc. It also can be expected, that the precipitating capability of salts is parallel to other physical and chemical properties, if it has been proved or it is most likely, that these properties are dependent on the water absorbing capability of the salts.[3,8]

The concepts by Robinson, Stokes, Gurney, Friedman, Collins and many others about specific ion hydration can be traced back to Hofmeister's conclusion:

> ... there are reasons to relate the lowering of vapour pressure, the lowering of the freezing point etc., to the existing attracting forces between salt and water particles, there is no obstacle to use the same explanation for the precipitation of colloids by salts. In my experiments, similar relations are detectable, and preceding considerations suggest the attractiveness of the same interpretation.[3,8]

Finally, Hofmeister also described the origin of different osmotic pressures found from different salt solutions, as it will be discussed in Sec. 2.1.1:

> ... the osmotic pressure of a defined volume of salt solution differs from the pressure of a gas volume containing the same number of molecules, this can originate from the dissociation capability of the salt solution on the one hand, but also from the special relations of salt particles to the water particles. Finally, it could also originate from both phenomena.[3,8]

In Hofmeister's papers, only an ordering of salts and not of isolated ions is given. Such a different ordering, usually known as 'Hofmeister series', appeared only many years later.

However, it should be mentioned that specific ion effects were found even several decades before Hofmeister. In 1847, Poiseuille[9] was probably the first who noted that some salts increase the viscosity of water, whereas others decrease it. Jones and Dole in 1929,[10] Cox and Wolfenden in 1934,[11] and several other groups further refined the specific ion effect on water viscosity. From these viscosity studies and in particular the Jones–Dole viscosity *B* coefficients, the expressions for 'water-structure maker' and 'water-structure breaker' were finally derived. They were first introduced in 1945 by Frank and Evans,[12] who showed the relationship between viscosity and entropy of dilution. There is a third concept introduced by A. Voet[13] [see also Eckfeldt[14]]: the ordering of ions according to their lyotropic numbers. It nicely correlates with ion effects on the swelling

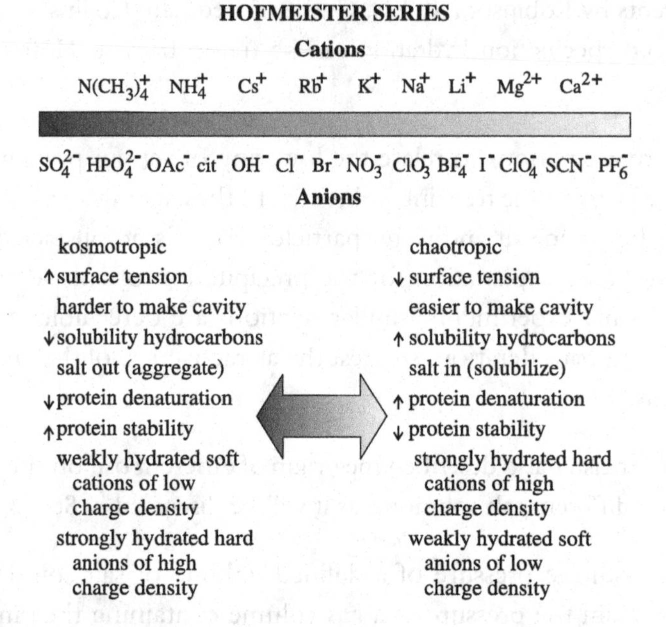

Fig. 1. A typical ordering of cations and anions in Hofmeister series.

of gels, viscosity of sols, zeta potentials of colloids, the heats of hydration of ions, etc. More recently, Pearson,[15] attempting to unify inorganic and organic reaction chemistry, ordered the ions according to their 'softness' and 'hardness'.

Today, these different series are getting more and more mixed. A typical ordering of ions in Hofmeister series is shown in Fig. 1. Of course, this is only an ordering based on macroscopic properties and simplified pictures that turned out to describe quite a lot of specific ion effects. We will see, in the next few chapters, to what extent they can still be used on a molecular level or where they must be modified or even abandoned. Nevertheless, what we can see here is that from left to right the cations are ordered from polarisable 'soft' cations to unpolarizable 'hard' cations. By contrast, for the anions, the series is just the other way round. This is still a mystery and will be further discussed in the following chapters. As will be demonstrated further on, there are also remarkable exceptions from this classification. For example, as illustrated in Chap. 6, the guanidinium ion is only very weakly hydrated, but nevertheless a strong denaturant due to its unique structure and aggregation behaviour.

As written in the introduction, numerous examples of specific ion effects are known, and nearly every day, new ones are published. It is almost impossible to give a complete list of them. However, most of them can be classified into a few categories, for example: ion effects in simple solutions and in complex mixtures, near flat and well-defined surfaces, or near macromolecules such as proteins. To further classify according to the respective systems, the effects can be subdivided into the experimentally observed quantities such as thermodynamics, transport properties, and kinetics, or into the methods used to deduce specific ion effects such as macroscopic probes, spectroscopy, scattering, etc.

In the next part of this chapter, we will follow the classification according to systems, in which different ions lead to different properties, depending on the composition of the solution. We will start with the simplest one, e.g. simple ions in water and their thermodynamic properties, and end with very complex ones e.g. ions in biological systems. We will not discuss the influence of ions on chemical reactions, as various outstanding papers exist on this topic, e.g. see Ref. 16.

Furthermore, the nucleophilic or electrophilic power of ions is often the decisive parameter. The softness or hardness of the ions is of course also important, but this parameter will be elucidated in the context of more basic physic-chemical parameters. Finally, our discussion will be restricted to aqueous salt solutions. Non-aqueous electrolyte solutions are described in other textbooks.[17]

In the last part, a review will be given on classical concepts, theories and simulations to interpret ion effects. We will try to elaborate to which extent the older approaches are still useful today, and where they must be replaced by new ideas.

In the following chapters, tremendous amount of data will be given together with appropriate references. More-or-less-complete data banks and general compilations of salt solution and ion properties are the most useful for any scientist dealing with such systems. Here, we should especially mention the impressive work by Yizhak Marcus[18–21] as well as the ELDAR (Electrolyte Data Bank Regensburg), the most comprehensive collection of experimental data on salt solutions in all types of solvents.[22] ELDAR is available as data book series and can also be consulted online within the DETHERM data bank system of the German DECHEMA Institute.[23]

2. Selected Experimental Evidence

2.1. *Simple Salts in Bulk Water*

2.1.1. *Free energy: Osmotic and activity coefficients*

In this section, we will consider simple salts in aqueous solutions, i.e. binary mixtures of one type of salt and water. In this case, it is easy to define the chemical activity $\mu^{\text{salt}}(p, T)$ — the change of the free energy G of the salt with salt concentration at constant temperature T and pressure p — in the following way:

$$(\partial G/\partial n_{\text{salt}}) = \mu^{\text{salt}}(p, T) = \mu^{\text{salt}\,\infty}(p, T) + RT \ln x_{\text{salt}} f_{\text{salt}} \\ = \mu^{\text{salt}\,\infty}(p, T) + \nu RT \ln x_{\pm} f_{\pm}. \quad (1)$$

Here, the stoichiometric coefficient of the salt is $\nu = \nu_+ + \nu_-$. So $\nu = 2$ for a 1:1 salt such as NaCl. As usual, $x_\pm$ is the molar fraction and $f_\pm$ is the mean activity coefficient of the salt. R is the gas constant and ∞ means infinite dilution.

$f_\pm$ can be converted into the molarity scale or the molality scale, yielding $f_\pm^{(c)} = y_\pm$ or $f_\pm^{(m)} = \gamma_\pm$, respectively. In the case of a binary system, these salt activity coefficients can be related to the molar osmotic coefficient φ of the solution via the Gibbs–Duhem equation:

$$\ln \gamma_\pm = (\varphi - 1) - 2 \int_0^m (1 - \varphi)/m^{\frac{1}{2}} \mathrm{d}m^{\frac{1}{2}}, \quad (2)$$

where m is the molality (moles per kilogram of solvent).

The water activity $a_{\text{water}} = x_{\text{water}} f_{\text{water}}$ is related to φ through the following:

$$\varphi = -\ln a_{\text{water}}/(\nu m M_{\text{water}}), \quad (3)$$

where M_{water} is the molar mass of water. The osmotic coefficient φ is linked to the osmotic pressure as shown below:

$$\Pi = m\varphi\nu RTM_{\text{water}}/V_{\text{water}}, \quad (4)$$

where V_{water} is the molar volume of water. It is difficult to measure directly the osmotic pressure, because of the lack of appropriate membranes due to the small size of the ions. Therefore, the activity of water [equivalent to the determination of the osmotic coefficient, see Eq. (3)] or of the salt is usually derived from vapour-pressure measurements according to the following relation:

$$\ln a_{\text{water}} = \ln(p/p^*) + (B_{\text{water}} - V_{\text{water}})(p - p^*)/(RT), \qquad (5)$$

where p and p^* are the vapour pressures of water in the salt solution and of pure water, respectively, and B_{water} is the second virial coefficient of water vapour.

It should be noted that various other known methods are applied to determine the activity coefficients of salt solutions. Here, the famous book by Robinson and Stokes,[24] initially published in 1959, is still a good reference. Other methods can be found in Ref. 25.

Figure 2 shows some activity coefficients of 'simple' salt solutions. Similar curves were obtained for many other systems (chlorides, iodides, nitrates, chlorates, perchlorates and many more), c.f. the critical compilation by Hamer and Wu.[26] For comparison, we also added the activity coefficients of one non-electrolyte, namely methanol, demonstrating that the non-idealities, i.e. the deviations from one, are much more pronounced for charged particles in solution.

In all cases, the activity coefficients of the electrolytes first sharply decrease, due to the strong electrostatic interactions between cations and anions. After a certain concentration of salt, a classical ion specific effect appears: the decrease is much smaller with further increase of salt concentration or even there is a change to a significant increase. Very roughly and qualitatively speaking, this phenomenon is attributed to the hydration of the ions and also due to the increasing repulsion between the hydrated ions. As can be seen in Fig. 2(a), for a given concentration, the values of the activity coefficients increase in the series $Cs^+ < Rb^+ < K^+ < Na^+ < Li^+ < H^+$. This behaviour is classically explained by an increase of the *effective* size of the ions in the same direction. Effective means that the first hydration shell is considered to be part of the ion so that the series is just opposite to the series of the sizes of the bare ions.

All these seem to be rather trivial today. However, to our knowledge, *there is not a single published work in which a prediction of these values can be found*. For more than 90 years, many scientists tried to *describe* or to *fit* these data rather than to calculate them without any adjusted parameters. To be precise, periodically publications appear in which the activity coefficients of one or two salts are more or less correctly predicted, mostly for NaCl and KCl. However, as we will see many times in this

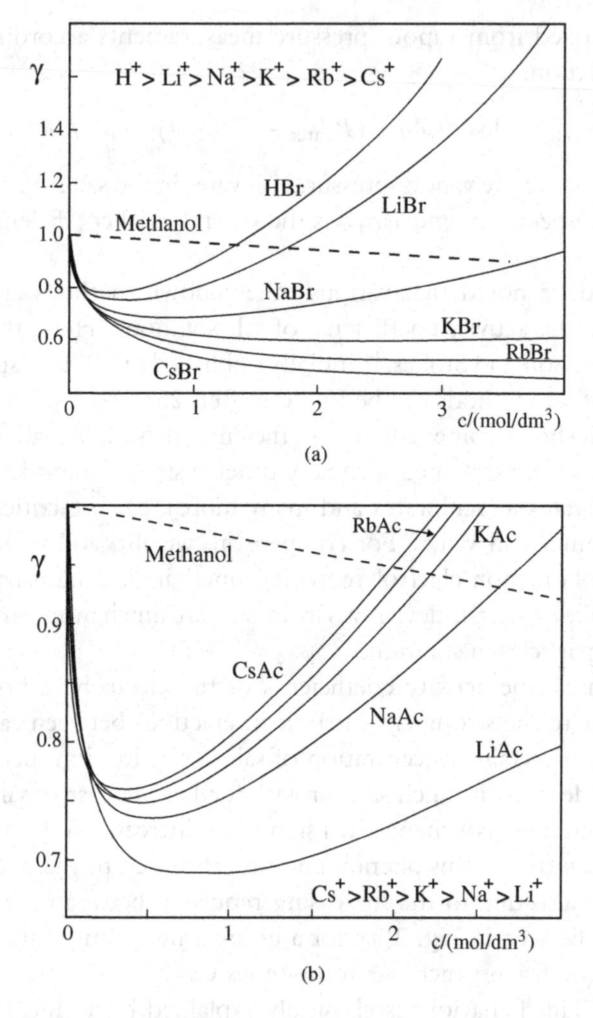

Fig. 2. (a) Activity coefficients of HBr and several alkali bromide solutions as a function of salt concentration. (b) Activity coefficients of several alkali acetate solutions as a function of salt concentration. For comparison, activity coefficients of methanol in water are also given.

book, these salts are not a crucial test, because they behave almost ideally, meaning that they can be described without the assumption of significant interactions beyond electrostatics.

In total, it is still a fact that over the last decades, it was easier to fly to the moon than to describe the free energy of even the simplest salt solutions beyond a concentration of 0.1 M or so. We will come back to the reasons in Sec. 2.3, where we present some of the most promising 'primitive' model approaches to describe the activity coefficients of aqueous electrolyte solutions.

Now the problem is even more complicated: not all anions show this typical cation series with activity coefficients increasing from caesium salts to lithium or hydrogen compounds. In Fig. 2(b), just the reverse series is found, and this is not only true for acetate, but also for hydroxide and formate. As early as 1941, Robinson and Harned[27] (see also Robinson and Stokes[24] pp. 423–425) discussed this reversal in the series of ion specificity. They proposed a 'localized hydrolysis', in which water shares one of its hydrogen atoms with a proton acceptor anion, such as hydroxide or formate: Na^+ - - - OH^- - - - H^+ - - - OH^-. Obviously, this effect would be more pronounced for lithium than for caesium. Li^+ would lose its water hydration shell and at the end the effective radius of it would be smaller than that of Cs^+, which would explain the reversal of the specific ion series.

However, since this work by Robinson and Harned, no other convincing paper appeared that could prove or — on the contrary — discard this assumption. An alternative, and as we will see, more realistic idea was forwarded by Lyklema in 2003.[28] He applied the simple rule of thumb 'like seeks like' inferred from specific ion effects at solid surfaces to qualitatively explain why the mean activity coefficients of CsI are lower than those of LiI at concentrations of 1 M. This is a simple consequence of the more sophisticated model of matching water affinities proposed by Collins, as will be discussed in subsequent chapters.

This is not the end of the complexity of ion effects related to their Gibbs energy, in particular with divalent ions. Magnesium and barium acetates show a cross-over of their respective activity coefficients at 1 M, not to speak of mixed salt solutions or transition metal ions such as Zn^{2+} or Cd^{2+}. The prediction of their activity coefficients is hopeless today and certainly yet for a long time.

There is no need to go to such 'complex', but in reality still simple solutions, compared to biological systems, in which ions are involved. Figure 3 shows the temperature dependence of the osmotic coefficients of 'very simple' aqueous sodium chloride solutions. Here, a very specific salt effect appears that does not exist for similar salt solutions such as LiCl, KCl and NaBr: there is a maximum in the φ values at around 50°C for

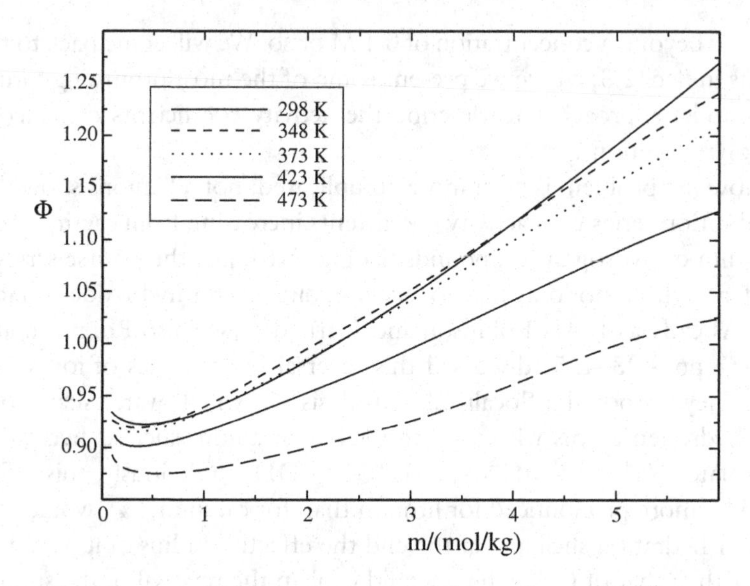

Fig. 3. The osmotic coefficients of aqueous NaCl solutions as a function of temperature. (After Gibbard *et al.*[29])

concentrations above 0.5 M. The whole temperature dependence is highly non-monotonic! And again, nobody could ever describe or at least explain this strange behaviour of NaCl solutions. Even a temperature-dependent fit requires numerous parameters and is today not satisfactorily done.

To conclude this section, let us have a look at the variation of the activity coefficient of water in simple electrolyte solutions (see Fig. 4). As can be seen, the water activity coefficient in some solutions is slightly increasing at medium salt concentrations, up to values higher than one. This means that a small tendency of demixing exists, which is, of course, overcompensated by the entropy gain and the favourable solution effect on the salt.

2.1.2. *Heat of solution*

The standard heat of solution of a salt at infinite dilution is the heat change that takes place when one mole of salt is completely dissolved in a very large excess of water, measured under standard conditions. It can be calculated as the sum of the reverse of the salt lattice enthalpy plus the sum of the hydration enthalpies of the ions.

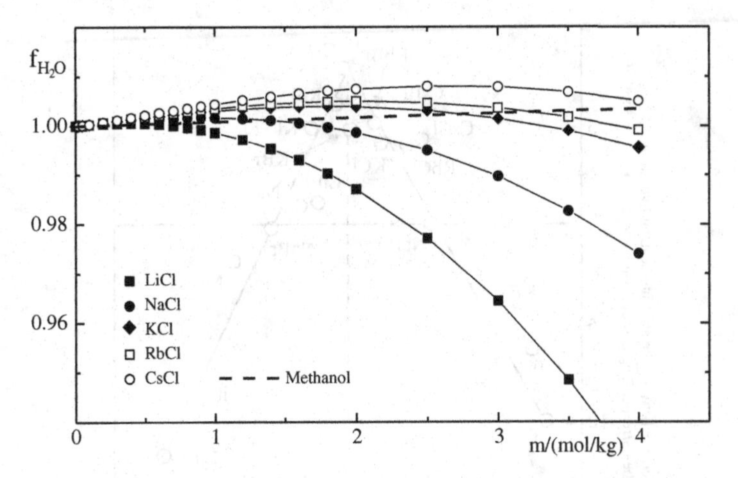

Fig. 4. Water activity coefficients in some simple salt solutions as a function of salt molality at 25°C. As in Fig. 2, the corresponding water activity coefficients in water–methanol mixtures are added for comparison. (After Hamer and Wu.[26])

The dissolution of some alkali halides yields cold solutions, whereas the dissolution of others yields warm solutions. This behaviour is illustrated in Fig. 5, where the enthalpies of solution are plotted against the differences in the hydration energy of each ion in the salt. This idea stems from Collins[30] who showed these striking differences in the heats of solution as an example of his 'concept of matching water affinities'. We will discuss this very valuable concept later on in Sec. 3.1. Here, it is important to see that even the sign of the heats of dilution can be different depending on the ion pairs. Obviously, pairs of two kosmotropic ions and pairs of two chaotropic ions produce cold solutions, whereas the opposite is true for pairs of one kosmotropic and one chaotropic ion, independently of the sign of the ions. This remarkable phenomenon must be related to different ways of hydration of the ions or the ion pairs if they form. To our knowledge, Collins' concept (Sec. 3.1) is the only one that provides a satisfactory explanation for it.

It will be a challenge for future theories and simulations to reproduce at least qualitatively this picture. The problem is, however, that the heat of solution is related to the temperature derivative of the Gibbs energy. Unless the latter is described satisfactorily, there is little hope to get the derivatives at least qualitatively right. The situation is of course worse for the description of heat capacity, a very important parameter in industry.

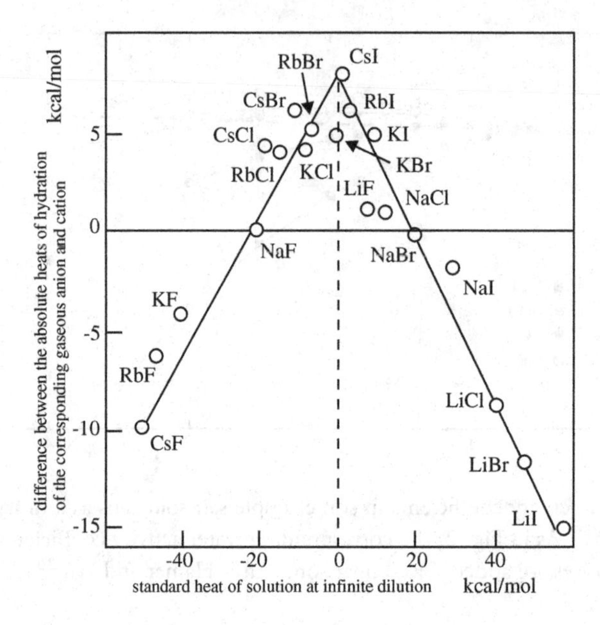

Fig. 5. Relationship between the standard heats of solution of a crystalline alkali halide (at infinite dilution) in kcal mol^{-1} and the difference between the absolute heats of hydration of the corresponding gaseous anion and cation, also in kcal mol^{-1}. (*Source*: Collins,[30] after Morris.[31])

The heat capacity is related to the second derivative of the Gibbs energy with respect to temperature. Here, even an acceptable fit is hard to get.[32]

Furthermore, care should be taken when trying to interpret quantities such as the heat of solution in terms of interactions in water. They also depend on the crystalline state, which is energetically different for every salt.

2.1.3. *pH*

Buffers are utilised to adjust a certain pH value in a given solution. Further added inert ions are not supposed to change the pH of a buffered system. However, this is not completely true, and this can have consequences, especially in biological systems, where small changes in pH may cause significantly different enzymatic activities.[33]

As shown in Figs. 6(a) and 6(b), even at a relative small concentration of 0.3 M of lithium chloride, the pH of the citric buffer, initially fixed at

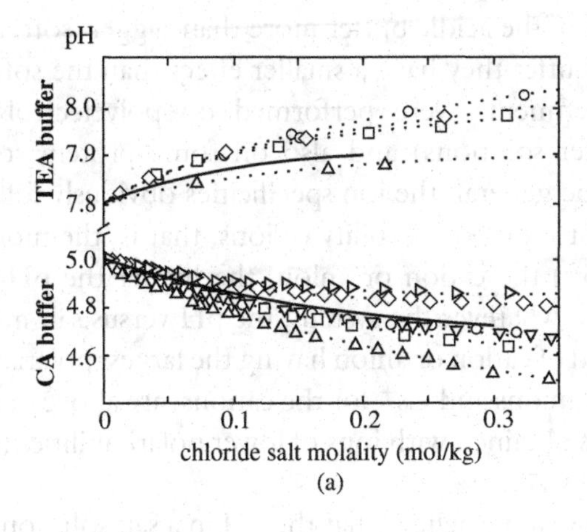

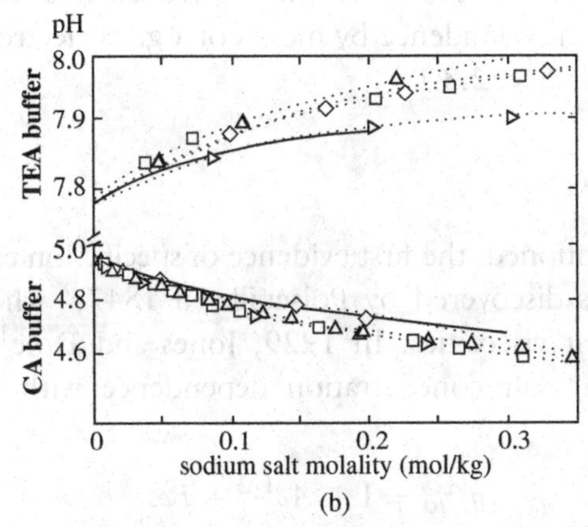

Fig. 6. (a) pH of the citric buffer ($c = 0.025$ M) and of the triethanolamine buffer ($c = 0.025$ M) versus chloride salt molality (mol/kg): (▷) $N(CH_3)_4Cl$, (◇) CholineCl, (○) CsCl, (▽) KCl, (□) NaCl; (△) LiCl. The full curve represents the pH calculated for both buffers from an extended Debye–Hückel equation for an ionic size parameter of $a_i = 4 \times 10^{-10}$ m. (b) pH of the citric buffer ($c = 0.025$ M) and of the triethanolamine buffer ($c = 0.025$ M) versus sodium salt molality (mol/kg): (△) NaSCN, (◇) NaBr, (□) NaCl; (▷) $NaNO_3$. The full curve represents the pH calculated for both buffers from the extended Debye–Hückel equation for an ion size parameter of $a_i = 4 \times 10^{-10}$ m. (After Voinescu *et al.*[33])

5.0, is actually at around 4.6. If the initial buffer pH is in the alkaline area, then the actual pH is always increased by added salt. The overall effect can be roughly estimated from a simple extended Debye–Hückel calculation. More interestingly, there is an ordering in the ion effects: small 'hard' ions decrease the pH of the acidic buffer more than bigger 'soft' ions, whereas for the alkaline buffer they have a smaller effect than the soft ions.

Similar experiments were performed on polyelectrolyte solutions (instead of buffer solutions) and also on some protein solutions. The results seem to be general: the ion specificities obviously follow a general trend according to the polarisability of ions, that is, the more polarisable (salting-in or soft) the cation or anion, the higher the pH through salt addition. Indeed, whatever the system, the pH versus salt molality curves of salts composed of cation or anion having the largest polarisabilities, such as SCN^- for the anions and Cs^+ for the cations, are above or at least never below the curves obtained with ions of lower polarisabilities (harder ions), such as Na^+ and Cl^-.

This is all true, if we believe that the pH in a salt solution can really be measured with some confidence by means of a glass electrode. This issue will be discussed in Sec. 2.3.1.

2.1.4. *Viscosities*

As previously mentioned, the first evidence of specific ion effects in aqueous solutions was discovered by Poiseuille in 1847,[9] when he studied systematically their viscosities. In 1929, Jones and Dole[10] proposed a description of the salt concentration dependence with the following equation:

$$\eta/\eta_o = 1 + Ac^{1/2} + Bc, \tag{6}$$

where η and η_o are the viscosities of the aqueous solution and pure water, respectively; A is an electrostatic term that is close to one, as long as the salt concentrations are not significantly higher than 0.1 M; c is the molarity and B is the so-called Jones–Dole viscosity coefficient. For equally charged ions, the specificity is thus expressed with this B coefficient. The viscosity of neutral solutes would best be described with $A = 0$.

In the review paper by Jenkins and Marcus,[34] a comprehensive collection of viscosity data and B coefficients can be found. Table 1 shows a few of these Jones–Dole coefficients.[30]

Interestingly, the B values of the ions are nearly additive for a given salt. Although no convincing theory exists so far to explain the concentration dependence of salt solution viscosities, it seems plausible to attribute the B values to ion–water interactions. According to Collins, negative values can be related to soft ions with low surface density. Here, water–water

Table 1. Jones–Dole viscosity *B* coefficients.

Cations	*B*	Anions	*B*
Mg^{2+}	0.385	PO_4^{3-}	0.590
Ca^{2+}	0.285	$CH_3CO_2^-$	0.250
Ba^{2+}	0.22	SO_4^{2-}	0.208
Li^+	0.150	F^-	0.10
Na^+	0.086	HCO_2^-	0.052
K^+	−0.007	Cl^-	−0.007
NH_4^+	−0.007	Br^-	−0.032
Rb^+	−0.030	NO_3^-	−0.046
Cs^+	−0.045	ClO_4^-	−0.061
		I^-	−0.068
		SCN^-	−0.103

Note: After Collins,[30] and references therein.

interactions are supposed to be stronger than ion–water interactions. Positive values signify strong hydration. Even within these two groups, the stronger (or weaker) the hydration, the higher (or more negative) the *B* coefficients.

It is surprising that the concept of hard and soft ions is as applicable to transport properties as it is to thermodynamic properties. There is no reason why a non-equilibrium quantity, related to hydrodynamics, is equally sensitive to ion–water interactions relative to water–water interactions.

In principle, other transport properties such as diffusion coefficients and electrical conductivity also show ion specificity. We will not discuss these properties here. The diffusion coefficients show only slight concentration dependence so that the specific ion effects do not appear as clearly as that with viscosity. Since conductance is a very complex phenomenon, where several effects such as the electrophoretic and the Debye relaxation effect interfere, it is difficult to infer predictions and general rules. The book written by Robinson and Stokes[24] is still a valuable introduction to this field, although a huge number of experimental and theoretical papers appeared ever since.[35,36]

2.1.5. *Mixtures of salts and other solutes*

In the preceding sections, it is shown that 'simple' aqueous solutions of one salt are far from being trivial to understand. Obviously, this is all the more true, when a third component is added.

More than half a century ago, Long and McDevit published an extensive review about activity coefficients of non-electrolyte solutes in aqueous salt solutions.[37] This paper is still very valuable today. It is shown there how solubility, phase distribution, and vapour pressure measurements can be used to determine the influence of ions on other solutes. Detailed examples are given for salt effects on hydrogen, oxygen, carbon dioxide and other gases, but also on liquid compounds, such as benzene. This classical example is shown in Fig. 7.

As expected, highly charged and hard ions show a significant salting-out effect, i.e. the activity coefficients of benzene are higher than one, whereas bigger, softer ions and their salts (CsI) slightly salt in benzene in water. Organic ions like $(CH_3)_4N^+$ have a significant salting-in effect.

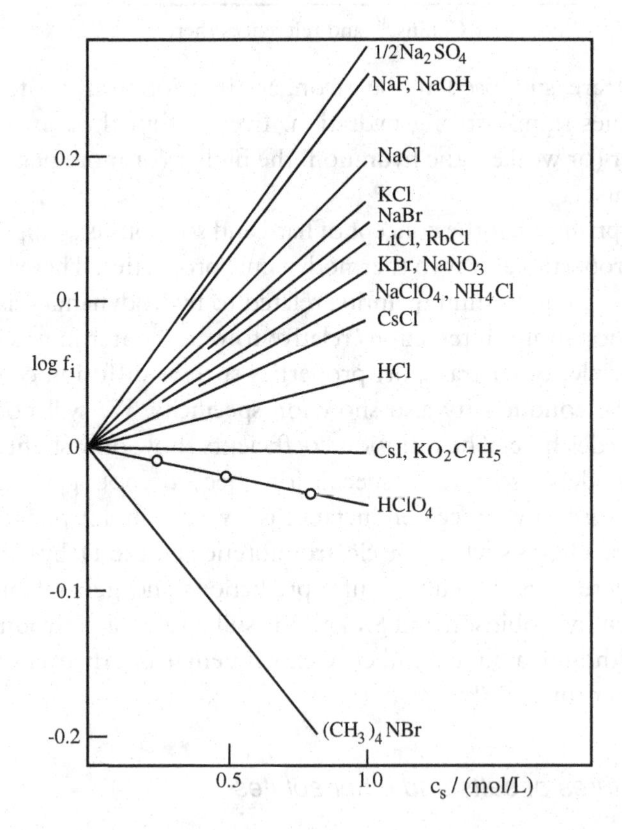

Fig. 7. The activity coefficients of benzene in different salt solutions. (After Long and McDevit.[37])

However, when looking in more details into the series, some unexplained inconsistencies appear: lithium has an unexpectedly small salting-out effect and perchloric acid is even a salting-in species. Nonetheless, in total roughly a Hofmeister series is found.

For a given non-electrolyte concentration, the activity coefficients of the non-electrolyte can be approximated by

$$\log f_{\text{non-electrolyte}} = k_{\text{salt}} c_{\text{salt}} \quad (7)$$

Table 2 shows some k_{salt} values for different salt effects on hydrogen, oxygen, and benzene.

It is interesting to note that the effects of cation and anion in the considered salts are almost additive, even at high concentrations of salt. More data on salting-in and salting-out of various aromatic substances can be found in Ref. 38.

Table 2. Salt parameters k_{salt} for hydrogen, oxygen and benzene in different salt solutions.

	Hydrogen		**Oxygen**		**Benzene**	
Salt	**k_s (obs.) l/mol**	**k_s (calc.) l/mol**	**k_s (obs.) l/mol**	**k_s (calc.) l/mol**	**k_s (obs.) l/mol**	**k_s (calc.) l/mol**
Na_2SO_4	0.278	0.46			0.548	1.33
NaOH	0.140	0.25	0.179	0.29	0.256	0.85
NaCl	0.114	0.12	0.141	0.15	0.198	0.42
NaBr			0.110	0.12	0.155	0.35
KCl	0.102	0.10			0.166	0.34
LiCl	0.076	0.09	0.100	0.11	0.141	0.31
RbCl					0.140	0.31
$NaNO_3$	0.100	0.09			0.119	0.31
KNO_3	0.070	0.07	0.100	0.08		
NaI					0.095	0.27
$KaClO_4$					0.106	0.26
CsCl					0.088	0.26
NH_4Cl					0.103	0.15
HCl	0.030	0.03		0.03	0.048	0.09
$HClO_4$					−0.041	−0.05
$(CH_3)_4NBr$					−0.24	−0.54

Note: After Long and McDevit.[37]

Table 3. Corrected optical rotation values for glucose and serine enantiomers in the presence of different sodium salts (maximum error is ±0.060). ρ are estimated ion polarisabilities in solution in Å^3.

	α				
Anion	**D-Glucose**	**L-Glucose**	**L-Serine**	**D-Serine**	ρ
I^-	51.649	−51.534	−6.450	6.550	7.46
SCN^-	52.046	−51.745	−6.440	6.601	6.47
NO^-	52.482	−51.976	−6.702	6.678	4.47
Br^-	52.388	−52.016	−6.681	6.691	5.06
Cl^-	52.654	−52.060	−6.882	6.894	3.73
CH_3COO^-	52.245	−51.846	−6.518	6.599	5.50
ClO_4^-	52.344	−51.906	−6.540	6.626	5.26
F^-	52.815	−52.228	−7.091	7.021	1.36
$H_2PO_4^-$	52.868	−52.312	−7.195	7.201	(4–5)
(H_2O)	52.894	−52.517	−7.589	7.495	

Note: After Lo Nostro *et al.*[41]

Very roughly speaking, the same series holds for both unpolar and polar non-electrolytes, although for the last ones, the ion sensitivity depends on the acid or basic character. Of particular interest are the ion specificities in amino acid solutions. Here several papers appeared in the last few years. They show still unexplained specific interactions between the partially dissociated organic solutes and the charged species.[39,40]

Apart from such thermodynamic considerations, other techniques are helpful to illustrate specific ion effects on non-electrolytes. This can be, for example, the optical rotation of a chiral component. This was done recently, and in Table 3 the influence of different ions on the optical rotation of D- and L-Glucose and of L- and D-Serine is given.[41] Obviously, the change in optical rotation roughly correlates with the supposed absolute ion polarisabilities, but not with a classical Hofmeister series, in which, for example, acetate would be located much closer to fluoride than to thiocyanate.

As far as mixtures of electrolytes are concerned, a vast literature exists. The reader is referred to the book of Robinson and Stokes[24] for more information on this topic. A more recent and extensive paper about mixed salt solutions is written by Pitzer and Kim.[42]

In such systems, it is very difficult to find an ordering of specific ion effects. The activity coefficients of mixtures can significantly differ from that of the respective pure solutions. The various ion–ion and ion–solvent interactions and the limited data sets make it difficult to find general rules.

It should be noted that in aqueous salt mixtures, it is not always easy to differentiate between the properties of both salts. The osmotic pressure is a measure of the water activity, but it can no longer be converted to the activity coefficients of either the salt or the third component. The Gibbs–Duhem equation relates the activities of all three components. The same is true for conductivity, density, etc. A way to solve the problem is to perform a second independent experiment or to find a probe that is only specific to the property of the third component, for example selective electrodes.

2.2. *Salts near 'Simple' Interfaces*

2.2.1. *Air-solution interface*

In the last section, we discussed some specific ion effects in bulk solutions, where no interfaces are involved. The probably simplest interface is the surface of an aqueous salt solution against air. Here the relevant property is the surface tension. Figure 8 shows how the surface tension of water is influenced by the presence of ions,[43] see also Ref. 44.

In Chap. 8, a more comprehensive data set including salt mixtures in water will be presented. Nevertheless, Fig. 8 shows the main features. Salts with soft, polarisable ions increase the surface tension of water less than salts composed of hard ions. However, the effect is small, which makes it difficult not only for precise measurements, but also for quantitative predictions with simulations and theories. Anions seem to change the surface tension more than cations. Whereas it is intuitively comprehensible that 'hard' ions tend to stay away from the surface, more than only weakly hydrated soft or hydrophobic ions, it was for a long time a mystery, why acids lower the surface tension. This means that protons should go to the surface, which is counter-intuitive, at least for a chemist. A plausible explanation is given by Jungwirth and co-workers, based on their molecular dynamics (MD) simulations.[45,46] It seems that for geometrical reasons, H_3O^+ fits better into the surface layer than into the bulk.

Besides surface tension, the surface electric potential is another macroscopic and evident example of specific ion effects at the surface of salt solution. Only few detailed discussions exist in the literature about this property, and one such example is shown in Fig. 9. Petersen and Saykally

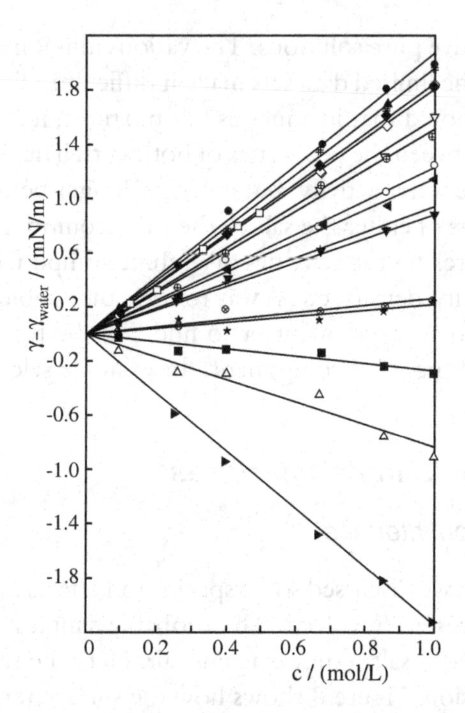

Fig. 8. Effect of electrolyte concentration on the change in surface tension relative to water for 1:1 electrolytes. Experimental error in data points is $\pm 0.1\,\mathrm{mN\,m^{-1}}$. HCl (■), LiCl (▲), NaCl (•), KCl (♦), CsCl (⊕), NaF (□), NaI (○), NH_4Cl (∇), NaBr (◇), HNO_3 (Δ), $(CH_3)_4NCl$ (+), NH_4NO_3 (◄), $HClO_4$ (►), $NaClO_3$ (▼), $LiClO_4$ (⊗), $NaClO_4$ (★), KOH (#). (After Weissenborn and Pugh.[43])

recently summarised the present knowledge and ideas on surface electric charge.[47] In Randles' extensive review,[48] the experimental difficulties are also discussed in detail. Due to these difficulties, which involve some more or less questionable thermodynamic assumptions, the data must be taken with care. Nevertheless, some trends seem to appear (see Fig. 9). In general, anions change the surface potential much more than cations do. Obviously anions go closer to the surface. The exception is hydrogen, which is now also known to go to the surface (see preceding paragraph). In agreement with surface tension, softer and more polarisable ions go closer to the interface. As a consequence, the surface tension increment is *smaller* for such ions, whereas the surface potential increment is *higher*.

However, in contrast to surface tension, where only the ion profiles are relevant, the surface potential is influenced not only by the different

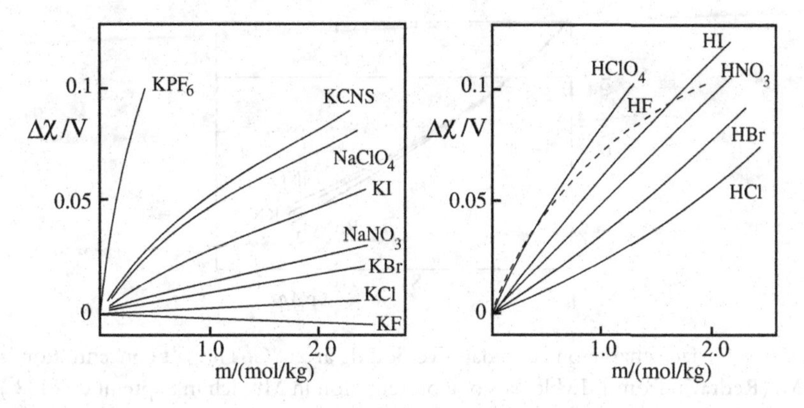

Fig. 9. Surface electric potential increments of various aqueous electrolyte solutions as a function of salt molality. (After Randles.[48])

layering of the ions near the surface, but also by the orientation of water dipoles. These water dipoles may slightly reorientate at the surface, at least in the ion hydration shells, and this reorientation may contribute to the change in the surface potential. Within a simple (primitive) model, in which water is taken as an average bulk with a dielectric constant only, it will be difficult to describe surface electric charges of salt solutions.

Today, the ion profiles near the solution surface are subject of intensive studies; see the recent review by Petersen and Saykally[47] and references given therein. Several of the following chapters will deal with these ion profiles, since they are considered as a key for the understanding of specific ion effects.

2.2.2. *Solution–solid interfaces*

Physicists dealing only with 'physical interactions' are surprised that ions can adsorb to surfaces that have the same sign of charges as the adsorbed ions. And if so, they ascribe this finding to a 'dispersion' interaction only, forgetting that 'chemical' interactions may be present.

In the present subsection, we will focus on more classical ion effects, i.e. the adsorption of cations on negatively charged solid surfaces. An excellent review on this subject was published in 2003 by Lyklema.[28] His discussion is essentially summarised here.

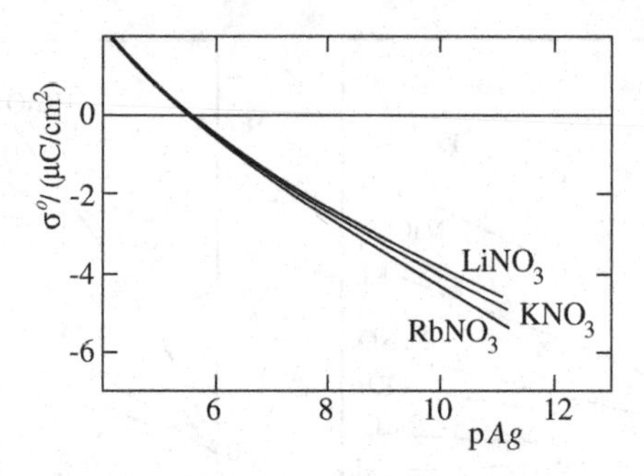

Fig. 10. Surface charge on colloidal silver iodide at 25°C. The salt concentration is 0.1 M. (Redrawn from J. Lyklema's oral presentation in Munich in September 2008.)

The reason why we classified this subsection under 'simple surfaces' is that only hard surfaces are considered without ion penetration and without significant curvature of the surfaces at a molecular scale.

Probably the most studied system is silver iodide, AgI, for which many reliable experimental data are available, such as the critical coagulation concentrations (ccc) in different salt solutions and the surface charges as a function of Ag^+ concentration (see Fig. 10).

According to this picture, the rubidium salt gives the highest surface charge, but at the same time the ccc is lowest for this salt ($LiNO_3$, 165 mM; KNO_3, 136 mM; $RbNO_3$, 126 mM). This apparent contradiction is resolved by assuming that Rb^+ is the most strongly adsorbed type of ion in the inner Helmholtz plane. The adsorbed cations attract iodide counterions, which in turn decreases the outer Helmholtz potential. For more detailed information, see Ref. 28.

From the experimental results, the chemical i.e. non-electrostatical part of the Gibbs energy of adsorption can be estimated, although the separation of $\Delta G^{\text{adsorption}}$ in $\Delta G^{\text{ad elect}}$ and $\Delta G^{\text{ad chem}}$ is not rigorous and somewhat arbitrary. Anyway, $\Delta G^{\text{ad chem}}$ is in the order of $-3\,kT$. The differences between the chemical Gibbs adsorption energies of the three salts are of the order of $0.5\,kT$. This is comparable to the differences in $-RT \ln f_{\pm}$ for these salts. Surprisingly, the specificities both in surface charge and ccc disappear above around 65°C.

Lyklema also considered the adsorption of ions to various oxides and to mercury. In contrast to all the other surfaces, the surface charge on mercury is of electronic nature and completely smeared out. A very weak attraction is found here and the ordering is as for AgI. For oxides the ordering of the counterion adsorption depends on the point of zero charge (pzc) of the oxides. If the pzc is low (i.e. the oxide is acidic), there is a preference for big ions, whereas the series is reversed for basic oxides (high pzc). However, the exact ordering depends not only on the pzc, but also on the material and even its preparation (surface treatment, crystal plane, etc.). In total it seems that the specific ion interactions depend both on the ions and on the specific groups and that they are highly water-structure mediated. We will come back to this point in the final chapter, because it is one of the main conclusions from the newest experimental and theoretical work in this field.

2.3. *Salts Near More Complex Interfaces*

2.3.1. *Again pH: The electrode interface*

Since the pioneering work by Harned[49] and later on by Schwabe and Suschke[50] about the use of glass electrodes, only few papers dealt with the question of whether concentrated salt solutions can falsify the measured pH, for example because of strong ion adsorption at the electrode. In a very careful study Kron *et al.* fixed this problem.[51] They carried out high-precision potentiometric titrations in cells with liquid junction using either glass or hydrogen electrodes. They considered various salt solutions up to several molar concentrations (see Fig. 11). The variation they found were always in the range pH $= 7.0 \pm 0.2$. This is significant, but it is difficult to deduce specific ion effects from these deviations. Very probably, ions adsorb at electrodes or interfere in chemical equilibria via the change of water activity, but surprisingly (and fortunately) these effects do not significantly disturb the pH measurements or adulterate the indicated pH values, neither for the hydrogen nor for the glass electrode. This may be different when further solutes, especially surface active ones, are present in the solution.

What is more, glass electrodes seem to work properly even in very complex systems such as reverse microemulsions.[52] This could be checked by comparing the pH values indicated by the electrode to the activity of a special enzyme that is very sensitive to tiny changes in pH. Both independent measurements suggested the same pH values. It is very improbable that ions falsify both experiments in the same way. Therefore it can be concluded that the glass electrode shows the right pH.[52]

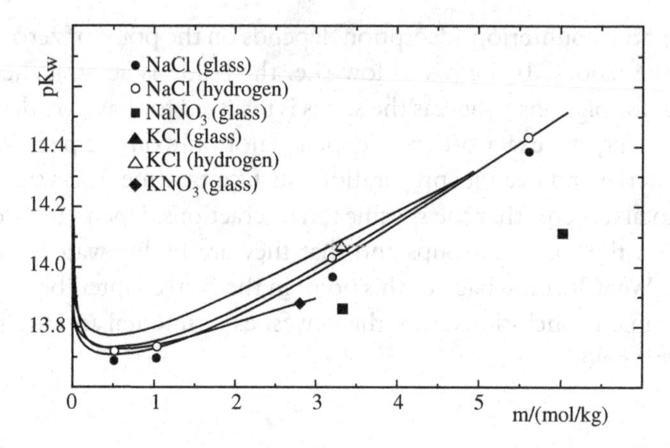

Fig. 11. Experimental and calculated values of the ion product of water, pK_W, for various salts at 25°C. The experimental data were obtained either with a hydrogen or a glass electrode and the lines are Pitzer fits. For details see the original paper. (After Kron *et al.*[51])

2.3.2. *Molecular sieves and chromatographic surfaces*

In contrast to electrodes and oxides with more or less smooth surfaces, the surface of chromatographic material is usually porous with special binding sites or cavities that allow for a separation according to the molecular weight (size-exclusion mechanism) or the chemical affinities.

Washabaugh and Collins[53] published a very interesting study on the retention of various salts (and also neutral molecules) in Sephadex G-10 columns. This material consists of epichlorohydrin cross-linked dextran in beaded form. Solutes with a weight smaller than 700 Dalton can penetrate the beads resulting in a slower elution than that of molecules with a higher molecular weight.

In Fig. 12, solutes with different molecular weights are given as well as their distribution coefficients K_d. $K_d = 0$ means that the species cannot enter the beads at all, whereas $K_d = 1$ signifies that all species follow the longer pathways through the beads. $K_d > 1$ cannot be explained by a gel sieving mechanism. There must be an additional retention of the solute, very probably due to adsorption. Obviously the most polarisable and most weakly hydrated ions absorb most. The concentration and temperature dependence of this adsorption phenomenon gives additional information.

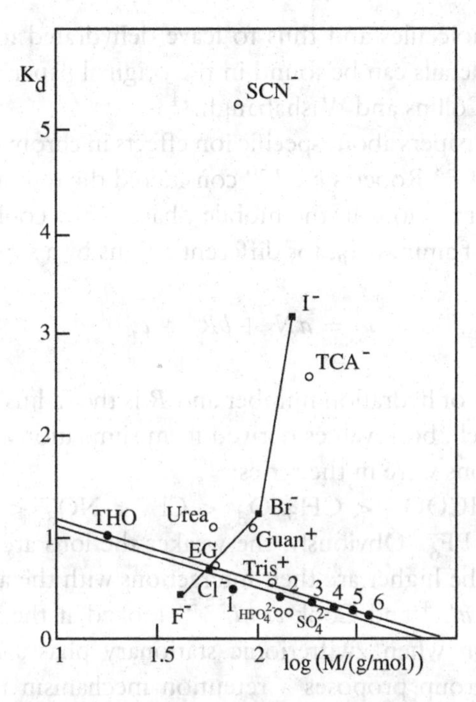

Fig. 12. Gel sieving chromatography of salts and solutes on Sephadex G-10. One-millimetre samples containing 0.1 M solute in 0.1 M NaCl plus $\sim 2\,\mu$Ci ^3H_2O, and 0.5% dextran were chromatographed on a Sephadex G-10 column (1.5×85.5 cm) at 30°C and a flow rate of 0.5 ml/min. The eluant was 0.1 M NaCl, pH 7.0. Anions were chromatographed as sodium salts; cations were chromatographed as chloride salts. The double line is the best estimate of ideal behaviour for solutes on Sephadex G-10. The relative elution position K_d is defined as $K_d = (V_e - V_o)/(V_i - V_o)$, where V_i is the included volume (measured with ^3H_2O), V_o is the excluded volume (measured with dextran), and V_e is the elution volume for a given solute. The points labeled 1–6 represent glycine and its homopolymers through hexaglycine. *EG*, ethylene glycol; *TCA*$^-$, trichloroacetate; *GUAN*$^+$, guanidinium; *TRIS*$^+$, protonated Tris. *THO*, [^{3}H]OH. (After Washabaugh and Collins.[53])

With increasing salt concentration, the chaotropes elute sooner, a salting-in effect, which is — according to the authors — related to an indirect water-structure mediated effect. As the temperature increases, chaotropes adsorb less strongly. This may be due to water hydrogen bonding becoming less strong with higher temperature. As a result, the tendency of loosely bound water (to the chaotropes) to form stronger hydrogen bonding to other water molecules and thus to leave dehydrated ions is less pronounced. More details can be found in the original paper and also in the review paper by Collins and Washabaugh.[44]

Several other papers about specific ion effects in chromatography have been published.[54–58] Roberts *et al.*[56] considered the retention of organic amines by different anions in the mobile phase. They could describe the retention factor of amines, k_p, for different anions by a simple equation:

$$k_p = aN_s + bR^2 + c, \tag{8}$$

where N_s is a sort of hydration number and R is the radius of the ion plus first hydration shell, both values derived from simulations. The retention power of the anions were in the series:

$H_2PO_4^- < HCOO^- < CH_3SO_3^- < Cl^- < NO_3^- < CF_3COO^- < BF_4^- < ClO_4^- < PF_6^-$. Obviously, the weaker the ions are hydrated (the softer they are), the higher are their interactions with the amines.

Umemura *et al.*[54] and Cook *et al.*[55,57] looked at the ion chromatographic separation when zwitterionic stationary phases are employed. The Australian group proposes a retention mechanism based on both an ion exclusion and a chaotropic interaction between ions and the opposite charge of the headgroup. The outer choline group of the *N*-tetradecylphosphocholine creates a Donnan potential. The ion pairing of the inner negative and outer positive part of the zwitterionic group with the counterions depends on the position of the ions in the Hofmeister series. Hard cations are more strongly attracted to the phosphonate group, whereas soft anions are more strongly attracted to the choline headgroup. If we make the reasonable (and often verified) assumption that choline is a soft group, whereas phosphate is a hard one, one finds again that 'like seeks like' explains this experimental result.

The magnitude (and even the sign) of the Donnan potential depends on the relative association of the ions with the two charges on the headgroup. For example, the hard Ce^{3+} ions strongly interact with the hard phosphonate, whereas the Hofmeister neutral chloride ion does not strongly interact with the trimethylammonium headgroup. As a result, the Donnan potential is very positive. By contrast, sodium interacts much weaker with the phosphonate group than Ce^{3+}, whereas ClO_4^- strongly screens the positive charge of the zwitterions. The overall result is a small or even negative Donnan potential (Fig. 13).

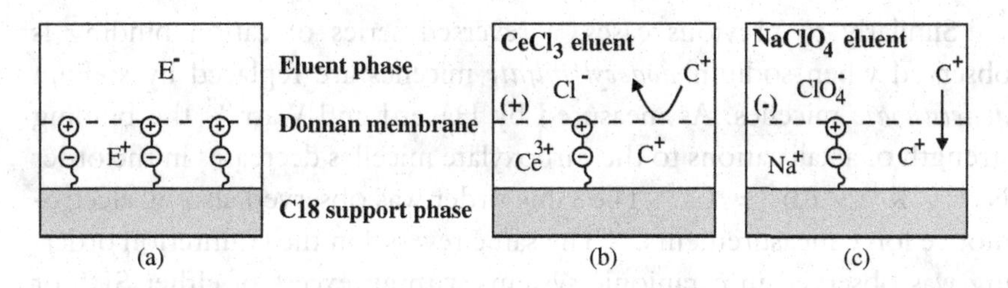

Fig. 13. Schematic representation of the proposed mechanism for the phosphocholine stationary phase system. (a) Establishment of the Donnan membrane, (b) use of $CeCl_3$ and (c) use of a $NaClO_4$ mobile phase. (After Cook *et al.*[57])

Harrison *et al.*[58] undertook a particularly interesting ion-pair chromatography study. In contrast to most other investigations in this field, they varied the nature of the ion-pairing agent (IPA). Further to a classical ammonium-based IPA, the tetrabutylammonium ion (TBA), they used tetrabutylphosphonium (TBP) and tributylsulfonium (TBS). They found that TBP is the most chaotropic ion, whereas the TBS is the least chaotropic. Again in agreement with the rule 'like seeks like', the most chaotropic anions associate most with TBP and least with TBS.

2.3.3. *Surfactant and polyelectrolyte colloids: Micelles, vesicles, and liquid crystals*

In systems of classical association colloids (made of surfactants and polyelectrolytes), specific ion effects are ubiquitous. Consequently, a huge number of papers deal with such effects. The Swedish groups of Ekwall and Lindman made numerous detailed studies of counterion binding to surfactants, polyelectrolytes and charged biomolecules (see Refs. 59–62). They also characterised in details the influence of ions to changes in phase diagrams. In a recent feature article, Romsted discussed extensively the subtle differences between chloride and bromide in amphiphile aggregation.[63]

Many other groups and their work could be cited. Schulz and Warr[64] reported the selective binding of alkali counterions to anionic SDS surfactant films by ion floating technique. The observed binding strength decreases from Cs^+ to Na^+. This specificity order agrees well with interfacial probe studies of alkali metal ion binding performed by He *et al.*[65] and Hafiane *et al.*[66]

Similarly to previous cases, a reversed series of cation binding is observed when sodium *dodecyl sulfate* micelles are replaced by sodium *dodecanoate* micelles. As measured by Haverd and Warr,[67] the binding strength of alkali cations to the carboxylate micelles decreases in the order $Na^+ > K^+ > Rb^+ > Cs^+$. The same order was observed also by electromotive force measurements.[68] The same reversal in the counterion ordering was observed in catanionic systems with an excess of either SDS or sodium dodecanoate. These systems can form spontaneously vesicles with increasing concentration of added salt.

Twenty years ago, Ninham[69–71] and Khan[72–74] reported on the unusual behaviour of acetate, hydroxide and formate double-chain surfactants. Didodecyldimethylammonium sulfate (DDAS) is almost insoluble in water. However, it swells in water, producing a lamellar liquid crystalline phase. On the other hand, didodecyldimethylammonium acetate and hydroxide are highly water soluble and form an extended micellar phase. Similar counterion effects have been observed for the swelling of lamellar phases. Many possible explanations for the observed phenomenon are given. Ninham *et al.* attributed these peculiarities to the high hydration of the headgroups. Morgan *et al.*[75] proposed that partial dehydration of the competing counterions dominates their exchange at the interface, while Robinson and Harned[27] suggested the formation of a solvent-shared ion pair (effect of the so-called 'localised hydrolysis'). However, no really convincing and general rule could be inferred.

The counterion selectivity can be extended to phosphate-containing systems. Here, the counterion selectivity for the phosphate headgroups depends strongly on the charge of the headgroup (single or double). For comparison reasons, only the single-chain hydrogen alkyl phosphate is considered. The selectivity order of alkyl phosphate micelle surfaces as determined from flotation experiments[67] resembles that of alkyl sulfate micelles; the alkali metal ion selectivity coefficients follow the order $Na^+ < K^+ < Rb^+ < Cs^+$. This order is found also for phosphoric acid resins. At medium pH (6–8.5), where hydrogen phosphate is present, the order of selectivity coefficients is $Cs^+ > Rb^+ > K^+ > Na^+ > Li^+$.[76] However, while in the case of alkylsulfates and alkylcarboxylates, the trend always follows the described series; different results are observed in the case of phospholipids. Experiments performed on the negatively charged phospholipid dioleoylphosphatidylglycerol (DOPG) show that the maximum in binding is found for K^+.[77] It is as if the negatively charged phosphate group behaves like an intermediate case somewhere between a kosmotropic and a chaotropic object. Nevertheless, the phosphate group in vesicle-forming lipids more often behaves as a hard group, in contrast to phosphate surfactants.[78] We will return to this classification later in this chapter.

Similar counterion specificity, as reported in the catanionic systems, is observed in the case of polyelectrolytes. Nevertheless, it should be kept in mind that ion effects do not depend only on the individual properties of the participating ionic group and its counterion (i.e. charge, size, charge distribution and polarisability), but also on the overall charge of the polyion, as well as on the possible cooperating binding sites. Furthermore, even at high dilution, a substantial number of counterions is forced into close proximity to the polyion by the long-range electrostatic forces,[79,80] so that there always exists a large number of ion pairs, for which solvation effects should be observable.

Strauss[81,82] used the dilatometric method to measure the volume changes that occur when polyelectrolyte solutions are mixed with solutions containing different specifically interacting counterions. Polyelectrolytes containing sulfate and carboxylate headgroups exhibit opposite series of counterion binding, similar to the series observed in experiments on surfactant micelles. That is, in the case of polysulfonates, the volume increase becomes smaller with increasing ionic radius, whereas in the case of carboxylates, this series is reversed. Dilatometric results clearly show specificities depending on both the polyion and on the metal ion which would not be expected on the basis of long-range electrostatic density of the polyion. The effects of the latter are governed predominantly by the linear charge density of the polyion, and the linear charge densities of the studied sulfonates and acrylates are the same. It follows that the interactions giving rise to the observed volume changes involve specific sites on the polyions. Similar results for interactions of alkali cations with polyacrylates and polysulfonates have been discussed also by other authors.[83–87] The reversal of the cation series in the presence of polyacrylates was attributed to a competition between hydrating water molecules and the anion for a given cation, which is qualitatively related to the charge distributions, polarisabilities, and the effective field strengths of the ions.[88,89]

MD simulations show similar results. Chialvo and Simonson[90] modelled a short-chain polystyrenesulfonate and found the interaction with Li^+ too weak to cause desolvation. Consistently, Lipar *et al.*[85] reported binding to polyanetholesulfonic acid to be strongest for Cs^+. MD calculations for cationic polyelectrolytes[91] with linked tetramethylammonium groups predict the same ordering of counter-cations ($I^- > Br^- > Cl^- > F^-$) as that observed for a cationic trimethylammonium headgroup.

More examples and a more detailed discussion can be found in Ref. 92. It is not astonishing that the huge amount of data is puzzling. However, as we will see in the following chapters, a general rule emerges. This is presented and discussed in detail in the concluding chapter.

2.4. *The Most Complex Interfaces: Biological Ones*

2.4.1. *Biological membranes*

Doubtlessly, lipid membranes are among the most important biological interfaces. They control transport of molecules in and out of the cell and are responsible for cell adhesion, fusion, etc. The ions in the aqueous medium around the membranes play a significant role for the surface potential, the dipole potential, the structure and dynamics of the lipids, the transition from micelles to vesicles, etc. Relevant references on these different aspects can be found in the paper by Garcia-Celma *et al.*[93] Furthermore, it is still mostly trial and erroneous to extract proteins out of membranes and to stabilise them at convient pH with detergents and added salts.

The question arises as to how ion specificities in lipid membrane systems can be explained. We will try to give an answer based on three different studies. The first was recently published by Petrache *et al.*[94] They measured the swelling of multilayers consisting of 1,2-dilauroyl-*sn*-glycero-3-phosphocholine (DLPC) in the presence of KCl and KBr as a function of salt concentration. As expected, the swelling is highly ion specific. Bromide is more strongly attached to the neutral layers than chloride. Consequently, the membranes become equally charged so that the repulsion between the layers increases. But surprisingly, this swelling increases with increasing salt addition even at high salt concentration. Increasing electrostatic screening would lead to the opposite effects. However, not only the repulsion is reduced due to electrostatic screening, but also van der Waals attraction, when the salt concentration increases. So, from their study, Petrache *et al.* could nicely infer the subtle balance between specific adsorption and electrostatic as well as van der Waals screening. Finally, why is bromide more adsorbed than chloride to the neutral lipid layers? The probable reason is that the positively charged choline headgroup is soft and therefore interacts more strongly with the softer bromide ion.

A similar conclusion can be drawn from the work by Garcia-Celma *et al.*[93] on specific anion and cation binding to lipid membranes investigated on a solid-supported membrane. In this study, defect-free lipid membranes werc deposited on a gold support. The supported membrane has a surface that is similar to that of a free-standing lipid membrane. A solid-supported membrane (SSM) represents a model system for a lipid membrane with the additional benefit of being mechanically so stable that solutions may be rapidly exchanged at its surface.

When solutions of different ionic composition are exchanged, a charge displacement can be registered via the supporting gold electrode that represents the relaxation of the different ions in their respective equilibrium positions. The charge displacements measured on the SSM have been used to obtain information about the relative distribution of cations and anions. Ions are characterised relative to a reference ion, which was Na^+ for cations and Cl^- for anions. A positive charge displacement after a cation exchange or a negative charge displacement after an anion exchange means that the average equilibrium position of the ions is closer to the underlying electrode than that of Na^+ (for cations) or of Cl^- (for anions) and/or that more ions reside in this position. Figure 14 shows charged displacements relative to NaCl for various salts and two different lipids.

Several observations can be made. First, independently of the type of lipid (zwitterionic or neutral or positively or negatively charged — the last two cases are not shown here), soft anions bind more strongly to the lipids than hard anions. By contrast, it is just the other way round for cations hard and highly charged cations are more attracted than soft ones. A possible explanation, e.g. for phosphatidylcholine (PC), is as following: PC contains a soft cationic headgroup (choline) and also an anionic group (phosphate) that is known to be hard. Therefore, it is plausible that hard cations are attracted by the hard phosphate and soft anions are attracted by the soft choline group ('like seeks like'). Maybe this interpretation is simplified, but it correctly predicts the experimental findings. Due to the ion specificity, it is difficult to speak about the total average behaviour of the salt. Choline chloride can be in total a kosmotropic solute, whereas choline alone is chaotropic.

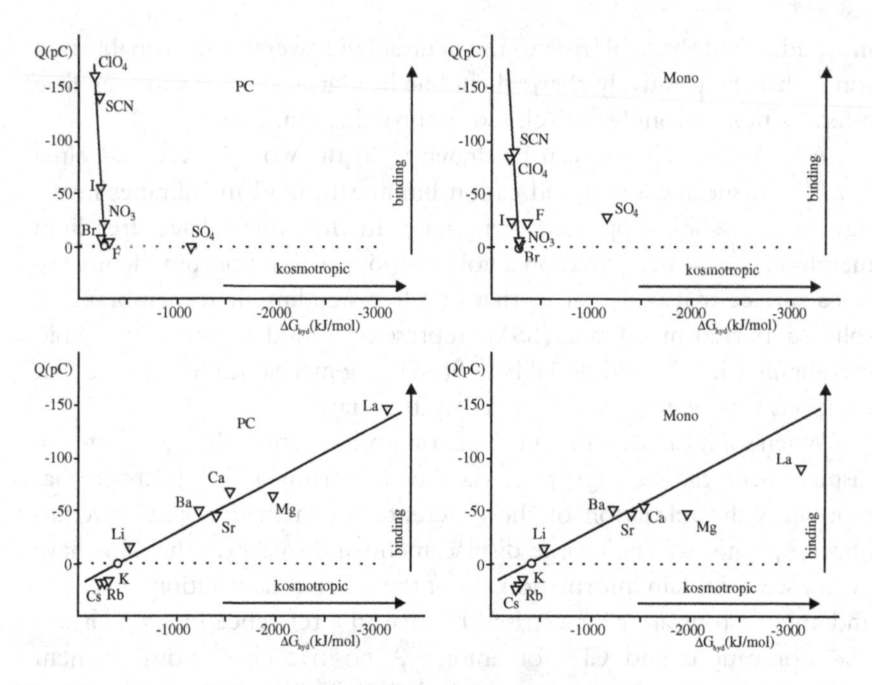

Fig. 14. Charge displacements of different anions and cations on a PC (phosphatidylcholine) and a Mono (monoolein) membrane. The open circle indicates the reference ions Na or Cl. Ions are classified according to their Gibbs free energies of hydration ΔG_{hyd}. (After Garcia-Celma *et al.*[93])

Up to now we discussed membranes that are in a highly ordered, dense state. What happens, if the lipid molecules are in a quasi-gaseous or at least liquid-expanded state so that water and ions can easily penetrate into the lipid layers? This case was extensively studied by Leontidis and coworkers.[95–97] For example, they spread Langmuir layers of PC on aqueous salt solutions at lipid concentrations that allow for expanded films. In this case, ion specificities also occur, but the results do not follow an ion-binding between the PC headgroups and the added salts. The authors[95] could show that the interpretation of Petrache *et al.*[94] is not sufficient and that probably chaotropic anions influence the structure at the interface, indirectly affecting the equation of state (osmotic pressure as a function of layer distance). The results can best be described by a partitioning model (see Fig. 1 in Chap. 2). This is an important finding, because it shows that Collins' idea of ion-headgroup binding is not always the dominating factor. Especially, when ions can easily penetrate into the lipid structure, partition can dominate. For further discussion, see Chap. 2.

2.4.2. *Protein 'surfaces'*

In his landmark papers, Hofmeister was the first to study the effect of salts on the solubility of protein mixtures, which were extracted from the egg white of hen. The terms 'salting-in' and 'salting-out' were created.

Over the following decades, other proteins were tested in this respect, and it turned out that mostly anions were responsible for protein precipitation. A reasonable interpretation was that the ions have a certain affinity for the proteins.[98]

However, more recent studies show that cations can also have a significant influence that is comparable to anions, on the crystallisation of proteins. It depends on the choice of the proteins. In a very interesting paper, Bénas *et al.*[99] examined the influence of a set of mono-, di-, and tri-valent cations on the the solubility of lysozyme. They found that these cations have as strong an effect on crystallisation as the anions, despite the fact that lysozyme is positively charged at neutral pH. Specific co-ion binding of cations, not only of anions, is found to be the origin of this phenomenon.

So the first rule that we can infer is: both cations and anions interact with proteins. Specificities occur in both cases. Most importantly, despite the electrostatic repulsions, co-ions can bind, sometimes significantly and very strongly. Collins called it 'electrostatics-defying charge interactions in biological macromolecules'. As a consequence, other forces, beyond electrostatic ones, must be dominant.

For example, anions can bind directly to electronegative pockets located around amino groups in nucleic acids.[100] Even more surprising are the phosphor-binding proteins, in which 'the ion-dipole interactions between the phosphate and dipolar groups compensate the ligand's isolated negative charges. Moreover, the surprise finding that the electrostatic surface in and around the cleft is intensely negative demonstrates the power of ion-dipole interactions in anion binding and electrostatic balance'.[101] It seems that phosphates buried inside the proteins are fixed by neutral amino acids, whereas closer to the protein 'surface' electrostatic interactions with cationic groups dominate.[102,a]

---

[a]These examples were presented by Kim D. Collins at a meeting on specific ion effects in Munich, September 2008.

All these results are very puzzling and complicated. As early as in 1911, this complication was already discussed in some details by Robertson,[103] he wrote:

> The influence of added salts of the alkalies and magnesium upon the precipitation of proteins by heavy metals varies with the concentration of salt employed (Pauli). At low concentrations (0.005 M) the salts inhibit precipitation in the order:
>
> $$SO_4^{2-} < Cl^- < C_2H_3O^- < NO_3^- < Br^- < I^- < SCN^-,$$
>
> while in high concentrations (4 M) they encourage precipitation (or coagulation) in the order:
>
> $$SO_4^{2-} > Cl^- > C_2H_3O^- > NO_3^- > Br^- > I^- > SCN^-.$$
>
> This is simply a particular instance of the general rule that the salts may act as precipitants and as coagulants at low and at high concentrations respectively, acting as solvents at intermediate concentrations. The heavy metal salts afford no exception to this rule. At low concentrations they precipitate, at higher concentrations they dissolve and at still higher concentrations they coagulate the proteins of egg white (Pauli). The concentration range throughout which the salt acts as solvent may be evanescent however, as it is in the case of silver nitrate acting upon egg-albumin.

Here a second rule appears: obviously, the dominating forces are concentration (and temperature and pH) dependent. Different effects dominate at low and high salt concentrations.

The subtleties concerning the balance between different effects emerge from the papers by Arakawa, Timasheff, and co-workers.[104–107] In most systems, proteins are indeed stabilised by salting-out ions and destabilised by salting-in ions, but sometimes preferential hydration does not stabilise the native structure of proteins, as would be supposed according to the Hofmeister series. Moreover, these effects seem to crucially depend on the salt concentration, the nature of the proteins (basic or acidic), the pH values, and so forth. It should be noted that these authors used techniques such as densitometry and inferred interaction parameters using relatively simple models. Thus, it seems difficult to make conclusions about general rules on molecular level. Curtis *et al.*[108,109] suggest an appealing interpretation of these apparently contradicting results. They forward a competition between Hofmeister and electroselectivity effects. The direct, lyotropic Hofmeister series concerns the structuring of the water molecules around the ions, which in turn leads to the stabilisation or destabilisation of proteins by salting-out and salting-in ions, respectively. By contrast, the electroselectivity series reflects the affinity of the ions for an anion-exchange resin. For monovalent ions, the electroselectivity series is suggested to be equivalent to the inverse Hofmeister series.

With these two opposing effects, the concentration dependence of ion effects can be explained, as well as the apparently contradictory behaviour of sulfate ions towards different proteins: if the sulfate interaction with the protein 'surface' is weak, sulfate strengthens the immediate water structure around the ion and acts as a stabilising salting-out ion, according to the direct lyotropic Hofmeister series. By contrast, if the sulfate ions can bind to the protein, the protein behaves like an anion-exchange resin and the sulfate ion must be classified according to the electroselectivity series. Indeed the electroselectivity series is often found to be the adequate description of ion–protein interactions. From the literature, it seems that at low salt concentrations, the electroselectivity series is prevailing, whereas at higher salt concentration, a competition is found between the two counteracting effects that are at the origin of the two series. In some cases, even a 'bellshaped' Hofmeister series was found.[110]

Yet another complication remains mysterious: hard anions are usually considered as salting-out and soft anions as salting-in, whereas for cations it is just the other way round — soft cations are salting-out and 'hard cations' salting-in. Again Collins' concept of matching water affinities could give an explanation. Proteins mainly have hard carboxylates as anionic groups, and soft amino ones as cationic groups. According to this model, soft cations would not interact with the proteins, but simply 'withdraw' water. Hard cations, by contrast, would bind more efficiently to the carboxylates, screen the electrostatics and salt the proteins in. For anions, it is just the opposite. This concept would also explain why sodium pumps are so important for living cells. In the inside, sodium would bind much more strongly than potassium to the carboxylate groups, which would lead to denaturation of the proteins.

In most papers where ion–protein interactions are studied, the interactions turn out to be highly specific. Chakrabarti[111] wrote a review in which he analysed the binding of sulfate and phosphate ions in 34 different protein structures. He listed in detail which amino acids are involved in the binding process and even gave the probable geometries of the amino acid–ion configurations and the hydrogen bonds between the ion and the amino acids, as well as between ions and the surrounding water molecules.

With so many specificities, it is clear that a simplified model of a uniformly charged protein 'surface' is not appropriate to describe the relevant interaction effects. On the other hand, some general features indeed exist. For example, knowing the amino acid residues that are accessible to bulk water and ions, it should be possible to predict the influence of each salt on protein stability at a given salt concentration. The term 'salt' includes also the buffer composition, in which the protein is immersed. Indeed, Kim *et al.*[112] showed in an elegant work that the enzymatic activity depends on the chosen buffer, even when the pH is the same for different buffers used.

A particular interesting and important protein is collagen. It is the main constituent of leather. For centuries, people use salts in the conversion process of animal hides to leather. The addition of salts 'pickling'[113] is necessary to prevent undesirable swelling of collagen in the preparation steps before tanning.

Still today, it is mostly an empirical process. Whereas chromium fixation is widely understood, this is not the case for the way how simple salts like NaCl, $CaCl_2$ or Na_2SO_4 work on a molecular level. What is known is that these salts can inhibit the swelling of collagen in acidic environment (NaCl) or stabilise the collagen structure at neutral pH ($CaCl_2$). Na_2SO_4 is even a very good stabiliser. Recently, Bulo *et al.*[114] could simulate these effects in a MD study. They measured the degree of swelling of the collagen fibril via the determination of the surface-accessible area. At low pH, where the pickling process was considered, anion effects prevail. Direct binding to the ammonium groups of arginine and lysine occur. These groups are soft, and according to Collins' matching water affinities, these groups should preferably bind to weak ions like SCN^- or I^- and less to Cl^-. Out of these ions, Bulo *et al.*[114] only simulated Cl^- and indeed they found only week binding. According to Collins, sulfate as a hard ion should not bind strongly. However, here the double charge is crucial, so that sulfate binds more strongly than Cl^-.

Komsa-Penkova *et al.*[115] measured the denaturation temperature of calf skin collagen type I as a function of added salt at pH 3. The results are shown in Fig. 15. Roughly three different domains can be distinguished. At very low salt concentration all salts are salting-in. The possible reason is that the charged ions screen the stabilising electrostatic interactions between close amino acid residues that have opposite charges. At higher concentrations the typical specific ion effects appear, mostly reflecting

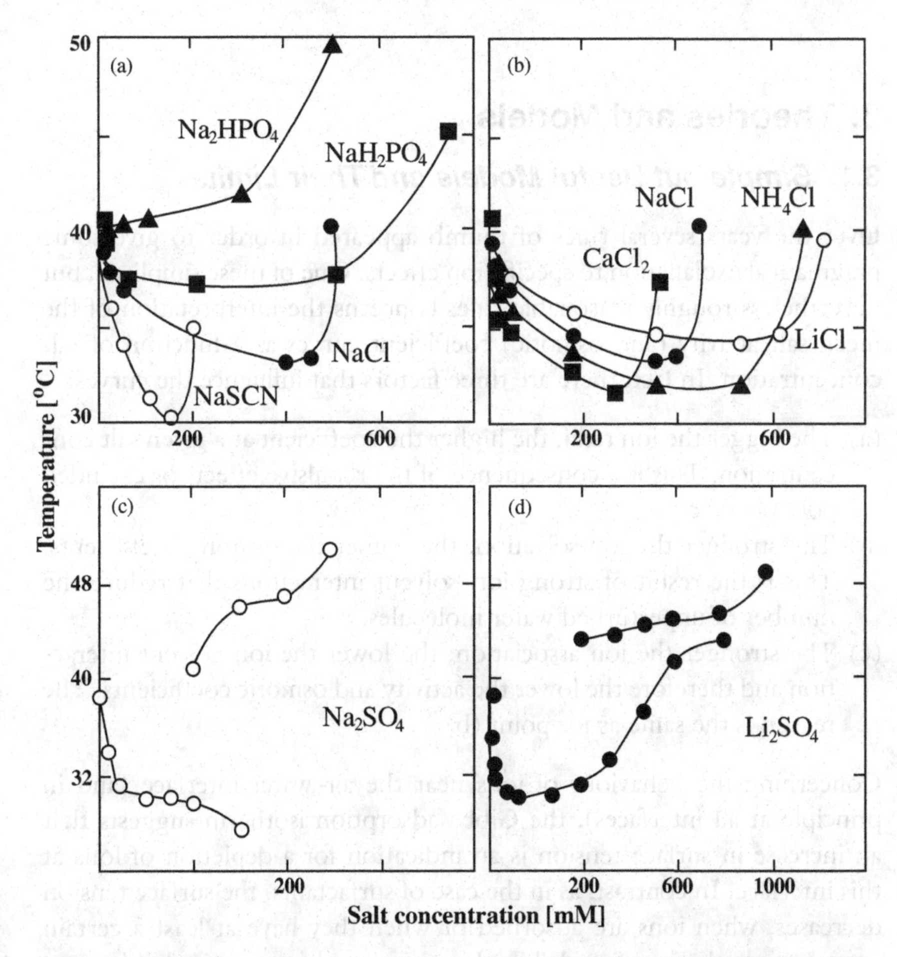

Fig. 15. Dependence of denaturation temperature of calf skin collagen type I on the salt concentration: (a) selection of sodium salts, (b) selection of chloride salts, (c) and (d) sulfates showing two different denaturation temperatures. (After Komsa-Penkova *et al.*[115])

direct ion binding to charged protein groups. And finally at still higher concentration, all salts become salting-out, simply because Hofmeister's 'water withdrawing' effect dominates, i.e. ion–water interactions are strongest for the hardest anions.

Of course, this is a somewhat simplified picture. In real systems these effects are always coupled. Nevertheless, it gives a clear hint of the molecular reasons for the Hofmeister series and elucidates that more than one effect must be taken into account to explain such series.

3. Theories and Models

3.1. *Simple but Useful Models and Their Limits*

Over the years several rules of thumb appeared in order to give some pragmatical explanation to specific ion effects. One of these simplistic, but nevertheless roughly reasonable rules concerns the interpretation of the mean salt activity (and osmotic) coefficient curves as a function of salt concentration. In fact, there are three factors that influence the curves:

(a) The bigger the ion radii, the higher the coefficient at a given salt concentration. This is a consequence of the repulsive effects or excluded volume effects.
(b) The stronger the ion solvation, the higher the osmotic coefficients. This is the result of strong ion–solvent interactions that reduce the number of unperturbed water molecules.
(c) The stronger the ion association, the lower the ion–solvent interaction and therefore the lower the activity and osmotic coefficients. The reason is the same as for point (b).

Concerning the behaviour of ions near the air–water interface (and in principle at all interfaces), the Gibbs adsorption isotherm suggests that an increase in surface tension is an indication for a depletion of ions at this interface. In contrast, as in the case of surfactants, the surface tension decreases, when ions are adsorbed or when they have at least a certain propensity to the interface. While this is rigorously correct in the thermodynamic limit (provided that the activity coefficients are taken into account correctly), caution must be taken if molecular interpretation is attempted. It is one of the great merits of Jungwirth and Tobias' work[116] to show that ion can indeed be *adsorbed* to interfaces, although the interfacial tension *increases.* At the beginning of our century, there was some confusion about this point. Today this apparent contradiction is well explained: the Gibbs adsorption isotherm is an integral over the total ion profiles from the surface until the pure bulk solution. However, this does not mean that in the very first molecular layers near the interface, depletion must occur. There can be a first (or second) adsorption layer followed by further layers that are depleted from ions. If this depletion prevails, a decreased surface tension can occur, despite adsorption of ions in the first molecular layers near the interface.

In the previous sections we repeatedly mentioned the general rule 'like seeks like'. But what is the reason? Kim D. Collins published a landmark paper[30] to elucidate and explain this phenomenon without using sophisticated theories. We quote here his main idea that he summarised in the following rule: *The law of matching water affinities: oppositely charged ions in free solution form inner sphere ion pairs spontaneously only when they have equal water affinities.*

The idea starts from the following simple reasoning: we can consider simple ions as a sphere with a point charge in the centre. When the ions are small, the surrounding water molecules are tightly bound (the ions are hard or kosmotropic), whereas when the ions are big, the hydration shell is only loosely bound (the ions are soft or chaotropic). The discrimination between both types comes from the relative strength of the ion–water interactions compared to water–water interactions, as illustrated in Fig. 16.

It turns out that this is a very valuable concept for the classification of ions. In the next step comes an explanation for the general rule 'like seeks like': two strongly hydrated small ions of opposite charge experience a very strong reciprocal attraction. Consequently, they can come together forming direct ion pairs and expelling the hydration spheres between them. In the case of weakly hydrated soft ions, the situation is very different, but the *result* is the same: the electrostatic attraction between them is much smaller than between kosmotropes, however, the hydration spheres are so loosely bound that the chaotropic ions can also form direct ion pairs expelling also the hydration water between them. Now comes the case of one hard and one oppositely charged soft ion — here the attraction by the soft ion is not strong enough that the hard ion loses its hydration shell. As a consequence, a soft/hard ion pair is always separated by water and cannot form

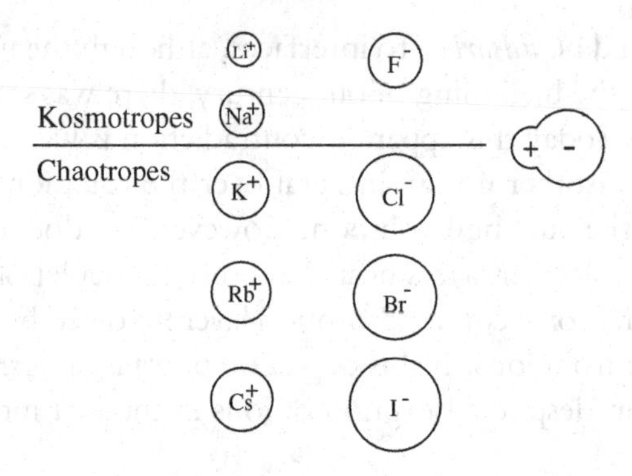

Fig. 16. Division of the group IA cations and the VIIA halide anions into [strongly hydrated] kosmotropes (water structure makers) and [weakly hydrated] chaotropes (water structure breakers). The ions are drawn approximately to scale. A virtual water molecule is represented by a zwitterion of radius 1.78 Å for the anionic portion and 1.06 Å for the cationic portion. In aqueous solution, Li^+ has 0.6 tightly attached water molecules, Na^+ has 0.25 tightly attached water molecules, F^- has 5.0 tightly attached water molecules, and the remaining ions have no tightly attached water molecules. (After Collins.[30])

strong ion pairs. This explanation of 'like seeks like' is illustrated in Fig. 17.

Collins' concept of matching water affinities turns out to be very general. For example it nicely explains the different heats of solution discussed in Sec. 2.1.2. In his original paper, Collins gives much more examples, e.g. the Jones–Dole viscosity coefficients of electrolyte solutions. Recently, Jungwirth and his group could confirm this concept by molecular dynamics simulation. Although such simulations are always based on interaction potentials that are not exactly known, the results seem to be sufficiently robust so that we can have some confidence in this coincidence between Collins' model and Jungwirth's approach.[45,117]

On the other hand, and as it is further discussed in Chap. 11, one should always keep in mind that this concept of matching water affinities is a very simplified one that should only be taken as a rule of thumb. In reality, there is certainly a subtle balance between different states of hydration, and like ions may simply tend to be a little bit more attracted than a combination of hard and soft ions.

So far only 'simple' ions are discussed. However, it would be useful to have a Hofmeister series also for charged headgroups that frequently

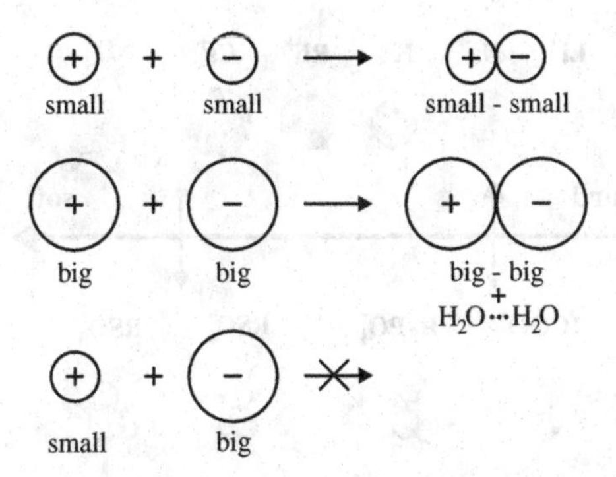

Fig. 17. Ion size controls the tendency of oppositely charged ions to form inner sphere ion pairs. Small ions of opposite sign spontaneously form inner sphere ion pairs in aqueous solution; large ions of opposite sign spontaneously form inner sphere ion pairs in aqueous solution; and mismatched ions of opposite sign do not spontaneously form inner sphere ion pairs in aqueous solution. A large monovalent cation has a radius larger than 1.06 Å; a large monovalent anion has a radius larger than 1.78 Å. (After Collins.[30])

occur in colloidal chemistry and biology. For this purpose Jungwirth and his group carried out simulations on alkyl sulfates, sulfonates, phosphates, and carboxylates and also on ammonium headgroups. Taking into account the results obtained by Harrison *et al.* (c.f. Sec. 2.3.2),[58] we can extend the expected interactions also to a series of cationic headgroups and negatively charged counterions. The following Fig. 18 shows the result.

Carboxylate turns out to be the hardest of all considered anionic headgroups, whereas sulfate and sulfonate are soft headgroups. The only and single parameter to distinguish them is their *charge density* and the resulting strength of the ion–water interaction compared with water–water interactions. Very probably, other effects may also play a role, such as the detailed geometry of the headgroups, the resulting water geometry and ion polarisabilities. However, Collins' law of matching water affinities together with the headgroup charge densities qualitatively explain all examples that were given in the preceding sections, especially those concerning micelles, polyelectrolytes, vesicles, etc., as discussed in Sec. 2.3.3.

As a conclusion of this section, we may note that now there is indeed a simple and sufficiently universal rule so that many specific ion effects in nature can be qualitatively predicted.

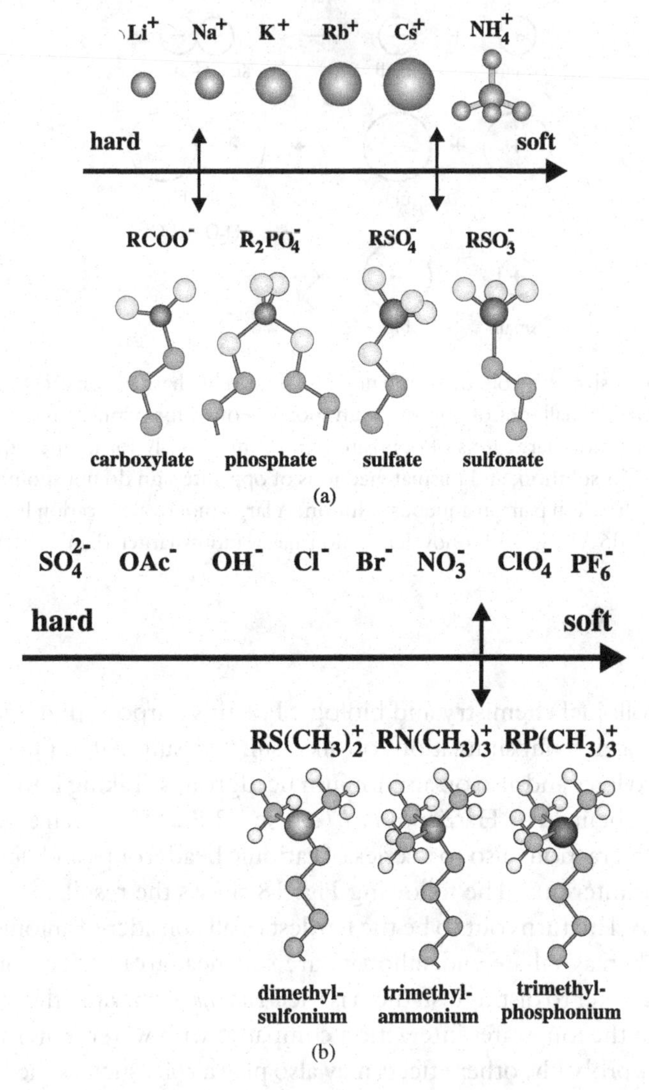

Fig. 18. 'Like seeks like': hard headgroups preferentially interact with hard counterions; soft headgroups preferentially interact with soft counterions.[92] (a) Cations with negatively charged headgroups, and (b) anions with positively charged headgroups.

3.2. *Primitive Model Calculations*

Properties of electrolytes have been modelled for roughly one century. The reason is that at the beginning of the last century, precise experimental data became available, challenging the theoreticians to describe them. However, the considerable success of the Debye–Hückel theory (and especially the universal limiting laws) and all further work based on it induced people to overestimate the electrostatic interactions and to neglect the solvent structure and properties. Furthermore, the model is limited to low concentrations and serious theoretical inconsistencies exist. Nevertheless, the so-called 'primitive models' still have their merits, as will be further discussed in Chap. 12.

Primitive models consider the solvent as a continuum with a dielectric constant or sometimes a frequency- or distance-dependent dielectric function.[17] The detailed structure of the water molecules and their possible geometrical rearrangement around ions are not or only very roughly taken into account. The ion specificity is considered through the charges and the radii of the ions. Since these two parameters define the charge density, they might be used to describe properly specific ion effects. However, since the ion–water interactions are modelled in a very crude way, the relative importance of ion–ion and ion–water interactions is difficult to estimate.

Therefore, so-called 'decorated' models are used. In a famous paper that appeared in 1997, Ninham and Yaminsky[118] proposed the consideration of dispersion forces in addition to electrostatic interactions and short-range repulsive interactions coming from the finite volume of the ions. To do so, they based their model on the Poisson–Boltzmann (PB) equation. The dispersion forces were derived from a Lifshitz-like approach. The merit of Ninham and Yaminsky's idea is to bring in quite naturally ion specificity and to explain also differences of ion behaviour near different interfaces, such as at the air–water and the water–oil interfaces. However, in the meantime it turned out that this approach is oversimplified. Especially, the neglect of the water geometry around ions and close to surfaces is a serious problem that can even lead to qualitatively wrong thermodynamic results.

Manciu and Ruckenstein[119] followed a pragmatical way: to describe the behaviour of ions at the air–water interface, they added a further empirical interaction term to the electrostatic interaction potential with parameters that reproduce the results from MD simulations. Of course, this is a valuable alternative for describing the data, but it does not give deeper insight in the origin of ion specificity.

Over the last decades, many more approaches appeared, all trying to reproduce specific ion effects, especially as they are reflected in

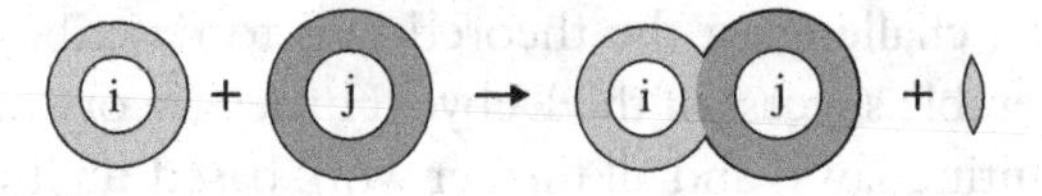

Fig. 19. Illustrating the effect of co-sphere overlap as the interionic distance gets small. The solvent displace by overlap is assumed to return to the normal bulk water state. The free energy change per mole in this process is A_{ij}. (After Ramanathan and Friedman.[123])

thermodynamic[120,121] and transport data, such as electrical conductivity, diffusion coefficients, etc.[35,36,122] Today, it seems that the most physical model is one of the oldest: the Friedman–Gurney (FG) model of electrolyte solutions.[123] This model, developed in the late 1960s, is in good agreement with Collins' concept of matching water affinities, as it was discussed in the preceding section. We briefly recall here the main features of the FG model.

In addition to the electrostatic and short-term interaction potential (and a negligible cavity potential), the ions i and j interact via a so-called Gurney potential $u_{ij}^{\mathrm{GURNEY}}(r)$. This term is

$$u_{ij}^{\mathrm{GURNEY}}(r) = A_{ij} V^{\mathrm{mutual}} / V^{\mathrm{water}}, \tag{9}$$

where V^{mutual} is a mutual volume function deduced by geometrical considerations and V^{water} is the molar volume of the pure solvent. A_{ij} is the molar free energy change of the co-sphere solvent returning to its normal state, when the two ions approach each other and part of the intermediate water is expelled, when the hydration spheres of the ions overlap. This process is illustrated in Fig. 19.

This means that ions can form close-contact ion pairs, if A_{ij} is significantly negative. They will not lose their hydration spheres when A_{ij} is zero or even positive. Therefore it can be expected that the description of thermodynamic properties of hard–hard and soft–soft ion couples require negative A_{ij} values and hard–soft combinations require non-negative ones.

In order to get these values, Ramanathan and Friedman[123] and later on Ramanathan, Krishnan and Friedman[124] used the hypernetted chain theory (HNC). This integral equation theory is supposed to be a somewhat better theoretical approach than PB at the primitive model level, but this is not really important in this context. What matters is that the following values were obtained for A_{+-} (the hydration overlap energy when a cation and an anion come together) in cal/mole (list not exhaustive):

NaF: −5; KF: 0; RbF: 12; LiCl: 100; NaCl: −14; KCl: −67;
RbCl: −84; LiI: 95; NaI: −25; KI: −82; CsI: −114;
Et_4NBr: −427; $Prop_4NCl$: −84; $Prop_4NBr$: −293;
$Prop_4NI$: −523; Bu_4NCl: 0; Bu_4NBr: −247.

It should be noted that the A_{ij} values are the only fitting parameters used to describe the osmotic coefficients as a function of concentration at 25°C. Due to the insufficiencies of both the interaction potentials and the primitive model HNC calculation, the values cannot quantitatively reflect the reality. But *qualitatively* the series is in nice agreement with Collins' interpretation of 'like seeks like'.

3.3. *Non-primitive Models and Simulations*

From the preceding sections, it seems evident that a real description of ion specificities in solutions can only be done if the geometry and the properties of water molecules are explicitly taken into account. Such models are called non-primitive or Born–Oppenheimer models. In the 1970s and 1980s, they were developed in two different directions. In particular, integral equation theories, such as the hypernetted chain (HNC) approach, were extended to include angle-dependent interaction potentials. The site–site Ornstein–Zernike equation with a HNC-like closure[125] and the molecular Ornstein–Zernike equation[126] are examples. For more information, see Ref. 17.

The second direction chosen to describe electrolyte solutions was the development of computer simulations such as Molecular Dynamics (MD) or Monte Carlo simulations.[17] Already 20 years ago, the detailed hydration sphere around ions was studied.[127] Of course, as with the integral equations, the result crucially depends on the chosen potential.

Today, such simulations are common and widely used, in contrast to integral equations theories that are much less popular, although new and interesting applications of HNC-like equations have been published recently.[128] The simulation codes can even be downloaded from freely accessible web sites and then adapted to specific problems such as specific ion effects.

The simulations of Jungwirth and Tobias at the beginning of this century[129–131] are landmark contributions to the understanding of ion effects at interfaces and initiated a Renaissance of this research topic. One of the reasons is that these authors study ion effects at interfaces and this topic is of considerable interest with lots of applications. The second and main reason is that their — at least qualitatively — explanation of surface tension goes well beyond the Gibbs adsorption equation.[116] Thanks to the work of Jungwirth and Tobias' group, it is now commonly accepted that soft ions can *locally* adsorb at the air–water interface, whereas *globally* depletion is observed. With other words, a certain propensity of ions to the surface is not in contradiction with an increase in surface tension, as already discussed (see Secs. 2.2.1 and 3.1).

Figure 20 is a typical example of ion profiles as obtained by Jungwirth and Tobias. Similar results were obtained by Dang *et al.*[132]

However, it is still a matter of debate, whether the potentials used by these authors are reliable. It cannot be excluded that this particular behaviour of the ions at interfaces is 'simply' a consequence of their charge density and only to a minor or negligible extent due to ion polarisability. It seems that reliable information about ion specificities at interfaces can also be obtained without taking into account ion polarisabilities.[133] In any case, the influence of the ions on the water structure at the interface is a major reason for the propensity of the ions to the interface, be it caused by polarisation effects or mainly size and charge density. More details are discussed in Chaps. 9–11.

To end this section, an interesting alternative simulation method should be mentioned.[134] The authors use the so-called Mercedes Benz (MB) model of water, a simple two-dimensional statistical mechanical model incorporated in a Monte Carlo simulation. Surprisingly, this simple model reproduces well several features of ion hydration and can even describe various experimental specific ion effects.

3.4. *Ab Initio* Calculations and Water Clusters Around Ions

There is a huge number of papers dealing with quantum mechanical calculations of hydrated ions. Still today, these calculations are limited in size due to restrictions in computer power. Therefore, a compromise is made. In order to have reliable interaction potentials, there is attempt

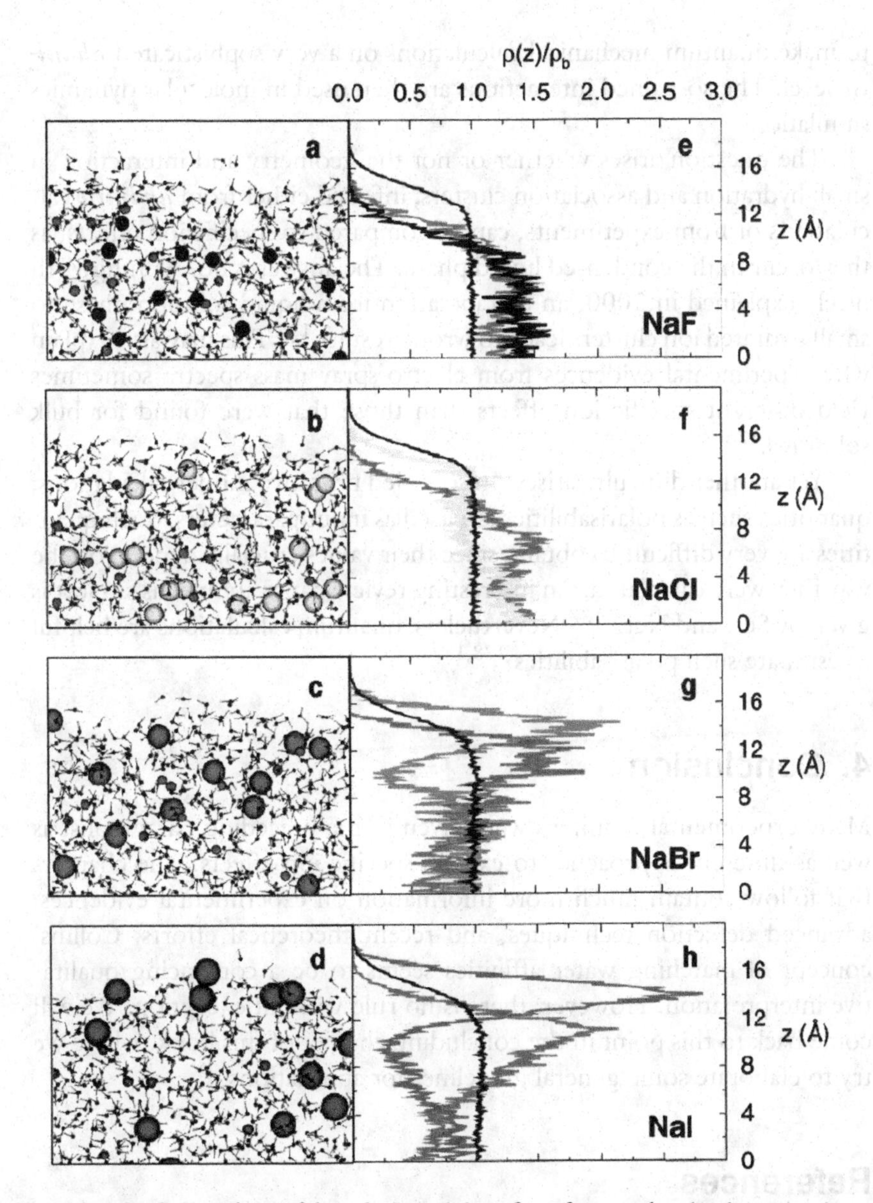

Fig. 20. (a–d) Snapshots of the solution–air interfaces from molecular dynamics simulations. Colouring scheme: water oxygen, blue; water hydrogen, grey; sodium ions, green; chloride ions, yellow; bromide ions, orange; iodide ions, magenta. (e–h), Number densities, $\rho(z)$, of water oxygen atoms and ions plotted vs distance from the center of the slabs in the direction normal to the interface (z), normalized by the bulk water density, ρb. The ion densities have been scaled by the water–ion concentration ratio of 48 for ease of comparison. The colours of the curves correspond to the colouring of the atoms in the snapshots. (Reprinted with Permission from Ref. 129. Copyright 2001 American Chemical Society.)

to make quantum mechanical calculations on a very sophisticated *ab initio* level. The so-gained interactions are then used in molecular dynamics simulations.

The question arises whether or not the geometry and interaction in small hydration and association clusters, inferred either from *ab initio* calculations or from experiments, can be compared to specific ion effects, as they occur in the condensed liquid phase. The answer is no. As Jungwirth nicely explained in 2000, an extrapolation from liquid solution phase to small hydrated ion clusters leads to wrong results.[135] This may also explain why experimental evidences from electro-spray mass spectra sometimes yield different specific ion effects than those that were found for bulk solutions.[136]

Yet another difficulty arises. In so-called primitive model calculations, quantities such as polarisabilities are used as input parameters. Such quantities are very difficult to obtain, since their values crucially depend on the way they were calculated. An interesting review paper about this subject is given by Serr and Netz.[137] Nevertheless, quantum calculations are helpful to estimate such polarisabilities.[138,139]

4. Conclusion

Many experimental examples were given in the preceding paragraphs, as well as different approaches to explain specific ion effects. The chapters that follow contain much more information on experimental evidences, advanced detection techniques, and recent theoretical efforts. Collins' concept of matching water affinities seems to be a convincing qualitative interpretation. However, there is no rule without exception. We will come back to this point in the concluding chapter of this book, where we try to elaborate some general guidelines for applications.

References

1. Lewith S. (1887) *Arch exp Path Pharm* **24**: 1–16.
2. Hofmeister F. (1887) *Arch exp Path Pharm* **24**: 247–260.
3. Hofmeister F. (1888) *Arch exp Path Pharm* **25**: 1–30.
4. Limbeck Rv. (1888) *Arch exp Path Pharm* **25**: 69–86.
5. Hofmeister F. (1890) *Arch exp Path Pharm* **27**: 395–413.
6. Hofmeister F. (1891) *Arch exp Path Pharm* **28**: 210–238.

7. Münzer E. (1898) *Arch exp Path Pharm* **41**: 74–96.
8. Kunz W, Henle J, Ninham BW. (2004) *Curr Op Coll Interf Sci* **9**: 19–37.
9. Poiseuille JL. (1847) *Ann Chim Phys* **21**: 76–110.
10. Jones G, Dole M. (1929) *J Am Chem Soc* **51**: 50–64.
11. Cox WM, Wolfenden JH. (1934) *Proc R Soc Lond* **A145**: 475–488.
12. Frank H, Evans MW. (1945) *J Chem Phys* **13**: 507–532.
13. Voet A. (1937) *Chem Rev* **20**: 169–179.
14. Eckfeldt EL, Lucasse WW. (1943) *J Phys Chem* **47**: 183–189.
15. Pearson RG. (1963) *J Am Chem Soc* **85**: 3533–3539.
16. Bunton CA, Nome F, Quina FA, Romsted LS. (1991) *Acc Chem Res* **24**: 357–364.
17. Barthel J, Krienke H, Kunz W. (1998) *Physical Chemistry of Electrolyte Solutions — Modern Aspects.* Steinkopff, Darmstadt, Springer, New York.
18. Marcus Y. (1997) *Ion Properties.* Marcel Dekker, New York.
19. Marcus Y. (1985) *Ion Solvation.* Wiley, Chichester.
20. Marcus Y, Hefter G. (2004) *Chem Rev* **104**: 3405–3452.
21. Marcus Y, Hefter G. (2006) *Chem Rev* **106**: 4585–4621.
22. Westhaus U, Sass R. (2004) *Z Phys Chem* **218**: 1–8.
23. http://www.dechema.de/en/detherm.html
24. Robinson RA, Stokes RH. (1970) *Electrolyte Solutions.* Second Revised Edition, Butterworth, London.
25. Pitzer KS (ed.). (1991) *Activity Coefficients in Electrolyte Solutions.* 2nd ed., CRC Press, Boca Raton, Florida.
26. Hamer WJ, Wu YC. (1972) *J Phys Chem Rev Data* **1**: 1047–1099.
27. Robinson RA, Harned HS. (1941) *Chem Rev* **28**: 419–476.
28. Lyklema J. (2003) *Adv Coll Interf Sci* **100–102**: 1–12.
29. Gibbard HF, Scatchard G, Rousseau RA. (1974) *J Chem Eng Data* **19**: 281–288.
30. Collins KD. (2004) *Methods* **34**: 300–311.
31. Morris DFC. (1969) *Struct Bond* **6**: 157–159.
32. Simonin JP, Bernard O, Papaiconomou N, Kunz W. (2008) *Fluid Phase Equilib.* **264**: 211–219.
33. Voinescu A, Bauduin P, Pinna C, Touraud D, Kunz W, Ninham BW. (2006) *J Phys Chem B* **110**: 8870–8876.
34. Jenkins HDB, Marcus Y. (1995) *Chem Rev* **95**: 2695–2724.
35. Bernard O, Kunz W, Turq P, Blum L. (1992) *J Phys Chem* **96**: 398–403.
36. Bernard O, Kunz W, Turq P, Blum L. (1992) *J Phys Chem* **96**: 3833–3840.
37. Long FA, McDevit WF. (1952) *Chem Rev* **51**: 119–169.
38. Xie WH, Su JZ, Xie XM. (1990) *Thermochim Acta* **169:** 271–286.
39. Tsurko, EN, Bondarev NV. (2004) *J Molec Liq* **113**: 29–36.
40. Tsurko EN, Neueder R, Kunz W. (2007) *J Solution Chem* **36**: 651–672.
41. Lo Nostro P, Ninham BW, Dilani S, Fratoni L, Baglioni P. (2006) *Biopolymers* **2**: 136–148.
42. Pitzer KS, Kim JJ. (1974) *J Am Chem Soc* **96:** 5701–5707.
43. Weissenborn PK, Pugh RJ. (1995) *Langmuir* **11**: 1422–1426.
44. Collins KD, Washabaugh MW. (1985) *Q Rev Biophys* **18**: 323–422.

45. Jungwirth P, Winter B. (2008) *Annu Rev Phys Chem* **59**: 343–366.
46. Petersen PB, Saykally RJ. (2005) *J Phys Chem B* **109**: 7976–7980.
47. Petersen PB, Saykally RJ. (2006) *Annu Rev Phys Chem* **57**: 333–364.
48. Randles JEB. (1977) *Phys Chem Liq* 7: 107–179.
49. Harned HS, Hamer WJ. (1933) *J Am Chem Soc* **55**: 2194–2205.
50. Schwabe K, Suschke HD. (1964) *Angew Chem* **76**: 39–49.
51. Kron I, Marshall SL, May PM, Hefter G, Königsberger E. (1995) *Monatshefte für Chemie* **126**: 819–837.
52. Bauduin P, Touraud D, Kunz W, Savelli MP, Pulvin S, Ninham BW. (2005) *J Coll Interf Sci* **292**: 244–254.
53. Washabaugh S, Collins KD. (1986) *J Biol Chem* **261**: 12477–12485.
54. Umemura T, Kamiya S, Itoh A, Chiba K, Haraguchi H. (1997) *Anal Chim Acta* **349**: 231–238.
55. Cook HA, Hu W, Fritz JS, Haddad PR. (2001) *Anal Chem* **73**: 3022–3027.
56. Roberts JM, Diaz AR, Fortin DT, Friedle JM, Piper SD. (2002) *Anal Chem* **74**: 4927–4932.
57. Cook HA, Dicinoski GW, Haddad PR. (2003) *J Chromat A* **997**: 13–20.
58. Harrison CR, Sader JA, Lucy CA. (2006) *J Chromat A* **1113**: 123–129.
59. Lindman B, Ekwall P. (1969) *Kolloid Z Z Polym* **234:** 1115–1123.
60. Lindblom G, Lindman B, Tiddy GJT. (1978) *J Am Chem Soc* **100**: 2299–2303.
61. Stilbs P, Lindman B. (1982) *J Mag Res* **48**: 132–137.
62. Fabre H, Kamenka N, Khan A, Lindblom G, Lindman B, Tiddy GJT. (1980) *J Phys Chem* **84**: 3428–3433.
63. Romsted LS. (2007) *Langmuir* **23**: 414–424.
64. Schulz JC, Warr GG. (1998) *J Chem Soc Faraday Trans* **94**: 253–257.
65. He Z, O'Connor PJ, Romsted LS, Zanette DJ. (1989) *J Phys Chem* **93**: 4219–4226.
66. Hafiane A, Issid I, Lemordant D. (1991) *J Colloid Interface Sci* **142**: 167–178.
67. Haverd VE, Warr GG. (2000) *Langmuir* **16**: 157–160.
68. Rosano HL. (1967) *J Coll Interf Sci* **24**: 73–79.
69. Brady JE, Evans DF, Warr GG, Grieser F, Ninham BW. (1986) *J Phys Chem* **90**: 1853–1859.
70. Ninham BW, Evans DF. (1986) *Faraday Discuss Chem Soc* **81**: 1–17.
71. Ninham BW, Evans DF, Wel GJ. (1983) *J Phys Chem* **87**: 5020–5025.
72. Kang C, Khan A. (1993) *J Colloid Interface Sci* **156**: 218–228.
73. Regev O, Khan A. (1994) *Progr Colloid Polym Sci* **97**: 298–301.
74. Regev O, Kang C, Khan A. (1994) *J Phys Chem* **98**: 6619–6625.
75. Morgan JD, Napper DH, Warr GG. (1995) *J Phys Chem* **99**: 94.
76. Ti Tien H. (1964) *J Phys Chem* **68**: 1021–1025.
77. Claessens MMAE. (2003) *Size Regulation and Stability of Charged Lipid Vesicles.* Doctoral dissertation, Wageningen University, The Netherlands.
78. Vlachy N, Drechsler M, Touraud D, Kunz W. (2009) *C R Chimie* **12**: 30–37.
79. Fuoss RM, Katchalsky A, Lifson S. (1951) *Proc Natl Acad Sci USA* **37**: 579–589.
80. Manning GS. (1969) *J Chem Phys* **51**: 924–933.

81. Strauss UP. (1974) In: Sélégny E (ed.), *Polyelectrolytes*. pp. 79–85. D. Reidel Publishing Company, Dordrecht-Holland.
82. Strauss UP, Leung YP. (1965) *J Am Chem Soc* **87**: 1476–1480.
83. Gregor HP, Frederick M. (1957) *J Polymer Sci* **23**: 451–465.
84. Dolar D. (1974) In: Sélégny E (ed.), *Polyelectrolytes*. pp. 97–113. D. Reidel Publishing Company, Dordrecht-Holland.
85. Lipar I, Zalar P, Pohar C, Vlachy V. (2007) *J Phys Chem B* **111**: 10130–10136.
86. Daoust H, Chabot MA. (1980) *Macromolecules* **13**: 616–619.
87. Hutschneker K, Deuel H. (1956) *Helv Chim Acta* **39**: 1038–1045.
88. Bungenberg de Jong HG. (1949) In: Kruyt HR (ed.), *Colloid Science Vol. II*. Ch. 9, pp. 335–384. Elsevier, New York.
89. Eisenman G. (1961) In: Kleinzeller A, Kotyk A. (eds.), *Symposium on Membrane Transport and Metabolism*, pp. 163–179. Academic Press, New York.
90. Chialvo AA, Simonson JM. (2005) *J Phys Chem B* **109**: 23031–23042.
91. Druchok M, Hribar-Lee B, Krienke H, Vlachy V. (2007) *Chem Phys Lett* **450**: 281–285.
92. Vlachy N, Jagoda-Cwiklik B, Vácha R, Touraud D, Jungwirth P, Kunz W. (2009) *Adv Coll Interf Sci* **146**: 42–47.
93. Garcia-Celma JJ, Hatahet L, Kunz W, Fendler K. (2007) *Langmuir* **23**: 10074–10080.
94. Petrache HI, Zemb T, Belloni L, Parsegian VA. (2006) *Proc Nat Acad Sci USA* **103**: 7982–7987.
95. Leontidis E, Aroti A, Belloni L, Dubois M, Zemb T. (2007) *Biophys J* **93**: 1591–1607.
96. Aroti A, Leontidis E, Dubois M, Zemb T, Brezesinski G. (2007) *Coll Surf A* **303**: 144–158.
97. Zemb T, Belloni L, Dubois M, Aroti A, Leontidis E. (2004) *Curr Op Coll Interf Sci* **9**: 74–80.
98. Carbonnaux C, Ries-Kautt M, Ducruix A. (1995) *Prot Sci* **4**: 2123–2128.
99. Bénas P, Legrand L, Ries-Kautt M. (2002) *Acta Cryst* **D58(10)**: 1582–1587.
100. Auffinger P, Bielecki L, Westhof E. (2004) *Structure* **12**: 379–388.
101. Vyas NK, Vyas MN, Quiocho FA. (2003) *Structure* **11**: 765–774.
102. Hirsch AKH, Fischer FR, Diederich F. (2007) *Angew Chem Int Ed* **46**: 338–352.
103. Robertson TB. (1911) *J Biol Chem* **9**: 303–326.
104. Arakawa T, Timasheff SN. (1982) *Biochem* **21**: 6545–6552.
105. Timasheff SN, Arakawa T. (1988) *J Cryst Growth* **90**: 39–46.
106. Arakawa T, Timasheff SN. (1990) *Biochem* **29**: 1924–1931.
107. Arakawa T, Timasheff SN. (1985) *Methods Enzymol* **114**: 49–77.
108. Curtis RA, Prausnitz JM, Blanch HW. (1998) *Biotechnol Bioeng* **57**: 11–21.
109. Curtis RA, Ulrich J, Montaser A, Prausnitz JM, Blanch HW. (2002) *Biotechnol Bioeng* **79**: 367–380.
110. Žoldák G, Sprinzl M, Sedlák E. (2004) *Eur J Biochem* **271**: 48–57.
111. Chakrabarti P. (1993) *J Mol Biol* **234**: 463–482.
112. Kim HK, Tuite E, Norden B, Ninham BW. (2001) *Europ Phys J E* **4**: 411–417.

113. Rabinovich D. (May 2008) The chemical reactivity control of collagen: The Hofmeister effect revisited, in *World Leather*, pp. 26–29.
114. Bulo RE, Siggel L, Molnar F, Weiss H. (2007) *Macromol Biosci* 7: 234–240.
115. Komsa-Penkova R, Koynova R, Kostov G, Tenchov BG. (1996) *Biochim Biophys Acta* **1297**: 171–181.
116. Jungwirth P, Tobias D. (2006) *Chem Rev* **106**: 1259–1281.
117. Vrbka L, Vondrášek J, Jagoda-Cwiklik B, Vácha R, Jungwirth P. (2006) *Proc Nat Acad Sci USA* **103**: 15440–15444.
118. Ninham BW, Yaminsky V. (1997) *Langmuir* **13**: 2097–2108.
119. Manciu M, Ruckenstein E. (2005) *Langmuir* **21**: 11312–11319.
120. Krienke H, Barthel J. (1998) *J Molec Liq* **78**: 123–138.
121. Kunz W, M'Halla J, Ferchiou S. (1991) *J Phys Cond Matt* **3**: 7907–7918.
122. Barthel J. (1985) *Pure Appl Chem* **57**: 355–367.
123. Ramanathan PS, Friedman HL. (1971) *J Chem Phys* **54**: 1086–1099.
124. Ramanathan PS, Krishnan CV, Friedman HL. (1972) *J Solution Chem* **3**: 237–262.
125. Andersen HC, Chandler D. (1972) *J Chem Phys* **57**: 1918–1929.
126. Fries PH, Patey GN. (1985) *J Chem Phys* **82**: 429–440.
127. Spohr E, Pálinkás G, Heinzinger K, Bopp P, Probst MM. (1988) *J Phys Chem* **92**: 6754–6761.
128. Chuev GN, Fedorov MV, Chiodo S, Russo N, Sicilia E. (2008) *J Comput Chem* **29**: 2406–2415.
129. Jungwirth P, Tobias DJ. (2001) *J Phys Chem B* **105**: 10468–10472.
130. Jungwirth P, Tobias DJ. (2002) *J Phys Chem B* **106**: 6361–6373.
131. Mucha M, Frigato T, Levering LM, Allen HC, Tobias DJ, Dang LX, Jungwirth P. (2005) *J Phys Chem B* **109**: 7617–7623.
132. Dang LX, Chang TM. (2002)*J Phys Chem B* **106**: 235–238.
133. Chaumont A, Schurhammer R, Wipff G. (2005) *J Phys Chem B* **109**: 18964–18973.
134. Hribar B, Southall NT, Vlachy V, Dill KA. (2002) *J Am Chem Soc* **124**: 12302–12311.
135. Jungwirth P. (2000) *J Phys Chem* **104**: 145–148.
136. Jakubowska A. (2008) *Chem Phys Chem* **9**: 829–831.
137. Serr A, Netz RR. (2006) *Int J Quant Chem* **106**: 2960–2974.
138. Parsons D, Ninham BW. (2009) *Coll Surf* **343**: 57–63.
139. Salanne M, Vuilleumier R, Madden PA, Simon C, Turq P, Guillot B. (2008) *J Phys Cond Matt* **20**: 494207/1–494207/8.

Chapter 2

Phospholipid Aggregates as Model Systems to Understand Ion-Specific Effects: Experiments and Models

Epameinondas Leontidis*

The Langmuir monolayers and bilayer stacks of zwitterionic phospholipids have been recently used as model systems to understand specific anionic effects. Simple thermodynamic experiments coupled to theoretical considerations and advanced spectroscopic methods have provided considerable insights. Inorganic monovalent anions were found to affect only the disorganised liquid-like phases of lipid monolayers. The perturbation of the surface pressure–area isotherm was treated as a problem of electrostatic ionic adsorption using a range of models, and the separate effects of sodium and various anions were assessed. A chemical potential of transfer of anions to the interface was obtained from the fits, and was found to correlate well with a function of the ionic size of the anions, suggesting that ionic specificity is largely a matter of size, and reflects the difference between ion–water and water–water interactions.

Osmotic stress experiments on lipid bilayers in the presence of sodium salts yield a perpendicular and a lateral equation of state (EOS). The attempt to fit these EOS using standard theory failed in the presence of large (0.5 M) concentrations of chaotropic anions, proving that the adsorption at the surface of the bilayers modifies their local structure considerably. Information about the changes at the interface propagates to the bulk, affecting the perpendicular EOS. Overall, these experiments revealed a strong need for better fundamental understanding of the effect of salts on the structure of lipid bilayers. In particular, more work is needed to understand the specific salt effect on the lipid headgroup region.

*Department of Chemistry, University of Cyprus, P.O. Box 20537, 1678 Nicosia, Cyprus.

1. Introduction

A major difficulty that one encounters upon considering ion-specific phenomena is that it is usually not possible to decouple the effects of cations from those of anions on the examined system. Another difficulty, which is of particular concern when one considers the effects of salts on *interfacial* phenomena — and the majority of ion-specific effects have to do with interfaces — is that close to an interface one can envision at least three different modes of interaction of the ions with the interface, or a combination thereof. Ions may interact locally with available 'interaction' sites (Fig. 1, left), and then one talks about local binding or ion pairing, a language that places emphasis on chemistry and is usually adopted by chemists. Alternatively, one can envision that an interface generates inhomogeneous fields (of electrostatic origin, of solvent density, of solvent polarisation, etc.), to which ions react in specific ways (Fig. 1, right). This picture places emphasis on the physics of collective interactions, and it is the usual approach adopted by physicists. The difference between the two views has an interesting mathematical implication. It means that when solving an equation to determine ionic distributions and other mean-field properties in an interfacial system, the chemical approach would concentrate on the boundary

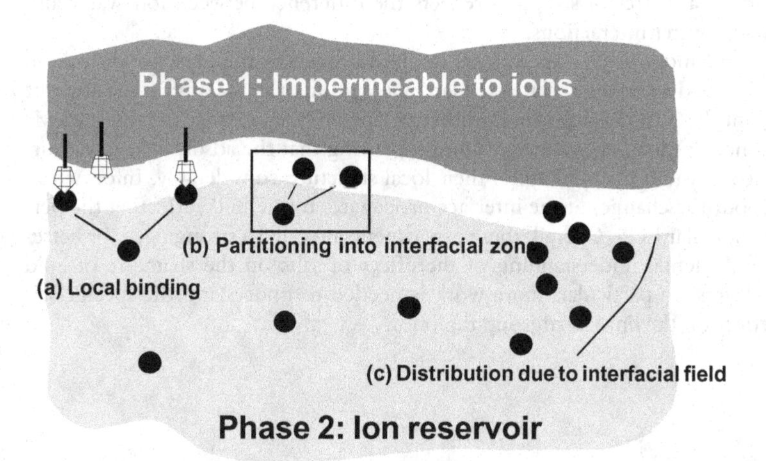

Fig. 1. Three modes of specific ion interaction at an interface between two phases. (a) Local chemical binding at available sites. (b) Partitioning into the interfacial zone. (c) Inhomogeneous ion distribution due to interfacial field effects.

conditions, while the physical approach would focus on the potentials of mean force.

There exist however even more ways of looking at ion-specific effects. Interfaces contain 'hazy' or 'grey' areas, with physical properties often quite different (and not always intermediate!) from those of the two phases that they connect. One could thus treat the interface as an 'inter-phase', i.e. as a different phase, in which the ions can distribute, and regard the ion-surface interaction as a partitioning phenomenon (Fig. 1, middle). Mathematically, this picture stands between the other two, although it is really a more complicated way of handling the boundary condition.

Which of the three pictures is valid in any particular situation? In most systems, in most phenomena, the manifestations of the three interaction modes would be almost equivalent, and it is then really hard for an experiment to reveal the underlying mechanism of ionic action. This is however a crucial piece of information. The literature is full of obscure ion-specific phenomena in systems that are not transparent enough to allow deeper understanding of mechanisms.[1,2] Simple, good model systems are needed that would allow a clear discrimination of the modes of ionic action. Measurements are needed that allow the juxtaposition of models with experiments. *Experimental observations not immediately amenable to theoretical treatment are not useful in the quest for understanding specific ion effects. On the other hand, abstract theory not capable of direct application to experimental data is also not particularly useful.*

Several years ago we postulated that surfactant aggregates could serve as such systems, especially if one could use the same lipid headgroup in a variety of different aggregate geometries. One would then always have the same chemical complexity, but would be able to use a very large variety of experimental methods, different for each aggregate type, to shed light on the nature of the ion–lipid interactions. As appropriate surfactants we proposed zwitterionic phospholipids with one or two hydrocarbon chains. These compounds offer several advantages: They are major components of biological membranes; working with them hopefully has useful biological implications. They are commercially available at very high purity. The structures that they form in water have been studied exhaustively over a period of 100 years, and large amounts of structural information are available. Lipids with the phosphatidylcholine (PC) headgroup were deemed particularly useful in the beginning. Both saturated and unsaturated double-chain PCs form bilayers and multilamellar vesicles when dissolved in water, while they form very well characterised insoluble monolayers when spread from an organic solvent on the surface of an aqueous solution (Fig. 2). In addition, the single-chain dodecylphosphocholine and its homologues form well characterised micelles in water (Fig. 2).

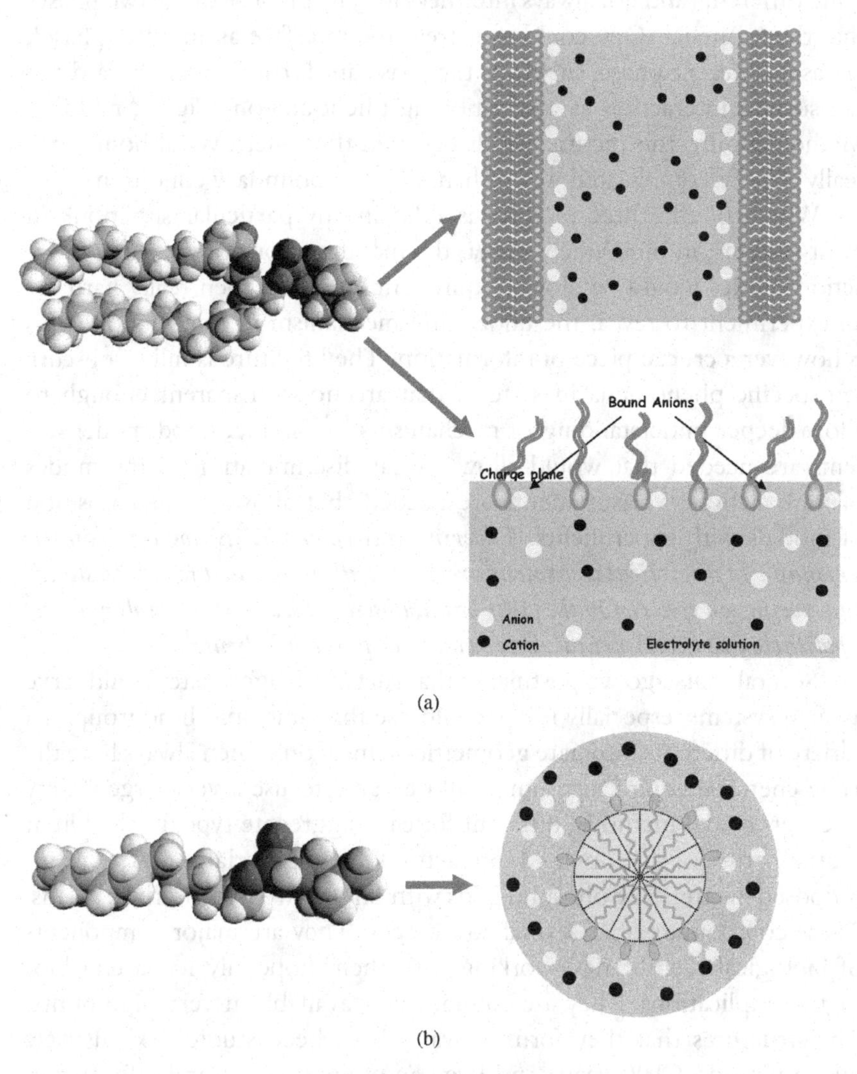

Fig. 2. (a) The DPPC molecule is a two-tailed phospholipid with saturated chains and can form bilayers in bulk water and Langmuir monolayers at the surface of aqueous solutions. (b) The DPC phospholipid is a one-chain analogue of DPPC and forms micelles in bulk water.

Before discussing some of the insights that have been gained with these systems, a note of caution is necessary. Soft matter interfaces are subject to considerable fluctuations, most of which are not present when one works with hard inorganic surfaces, or even with rigid colloids such as globular proteins. Especially in the case of 2-D organised lipid structures, there exist fluctuations in directions both perpendicular and lateral to the plane of lipid organisation (Fig. 3).

Perpendicular to the interface one observes positional (protrusion)[3] and orientation (tilt angle) fluctuations of the lipids, but also fluctuations in the water density and structure, which is non-homogeneous in any case. These types of fluctuations inhibit several advanced experimental techniques that attempt to obtain ionic profiles close to the interface.[4] Lateral fluctuations may be less of a problem. The scale of the well-known capillary waves is usually much larger than the molecular scale,[5] so that at the molecular level the curvature of the monolayer can be ignored. Lipid monolayer density fluctuations are short-lived. Fluctuations that occur because of lipid clustering in the case of multiple lipid binding to cations may become an issue.

In the next pages we take up phospholipid bilayers, and Langmuir monolayers as model systems and briefly discuss what fundamental

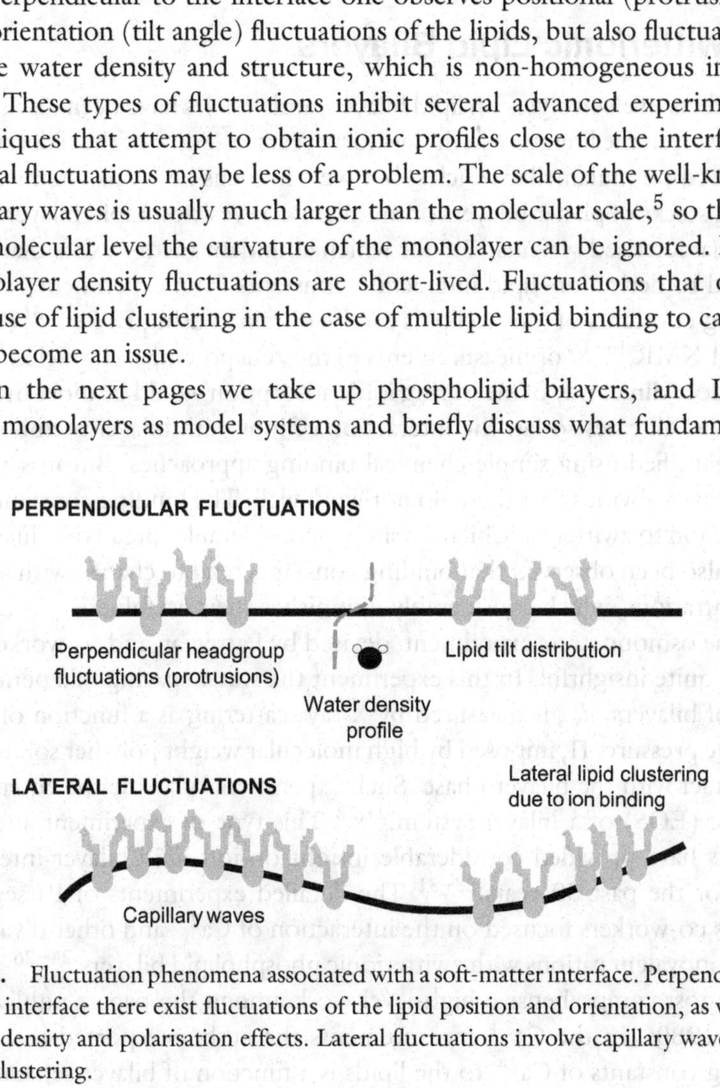

Fig. 3. Fluctuation phenomena associated with a soft-matter interface. Perpendicular to the interface there exist fluctuations of the lipid position and orientation, as well as water density and polarisation effects. Lateral fluctuations involve capillary waves and lipid clustering.

information can be gleaned from these systems. The emphasis is always on quantitative, 'mechanistic' information and not on qualitative or simply descriptive approaches that abound in the literature. Phospholipid micelles will be left out from this contribution, since the literature is still far from comprehensive and our own work in this area is still incomplete. The final chapter presents a comparative discussion and a summary.

2. Zwitterionic Lipid Bilayers

The effects of electrolytes on lipid bilayers and vesicles have been discussed extensively in the literature in the last forty years.[6,7] Some of the older work was based on unilamellar vesicles, but we will focus here on systems containing stacks of planar bilayers or multilamellar vesicles. Lipid bilayers are well-characterised systems,[8–13] for which abundant theory exists that can be used to derive ion–lipid interaction constants from experimental data. Binding constants of ions to lipids can be obtained using EPR,[14] ^{31}P, ^{1}H and ^{2}H NMR,[15–18] or measurements of the zeta potential of vesicles,[19,20] but a more direct way of measuring ionic adsorption would be more useful. In almost all cases so far, the interaction of ions with bilayer membranes was quantified using simple chemical binding approaches. But in several cases it was obvious that these do not work well. The binding constants of a single ion to zwitterionic bilayers show a considerable spread of values.[6,7] It has also been observed that binding constants tend to change with ionic concentrations, which considerably diminishes their usefulness.

The osmotic stress experiment, devised by Parsegian and co-workers,[8] can be quite insightful. In this experiment the water spacing of a periodic stack of bilayers, d_W, is measured by X-ray scattering as a function of the osmotic pressure, Π, imposed by high molecular weight polymer solutions in contact with the bilayer phase. Such experiments provide the equation of state (EOS) of a bilayer system.[21,22] This type of experiment and its variants have provided considerable insight on ion–lipid bilayer interactions for the past 30 years.[23–31] The detailed experiments of Parsegian and his co-workers focused on the interaction of Ca^{2+} and other divalent and monovalent cations with zwitterionic phospholipid bilayers.[24–26] This is the most comprehensive high-level work among the papers published in the 1980s. In the Ca^{2+} measurements the authors reported that the binding constants of Ca^{2+} to the lipids is a function of bilayer separation and ionic strength, and they obtained indications that the Hamaker constant is affected by the ions.[24] In addition they found that Ca^{2+}-binding is affected by the presence of Na^+ ions. They did not however examine the possibility of one Ca^{2+} binding to several lipids simultaneously, which is probably the preferred mode of interaction of Ca^{2+} even with zwitterionic lipids, as suggested by older experiments[15,16] and by recent computer simulations.[32] Cunningham *et al.*[27,28] and Tatulian *et al.*[29] found evidence that salts with large chaotropic anions, such as SCN^- and ClO_4^- may actually affect the structure of the bilayers at high concentration and lead to lipid interdigitation.

Petrache *et al.*[30,31] recently re-examined salt effects on lipid bilayers using the osmotic stress method. They worked at ambient temperature, using the zwitterionic lipid DLPC and KCl and KBr as the electrolytes, over a concentration range from 0.01 M to 1.25 M. They could fit their $\log\Pi$ versus d_w curves for all salt concentrations using the standard theory (see below) with chemical binding of ions on lipids and the charge-regulation model. The only requirement was that the Hamaker constant that determines the strength of the van der Waals attraction between two opposing bilayers must be reduced up to 70% upon increasing salt concentration, in rough agreement with existing theory.[33] This important investigation is not without its limitations. Potassium and chloride are expected not to interact with uncharged lipid bilayers, while Br^- is only a weak chaotrope, as indeed found by these researchers, who reported a binding constant of 0.22 M^{-1} for bromide on DLPC bilayers at 25°C.

We undertook a more extensive investigation of the effect of anions on bilayers, using also a large concentration range (0.05–0.5 M), but more chaotropic anions (NaI and NaSCN gave the most interesting results).[34] We chose Na^+ instead of K^+ as the common cation of all salts, which might create problems, since Na^+ definitely interacts more strongly with zwitterionic lipids than K^+[6,7] (see also the very recent computer simulations of Refs. 35–37). In addition we focused on both the lateral equation of state (EOS) of the bilayers and the perpendicular EOS. The former provides the area per lipid at the bilayer–water interface, a_L, as a function of the imposed osmotic pressure, Π, while the latter is usually the only measured EOS and provides the interbilayer distance, d_w, as a function of the osmotic pressure. The lipid DPPC was used in our work. Since this exhibits a gel-to-liquid phase transition at 41°C, the experiments were performed

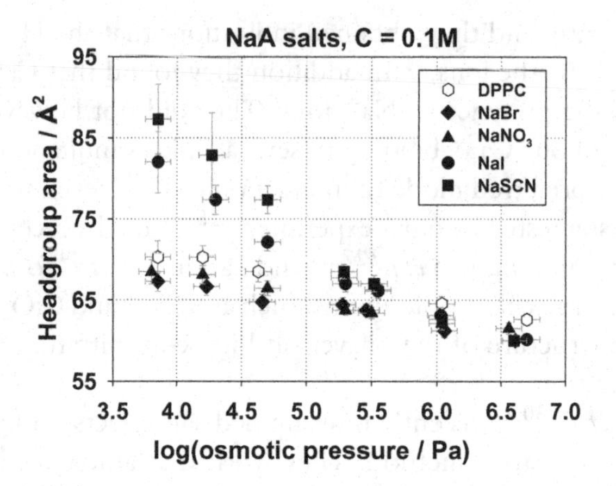

Fig. 4. Headgroup area, A, versus the osmotic pressure, Π, exerted on DPPC bilayers in the presence of NaA solutions of concentration 0.1 M (from Ref. 34 with permission).

at 50°C, which was both an advantage (relatively faster equilibration times) and a disadvantage (larger bilayer fluctuations). Some of the key results are summarised in Figs. 4 and 5 below. Figure 4 shows the dependence of the headgroup area on the osmotic pressure exerted on the lipid bilayers in the presence of 0.1 M NaA salt solutions, and proves that the headgroup area is affected in a significant way by the type of anions used. At small osmotic pressures, the chaotropic anions I^- and SCN^- have the strongest effect on the headgroup area. Br^- and NO_3^- dehydrate the DPPC headgroup to some extent, and it appears that the dehydration effect is stronger than ion–lipid association for these less chaotropic ions. In addition, as the osmotic pressure increases, the headgroup area in the presence of NaI and NaSCN decreases in a faster way, and at high pressures a_L becomes roughly independent of the salt present, indicating that the dehydration of the headgroups with pressure plays a more important role than ion–lipid association.

The $\log\Pi - d_w$ curves for DPPC in pure water and in NaBr, $NaNO_3$, NaI and NaSCN salt solutions of concentrations 0.1 M are presented in Fig. 5. The water bilayer separation d_w for the same osmotic pressure increases when salts are present. The change of d_w is more pronounced at small osmotic pressures. The experimental curves appear to converge at high osmotic pressures (low d_w), implying that the hydration forces that

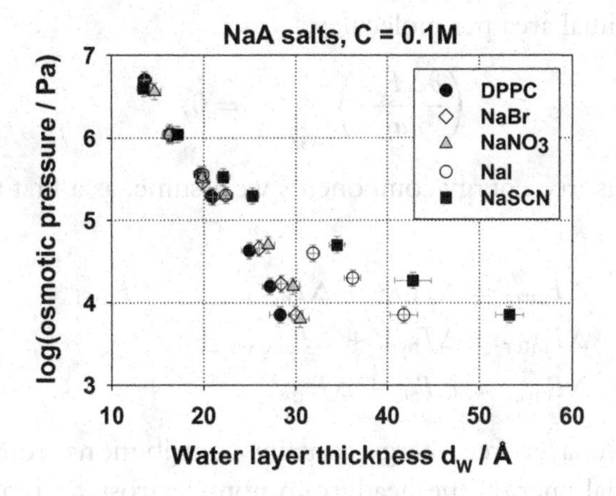

Fig. 5. $\log\Pi - d_w$ (perpendicular EOS) curves of DPPC in the presence of 0.1-M solutions of NaBr, $NaNO_3$, NaI and NaSCN (from Ref. 34 with permission).

dominate the interactions at these distances do not depend strongly on the salt presence. The increase of the water bilayer separation depends on the type of anion in the sodium salts, with SCN^- having the strongest effect on d_w, and Br^- the smallest. The effect of the anions on d_w follows the Hofmeister series. NaCl and NaBr appear to have a much smaller effect on DPPC at 50°C, compared to that of KCl and KBr observed by Petrache *et al.* on DLPC bilayers at lower temperatures.[31]

Fitting both the perpendicular and the lateral EOS is a difficult undertaking, and was based on the following thermodynamic formalism.[38] We write the total free energy difference *per mole of lipid* for the formation of the bilayer system as an *ad hoc* sum of intra-bilayer, inter-bilayer and cross terms:

$$\Delta F_{\text{tot}} = \Delta F_{\text{intra}} + \Delta F_{\text{inter}} + \Delta F_{\text{cross}}. \tag{1}$$

The osmotic pressure is obtained from this free energy difference via:

$$\frac{1}{a_L N_{AV}} \left(\frac{\partial \Delta F_{\text{tot}}}{\partial d_w} \right)_{a_L, T, n_L} = -\Pi_{\text{tot}}, \tag{2}$$

where n_L is the total number of moles of lipid. Also starting from ΔF_{tot} and minimising with respect to a_L at fixed d_w, T, n_L, we obtain the lateral EOS of the bilayer system, assuming that the bilayers are laterally free to adopt the optimal area per molecule:

$$\left(\frac{\partial \Delta F_{\text{tot}}}{\partial a_{\text{L}}}\right)_{d_{\text{w}},T,n_{\text{L}}} = 0. \tag{3}$$

For the various free energy components we assume, as a first approximation, that

$$\Delta F_{\text{intra}} = \Delta F_{\text{L/W}} + \Delta F_{\text{head-rep}} + \Delta F_{\text{conf}}, \tag{4}$$

$$\Delta F_{\text{inter}} = \Delta F_{\text{hydr}} + \Delta F_{\text{und}}, \tag{5}$$

$$\Delta F_{\text{cross}} = \Delta F_{\text{ele}} + \Delta F_{\text{vdw}}. \tag{6}$$

The intra-bilayer free energy contains contributions from the lipid–water interfacial energy, the headgroup non-electrostatic repulsion, and the conformational entropy of the tails (see below). We initially assume that these depend only on a_{L}, which is a good approximation in the absence of salts. However, in the presence of salts these terms may depend on d_{w}, through the ionic adsorption taking place, as will be discussed below. The hydration and undulation forces are considered to be pure inter-bilayer interaction terms, depending only on d_{w}. This is a usual assumption in the literature. Electrostatic and dispersion interactions depend on both d_{w} and a_{L} and are viewed as cross terms. If the in-plane free energy terms of Eq. (4) do not depend on d_{w}, they do not contribute to the osmotic pressure, and Eq. (2) leads to the usual expression for the osmotic pressure, with the four terms (hydration, undulation, dispersion, electrostatics) currently used by most investigators,

$$\Pi_{\text{tot}} = \Pi_{\text{hydr}} + \Pi_{\text{und}} + \Pi_{\text{vdw}} + \Pi_{\text{ele}}. \tag{7}$$

Standard equations were used for the hydration, undulation and van der Waals contributions[11,13,22–26,29,31]:

$$\Pi_{\text{hydr}} = P_0 e^{-d_{\text{w}}/\lambda}, \tag{8}$$

$$\Pi_{\text{und}} = \frac{3\pi^2 (k_{\text{B}}T)^2}{64\kappa_{\text{c}}} \frac{1}{d_{\text{w}}^3}, \tag{9}$$

$$\Pi_{\text{vdw}} = -\frac{H}{6\pi}\left[\frac{1}{d_{\text{w}}^3} + \frac{1}{(d_{\text{w}} + 2b_{\text{L}})^3} - \frac{2}{(d_{\text{w}} + b_{\text{L}})^3}\right]. \tag{10}$$

The parameters involved in these equations are the hydration decay length, λ, the hydration coefficient, P_0, the bending rigidity, κ_{c}, and the Hamaker constant, H. Their optimal values can be set by fitting the perpendicular EOS of DPPC bilayers *in pure water*. The electrostatic contribution was computed either by assuming localised ion binding, or by assuming that anions only can partition within the layer of lipid headgroups. In the former case the fitting parameter is the usual chemical anion–lipid binding constant, in the typical framework of the charge-regulation model. In the latter case one defines a partitioning chemical potential for the anions between bulk water and the lipid layer.[38] In any case, the electrostatic component of the osmotic pressure is given by standard double layer theory:

$$\begin{aligned} \Pi_{\text{ele}} &= \Pi_{\text{osm,med}} - \Pi_{\text{osm,ref}} \\ &= k_{\text{B}}T \sum_i \{C_{i,\text{med}} - C_{i,\infty}\} \\ &= k_{\text{B}}T \left\{ \sum_i C_{i,\infty} \left[\exp\left(-\frac{z_i q_{\text{e}} \varphi_{\text{med}}}{k_{\text{B}}T} \right) - 1 \right] \right\}, \end{aligned} \tag{11}$$

where $C_{i,\infty}$ is the ion concentration at the salt reservoir, and φ_{med} is the electrostatic potential in the centre of the water slab between two adjacent bilayers. The most interesting results of this fitting exercise were the following: in the absence of electrolytes, it is easy to fit the perpendicular EOS for a range of values of the parameters of the underlying models for the hydration, undulation and van der Waals forces, which fall within the range observed in several other similar systems.[9–11] However, in the presence of high concentrations of sodium salts of strongly adsorbing anions (e.g. NaI and NaSCN), the observed repulsion is extremely strong. No reasonable fiddling with the parameters of the various forces is sufficient to fit these data. Even if the Hamaker constant is reduced to zero (!), the binding constant of ions to lipids must be increased to unnaturally high values to obtain a good fit (Fig. 6).

This inadequacy of the theory, irrespectively of whether a binding or a partitioning electrostatic model is used, goes far beyond the observations of Petrache *et al.* that the Hamaker constant should be modified in the presence of salts,[31] or those of Manciu and Ruckenstein, that the hydration or the undulation force may also depend on the salt.[39] Our results imply that the electrolytes are affecting the actual water–lipid interface, in a way that cannot be accommodated by the existing theory.

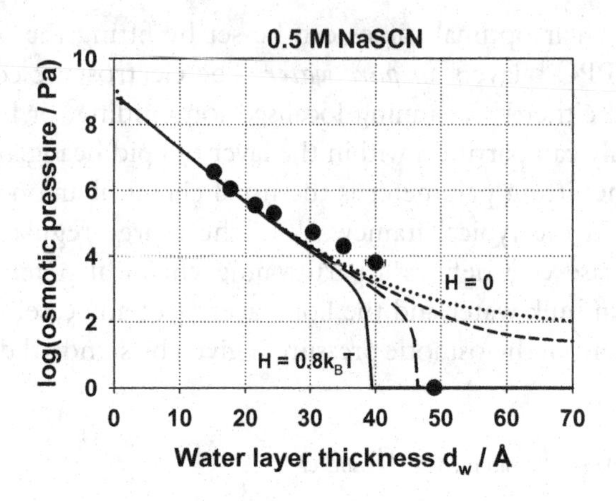

Fig. 6. Fitting curves obtained with the electrostatic binding model for DPPC in the presence of 0.5 M NaSCN, setting $K_A = 3.5\,\text{M}^{-1}$ and allowing the Hamaker constant to decrease and eventually drop to zero. The values for H are $0.8k_BT$ (solid line), $0.4k_BT$ (long-dashed line), $0.2k_BT$ (short-dashed line), and zero (dotted line). The experimental points are included for comparison (from Ref. 38 with permission).

The lateral EOS model is based on the intra-bilayer terms, modelled in a standard way:

$$\Delta F_{L/W} = \gamma N_{AV}(a_L - a_0), \tag{12}$$

$$\Delta F_{\text{head-rep}} = -\Re T \ln\left(1 - \frac{a_0}{a_L}\right), \tag{13}$$

$$\Delta F_{\text{conf}}(\text{kJ mol}^{-1}) = -55.88 + 2957.0 \exp(-10.07 a_L) + 2.981 a_L, \tag{14}$$

where

$$a_0 = \tilde{a}_0 + x_b \Delta a_i, \tag{15}$$

with γ the interfacial tension of the lipid–water interface, x_b the percentage of lipids with bound ions, $\tilde{a}_0$ the 'pure' molecular area of a lipid headgroup and Δa_i an ionic contribution to the headgroup area, based on the cross section of a hydrated lipid ion. There is nothing certain about the above equations. γ in Eq. (12) is in fact poorly known,[13,40] there is no consensus about the actual mathematical form of Eq. (13),[38,41] and Eq. (14) is just a numerical fit to theoretical results obtained in the absence of salts.[41] It is no surprise that these equations cannot fit the a_L versus $\log\Pi$ curve for various

NaI and NaSCN concentrations, or the a_L versus salt concentration at a specific osmotic pressure. The latter actually passes through a maximum for NaSCN, a singular behaviour that illustrates that more than one conflicting salt effects may act at the water–lipid interface.[38] However, Eqs. (12) to (15) provide a potential explanation about the high repulsion observed in the perpendicular EOS: If x_b depends on the inter-bilayer distance, d_W, the 'innocent' intra-bilayer terms would actually contribute to the osmotic pressure! For example, Eq. (12) would provide a term:

$$\Pi_{st} = \gamma \frac{\Delta a_i}{a_L} \left(\frac{\partial x_b}{\partial d_W} \right)_{a_L, T, n_L}. \qquad (16)$$

It turns out that such terms have the correct order of magnitude and the expected behaviour, which can be verified by using electrostatic models to compute the derivative in Eq. (16).[38] It is thus highly plausible that the strong interfacial adsorption of the chaotropic anions both affects the intra-bilayer free energy and contributes to the osmotic pressure in ways that have not been quantified until now. The work described here demonstrated that the level of existing theory for lipid bilayers and for the structure and thermodynamics of the water–lipid interface in the presence of ions is very unsatisfactory. Further progress with this model system can only be achieved if more theoretical work is carried out, probably assisted by computer simulation, to elucidate what happens at these important interfaces in the presence of electrolytes.

3. Zwitterionic Lipid Monolayers at the Air–Water Interface

Langmuir monolayers of phospholipids have been used for a very long time as membrane-mimetic model systems. It was thus highly surprising that until our own work, no detailed, systematic investigation about the effect of simple ions on zwitterionic lipid monolayers was published. This is all the more surprising because monolayers provide a considerable number of advantages as model systems: (a) They are the next simplest model beyond the free air–water surface. Interfacial lipid–ion binding interactions are possible. It is also possible to easily *tune the interfacial density of interaction sites* by simply changing the surface pressure of the monolayer. (b) Since they often exhibit *two-phase coexistence*, they allow the possibility of checking ion–lipid interactions with two different lipid arrangements simultaneously. (c) With zwitterionic phospholipids one avoids the strong Coulombic interactions (advantage common with bilayers). The geometry of the monolayer is ideal, as an electrostatic problem must be solved to infinite distance from the interface, and one does not need to worry about depletion of the salt reservoir. (d) Surface pressure (the main measurable quantity) is measured in the direction *lateral* to thc interface, so it may not suffer from uncertainties induced by fluctuations. (e) Lipid monolayers are very well characterised systems. Very powerful, novel experimental methods (spectroscopy and scattering of all kinds) allow a much more detailed molecular-level view of what is happening at a monolayer.

Some years ago, we undertook a detailed experimental investigation of the effect of sodium salts on the interfacial properties of Langmuir monolayers of DPPC at room temperature.[42] The fundamental result shown in Fig. 7 is that the equation of state of the monolayer (surface pressure versus lipid molecular area) is modified in the presence of salts in the aqueous subphase, moving to higher pressures at the same area, or higher areas at the same pressure. The effect becomes stronger with increasing salt concentration and is highly salt-specific, following closely the anionic lyotropic series (in Fig. 7 it is clear that NaSCN produces a much stronger effect than NaBr).

Application of advanced microscopic and spectroscopic methods, such as Brewster Angle Microscopy (BAM), Infrared Reflection–Adsorption Spectroscopy (IRRAS) and Grazing Incidence X-ray Diffraction (GIXD) at the monolayer revealed that when two lipid phases coexist, only the disordered liquid-expanded phase is affected by the ions. Table 1 summarises GIXD results for the crystal cell parameters of the ordered liquid-condensed phase. These results prove that the ordered DPPC phase is not strongly affected by salts even at high salt concentrations.

They also suggest that for the purpose of extracting quantitative information, one should focus on the liquid-expanded phase (LE). A closer look at the DPPC monolayer EOS in the range of 80 to 90 Å^2 per molecule, where the monolayer is in the pure LE phase, reveals that the electrolytes displace the EOS in an almost parallel fashion and the degree of displacement depends on anion type (in this series of experiments) (Fig. 7).

It is this set of data in the LE phase that we set out to model using a variety of mean-field models.[43] Modelling concentrates on the LE phase and specifically at 85 Å^2 per molecule, since the effect is roughly independent

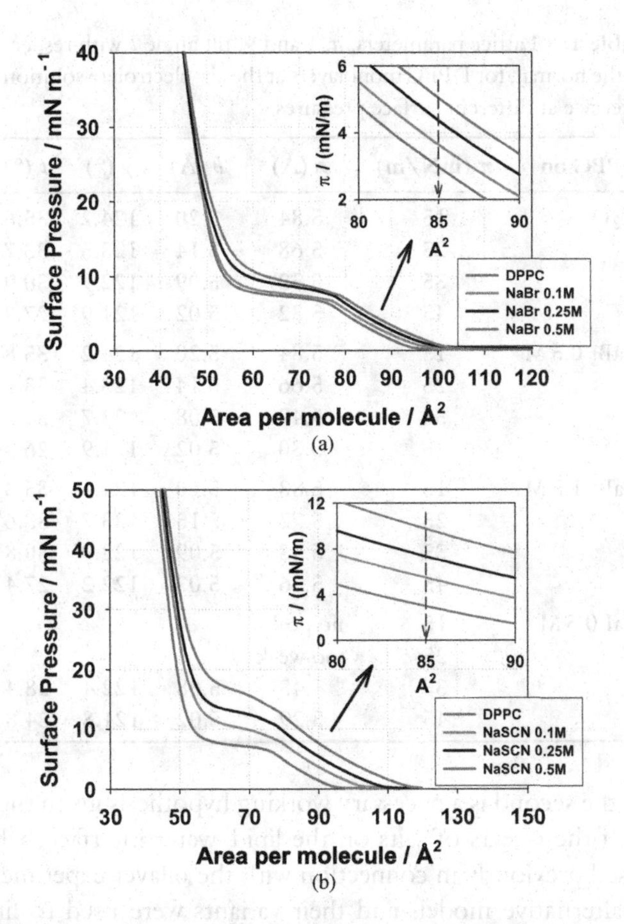

Fig. 7. DPPC monolayer equation of state (EOS) at 295 K over (a) NaBr solutions, and (b) NaSCN solutions of various concentrations.

of the headgroup area. It must be noted here that this molecular area is much larger than those examined in the bilayer systems, or found in bilayer computer simulations, which range between 55 and 70 Å^2 per molecule. This of course means that the monolayer interface is much more 'open' at the LE phase. The two major assumptions for the modelling effort are: (a) Since an anion effect is observed in Fig. 8, the modelling assumes that sodium does not interact with the lipid monolayers in any specific way. (b) The observed effect can be explained as specific ion adsorption at the monolayer, and not as a change in lipid tilt angle or headgroup conformation, induced by the electrolytes. While the first assumption can (and will)

Table 1. Lattice parameters, a, b and γ, tilt angle t with respect to the normal, for DPPC monolayers at the air–electrolyte solution interface at different surface pressures.

DPPC on	π (mN/m)	a (Å)	b (Å)	γ (°)	t (°)
H_2O	15	5.84	5.20	124.2	35.6
	25	5.68	5.14	123.5	33.7
	35	5.52	5.09	122.9	30.9
	45	5.32	5.02	121.9	27.2
NaBr 0.5 M	15	5.84	5.20	124.2	35.8
	25	5.66	5.14	123.4	33.2
	35	5.49	5.08	122.7	31.2
	45	5.30	5.02	121.9	26.4
NaBr 1.5 M	15	5.83	5.20	124.1	35.3
	25	5.72	5.15	123.7	33.6
	35	5.53	5.09	122.9	30.8
	45	5.36	5.03	122.2	27.4
NaI 0.5 M	15	no peaks			
	25	too weak			
	35	5.45	5.08	122.4	28.4
	45	5.29	5.02	121.8	24.8

be relaxed, the second is a necessary working hypothesis given the general ignorance of the effects of salts on the lipid–water interface, which was also discussed previously in connection with the bilayer experiments.

Three alternative models and their variants were used to fit monolayer data, such as those of Fig. 7.[44] The philosophy of these models was outlined in Fig. 1. Onc is a typical binding model based on the charge regulation approach; the second is an anion partitioning model, which treats the lipid–water interface as a separate phase; and the third is a mean-field model which includes dispersion forces in the potential of mean force between ions and monolayer, in the spirit of much recent work on the subject.[45–47] Details of the calculations with the binding model,[38,43,44] the partitioning model[38,43,44] and the dispersion-force model[44] are given in our previous publications. The binding model uses ion–lipid 1:1 binding constants as adjustable parameters, the partitioning model contains anionic partitioning chemical potentials, U_-, and the dispersion-force model contains dispersion force constants, B_-. Just for illustration purposes we write the Boltzmann relation for the ions close to the monolayer interface for the partitioning model:

$$C_i(x) = \begin{cases} C_{i,\infty} \exp\left(-\frac{z_i q_e \psi(x)}{k_B T}\right), & \delta < x \\ C_{i,\infty} \exp\left(-\frac{z_i q_e \psi(x)}{k_B T} - \frac{U_i}{k_B T}\right), & 0 < x < \delta, \end{cases} \tag{17}$$

and for the dispersion-force model in its simplest formalism adopted in the present work:

$$C_i(x) = C_{i,\infty} \exp\left(-\frac{z_i q_e \psi(x)}{k_B T} - \frac{B_i}{x^3}\right). \tag{18}$$

The partitioning chemical potentials, U_i, and the dispersion force coefficients, B_i, are treated as adjustable parameters, although the latter can in principle be estimated from spectroscopic data.[33,45–47] These two models are quite different in terms of their details, but also in terms of principle. The dispersion-force model treats the ion–lipid interface as a mathematical discontinuity and modifies the ion–lipid 'wall' PMF, while the partitioning model promotes an alternative, 'Swiss-cheese-like' picture of the ion–lipid interface. It is worthwhile to mention here that this penetration of anions within the lipid interface was observed in recent computer simulation work.[48]

In Fig. 8 it can be seen that the same data (DPPC monolayer surface pressure increments in the presence of various salt concentrations at 85 Å^2 per molecule) can be fitted equally well by the partitioning and the dispersion-force model, but not by the simple local binding model.

The first important piece of information from this model system is that the interaction of anions with lipids is not of the local chemical binding type. The fact that the other two models are successful is gratifying but not in itself surprising. Both partitioning chemical potentials and dispersion-force constants are roughly proportional to the ionic volume, so they provide a good representation of ionic size. The fitted U_- and B_- parameters are presented in Table 2 and their relation to the ionic volumes in solution is presented in Fig. 9.

The conclusion is that the specific ion effect on the EOS of the DPPC monolayers stems not from direct anion–lipid binding but from specific ionic adsorption driven by weaker intermolecular forces of the dispersion or hydration type.

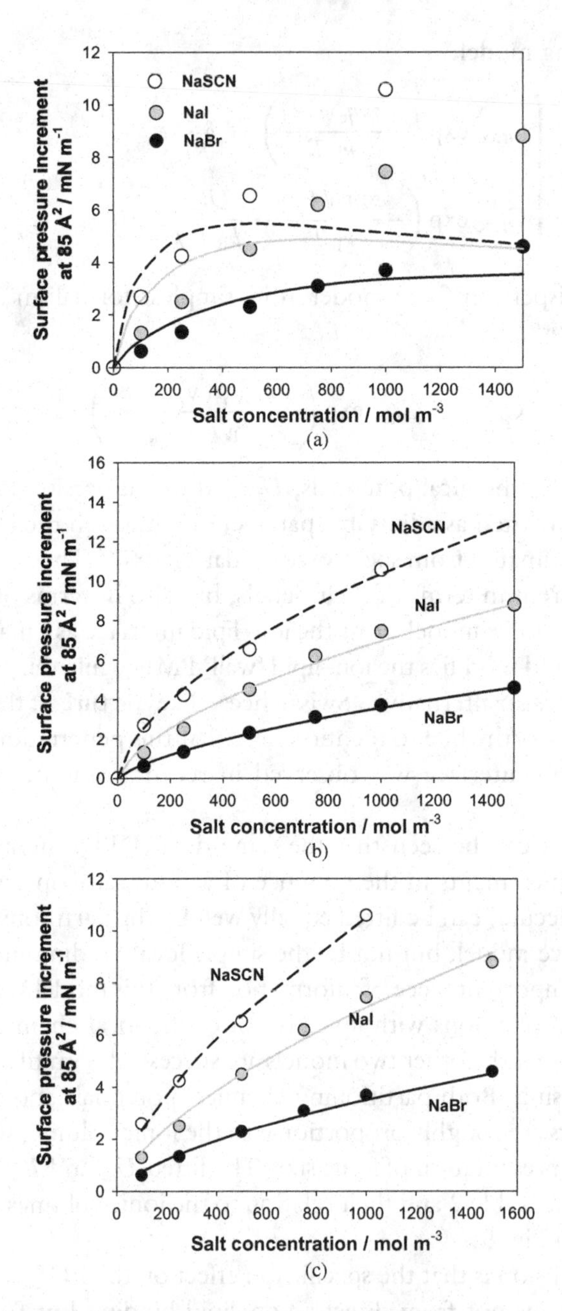

Fig. 8. Best fits of surface pressure increments for DPPC monolayers over electrolyte solutions at 85 Å^2 per molecule calculated using (a) the binding model, (b) the ion penetration model, and (c) the dispersion-force model. Results for NaBr, NaI and NaSCN are shown.

Table 2. Fitted anionic parameters for the partitioning model (assuming that the penetrable lipid layer width is equal to 10 Å), and the dispersion model (assuming $B_{Na} = +3 \times 10^{-50}$ J m^3).

Ion	U_-/k_BT Assuming no Na^+ binding to the lipids	U_-/k_BT with Na^+ binding to the lipids	$B_-/10^{-50}$ J m^3
Cl^-	−0.70 ± 0.10	−0.20 ± 0.05	−14.0 ± 0.4
CH_3COO^-	−1.40 ± 0.05	−0.95 ± 0.10	−16.5 ± 0.5
Br^-	−1.78 ± 0.05	−1.45 ± 0.05	−17.5 ± 0.5
NO_3^-	−2.50 ± 0.05	−1.85 ± 0.05	−18.5 ± 0.5
ClO_3^-	−2.90 ± 0.05	−2.50 ± 0.10	−19.8 ± 0.4
I^-	−3.15 ± 0.10	−2.95 ± 0.10	−20.9 ± 0.5
BF_4^-	−3.30 ± 0.10	−2.80 ± 0.05	−21.0 ± 0.3
ClO_4^-	−3.70 ± 0.05	−3.20 ± 0.05	−21.7 ± 0.4
SCN^-	−4.23 ± 0.10	−3.90 ± 0.10	−23.4 ± 0.4
PF_6^-	−4.50 ± 0.10	−4.05 ± 0.10	−24.6 ± 0.2

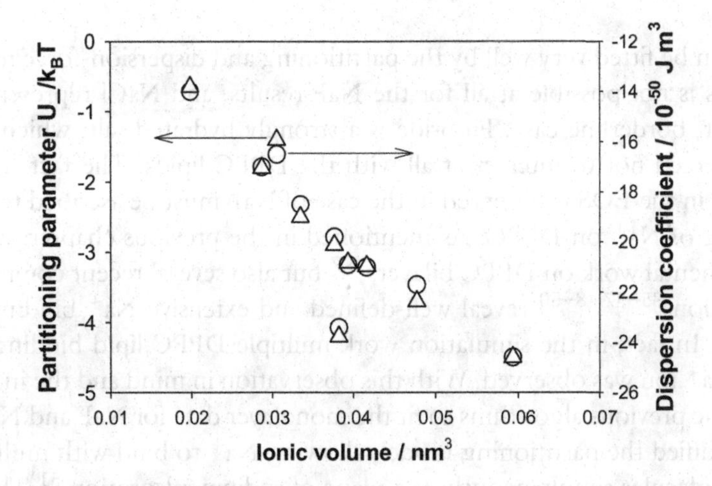

Fig. 9. Correlations of the optimal partitioning parameters and dispersion coefficients of Table 2 with the partial limiting ionic volumes.

The usefulness of lipid monolayers as model systems goes beyond the previous conclusion however. Figure 10 shows the DPPC surface pressure increments at 85 Å^2 per molecule and 295 K as a function of salt concentration for NaF, NaCl, NaBr and NaI. While the NaBr and NaI

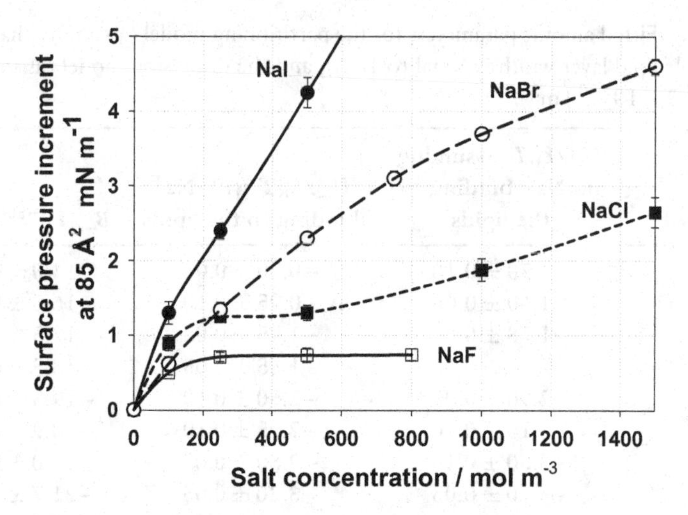

Fig. 10. Surface pressure increments of DPPC monolayers at 295 K and 85 Å^2 per molecule, over solutions of NaF, NaCl, NaBr and NaI.

lines can be fitted very well by the partitioning and dispersion-force models, this is not possible at all for the NaF results, and NaCl represents a difficult, borderline case. Fluoride is a strongly hydrated salt, which can be expected not to interact at all with the DPPC lipids. The fact that a change in the EOS is observed in the case of NaF must be ascribed to the binding of Na^+ on DPPC. As mentioned in the previous chapter, older experimental work on DPPC bilayers,[6,7] but also several recent computer simulations[35–37,48–50] reveal well-defined and extensive Na^+ binding to DPPC. In fact, in the simulation work multiple DPPC lipid binding on each Na^+ ion was observed. With this observation in mind and the inability of the previous algorithms to fit the monolayer data for NaF and NaCl, we modified the partitioning model, allowing Na^+ to bind with multiple lipid molecules simultaneously at a plane of sodium adsorption.[51] This is a highly idealised attempt to describe what is observed in recent simulations. This extended partitioning model can indeed fit practically all the EOS data. After fitting the NaF results with specific values for the sodium–lipid binding constants, we keep these fixed and recalculate the anionic partitioning chemical potentials, U_{-}. These new values are also listed in Table 2; they are roughly 0.5 $k_B T$ energy units smaller than those obtained from the original partitioning model, which ignores sodium binding to the lipids. This modelling exercise illustrates the usefulness of DPPC monolayers as model systems, and proves that both ions of an electrolyte play significant and distinct roles in these soft interfaces. Sodium ions bind to the lipids through a complexation mechanism with carbonyl groups, while large polarisable anions can in fact partition within the 'open' lipid monolayer and enhance the sodium effect. Similar conclusions were recently drawn by Dzubiella,[52] who examined salt effects on α-helix formation by oligopeptides. This coincidence shows that we are indeed approaching a much better understanding of specific salt effects on biological and physicochemical systems, as will be discussed in the final chapter.

The partitioning chemical potentials of Table 2 are quantitative measures of the specific ion effect and can be correlated to several Hofmeister parameters that are used to create scales for ionic specificity. Thus the U_- parameters of Table 2 correlate well with viscosity B coefficients,[53] von Hippel chromatographic parameters,[54] lyotropic numbers[55] and other measures of ionic specificity. Rather than displaying such correlations here, we have opted to discuss more deeply the nature of these partitioning chemical potentials. The chemical potential for partitioning of an ion between two solvents contains the following contributions[56–58]: (a) There is one contribution from the changed solvent polarisation, which for ions can be roughly approximated by a Born-type expression, although a huge literature exists on its modifications. But since the actual dielectric environment of an ion immersed between the lipid headgroups or in contact with the hydrophobic part of the monolayer is not known, we simply write:

$$\Delta G_{\text{pol}} \approx -\frac{(z_i q_e)^2}{8\pi\varepsilon_0(R_i + R_w)}\left(\frac{1}{\varepsilon_w} - \frac{1}{\varepsilon_{\text{lm}}}\right) \propto +\frac{1}{R_i + R_w}. \tag{19}$$

This is a term inversely proportional to the hydrated ion radius that involves the supposedly tight-held first solvation shell. The charge is omitted, since in this work we consider only monovalent anions. It is tacitly assumed that all ions feel 'a similar' dielectric environment within the monolayer. (b) A contribution from the cavity term that involves nonpolar ion-solvent interactions. This is consistently modelled as proportional to solute area or volume.

$$\Delta G_{\text{cav}} \approx (\gamma_{\text{lm}} - \gamma_w)4\pi(R_i + R_w)^2 \propto -(R_i + R_w)^2, \tag{20}$$

where again it is assumed that all ions sample roughly the same cavity environment in bulk water and at the interface. With these very crude approximations, we may conclude that the partitioning chemical potentials derived from our fitting should behave in the following way:

$$U_{-} \approx \Delta G_{\text{pol}} + \Delta G_{\text{cav}} \propto -c_1(R_{\text{i}} + R_{\text{w}})^2 + \frac{c_2}{R_{\text{i}} + R_{\text{w}}}. \tag{21}$$

According to this simple-minded approach, the partitioning chemical potentials should depend only on the ion size, although not in a simple fashion. The ratio of the unknown constants c_1 and c_2 can be fixed by assuming that the two opposing terms balance each other exactly for one particular anion. We choose this to be Cl^-, which is often thought to be the 'neutral' anion in the lyotropic series.[59] Using thermochemical ionic radii collected by Roobottom,[60] we find the ratio c_2/c_1 to be roughly equal to 29.2 Å^3. In Fig. 11 we plot the partitioning chemical potentials of Table 2 versus the ionic size function of Eq. (21).

The correlation is remarkable given the simplicity of the model, superior to correlations of the U_{-}s with the free energy of hydration, the polarisability, the viscosity B coefficient or the lyotropic number (not shown). Ionic specificity does exist however, beyond that covered by Eq. (21), as exemplified by the two important outliers, acetate and thiocyanate (we believe that the deviation of BF_4^- may be mostly due to a nonprecise value of the ionic radius). Acetate and thiocyanate possess significant dipole

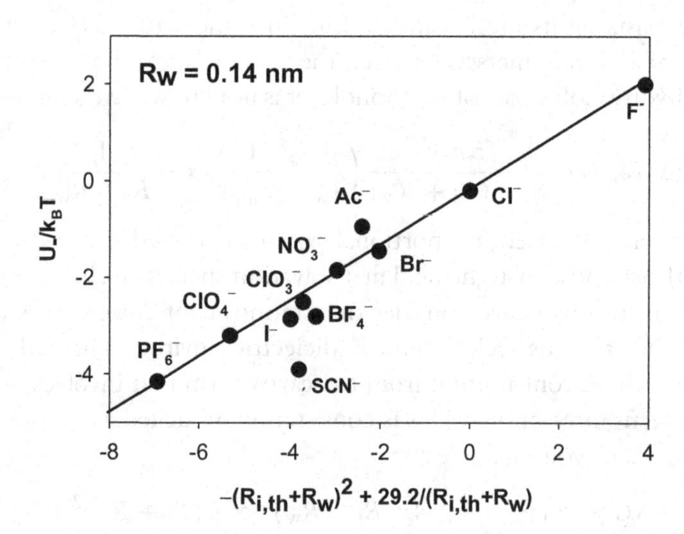

Fig. 11. Plot of the ionic partitioning parameter, U_{-}, versus the function of ionic size defined in Eq. (21).

moments and have large asymmetric polarisabilities.[61] In addition, these ions have a distinctly 'linear' shape, and acetate has a hydrophobic methyl group, which creates a local hydration asymmetry for the ion in bulk water,[62] and would force it to adopt a very special orientation in the lipid interface considered here. SCN^- was also recently reported to prefer an orientation either parallel to the surface at the air–water interface[63] or tilted with respect to the surface by roughly 40° (see the Vibrational Sum Frequency Generation (VSFG) Spectroscopy results of Ref. 64). It thus seems that the extra specificity of some simple ions, that goes beyond the effect of size (or charge-density), may be the result of a nonspherical charge distribution.

To conclude this chapter, the DPPC monolayers appear to be an extremely fruitful model system. The ion–lipid interactions in these systems can be studied theoretically and experimentally in great detail. The picture so far is one of sodium binding to the lipids and polarisable anions partitioning within the lipid monolayer in a way dictated mostly by their sizes, although shape (or charge distribution) effects can also play a role. We suggest that this picture may be valid for a large number of other systems as well, judging by the similarity of results observed in recent computer simulations of bilayers and oligopeptides.

4. Summary — Putting Everything Together

Phospholipid aggregates were shown to be useful model systems to understand ion-specific behaviour relevant to membranes. The bilayers still leave many open problems. The theoretical attempts highlighted here, based on the high-quality work with the osmotic stress method, prove that we still need considerable improvements in understanding and theory to advance. It is amazing that the phospholipid–electrolyte solution interface, so important a membrane-mimetic system and with considerable biological ramifications, is still very incompletely understood. What happens at the bilayer surface in the presence of electrolytes is just starting to be unveiled with a combination of novel experimental tools and computer simulation. Predictive theory lags far behind. The recent work of ourselves and others has highlighted several issues: The van der Waals interaction between bilayers is strongly affected by electrolytes in specific ways,[31] while the intra-bilayer free energy is affected by chaotropic anion adsorption and affects the inter-bilayer forces.[38] The multiple lipid–sodium binding observed in recent computer simulations of bilayers may not be an equally important issue in this model system, since the DPPC EOS at 50°C is hardly influenced by salts such as NaCl and NaBr. It is important to work simultaneously on the lateral and the perpendicular bilayer EOS, since we have obtained several indications that these two affect each other.

Langmuir monolayers are 'cleaner' model systems in several respects, since one is there confronted only with a lateral EOS. A fairly detailed picture has emerged from our recent work. Sodium salts appear to interact in a two-fold way with the DPPC monolayers. Sodium binds to many lipid molecules in a complexation mode predicted by recent computer simulations. Large chaotropic cations appear to be able to penetrate into the LE phase of the monolayer, which has a considerably open structure, but they cannot affect the ordered LC phase. Both partitioning and dispersion-force models, implicitly assuming sodium exclusion, can in fact fit the lateral EOS of DPPC at ambient temperature quite well. *These successful but very different models do not by necessity provide the correct actual molecular details at the lipid–water interface.* We have argued that their success stems from the fact that the fitting parameters reflect well the size of the ions, which is the major determinant of ion specificity. That this is indeed the case is supported by the correlation between the fitted anion partitioning chemical potentials into the lipid monolayer and a function of the ionic size that accounts for hydration and cavity interactions. Some recent investigations suggest that this partitioning picture may well describe the behaviour of ions at several interfaces, where strong local binding is absent or does not play a role. Iso and Okada explained the effect of salts on zwitterionic micelle mobility using a partitioning model,[65] while Kabalnov *et al.* explained salt effects on non-ionic microemulsions in this way.[66] Pegram and Record have recently applied the principle of ionic partitioning with considerable success to explain the effects of electrolytes on the surface tension of water[67,68] and on the folding and crystallisation of proteins.[69] The ideas of excess hydration and preferential exclusion of salts from polymer membranes discussed by Parsegian and Rau are very close to this partitioning picture.[70]

What is learnt about ion specificity from the work described in this contribution? Combined with current important work comprising experimental, theoretical and computational methods, our belief is that ion specificity is determined primarily by the ionic size and to second-order by

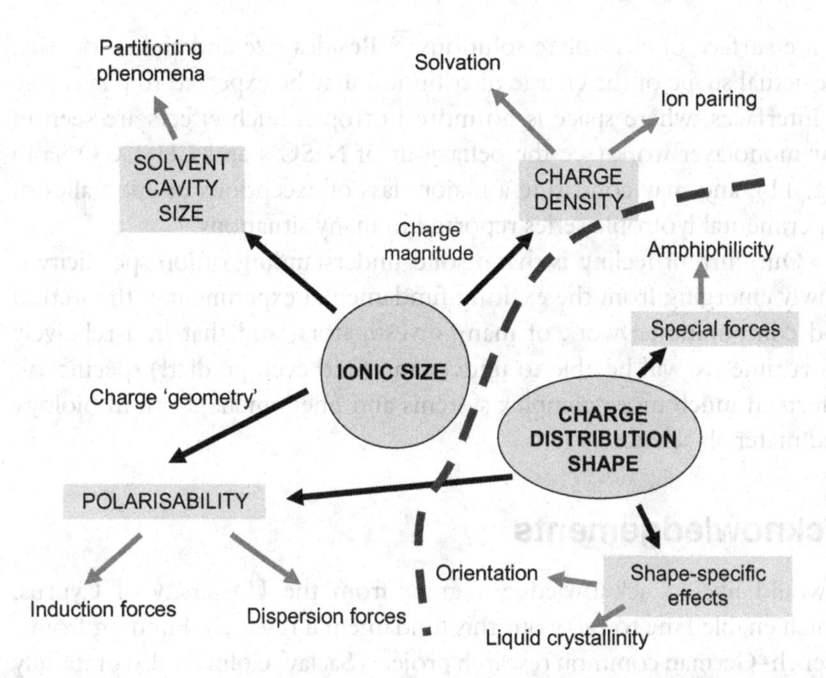

Fig. 12. Ionic parameters that directly determine ion specificity and derived ionic properties, which play important roles in several ion specific phenomena: A chart of interconnections.

the shape of the electron density of an ion. The situation is rather complex regarding ionic size, and is depicted in Fig. 12.

According to this scheme, ionic size is synonymous to charge density, when ions of the same charge are considered. Charge density is the major ionic property in processes where ion-pairing plays an important role. This is the case in many biological systems, as pointed out in the seminal work of Collins,[59,71] and supported by recent computer simulations from Dzubiella[52] and the group of Jungwirth.[72] In addition to this, ionic size determines the cavity size of the solute, which is always important, especially when ion-pair formation is not dominant. There are situations where both effects of the size must be considered, as in the case of our monolayer work, or in the case of the interaction of salts with PNIPAM reported by the group of Cremer.[73,74] Finally, the size is the major determinant of the molecular polarisability, which fine-tunes intermolecular interactions, and may play unexpectedly important roles in a situation of a near-balance of forces, such as that encountered for ions at the surface of electrolyte solutions.[75] Besides size and charge density, the actual shape of the charge distribution may be expected to play a role at interfaces, where space is no more isotropic. Such effects are seen in our monolayer work (see the behaviour of NaSCN and CH_3COONa in Fig. 11), and may constitute a major class of exceptions or anomalies of experimental lyotropic series reported in many situations.

Our current feeling is that a solid understanding of ion specificity is slowly emerging from the exciting fundamental experimental, theoretical and computational work of many investigators, and that in a relatively short time we will be able to understand (and even predict!) specific salt effects in much more complex systems and phenomena, both in biology and materials science.

Acknowledgements

I would like to acknowledge a grant from the University of Cyprus, which enabled me to carry out this fundamental research. Funding from a French–German common research project (Saclay–Golm) is also gratefully acknowledged. I am grateful to Dr Andria Aroti for all the experimental work presented here, to Prof Thomas Zemb, and Dr Monique Dubois for the collaborative work on bilayers, to Prof Luc Belloni (Saclay) for the theoretical work behind the partitioning model, to Dr Gerald Brezesinski (Golm) for the IRRAS, GIXD and BAM measurements on DPPC monolayers.

References

1. Collins KD, Washabaugh MW. (1985) The Hofmeister effect and the behaviour of water at interfaces. *Quart Rev Biophys* **18**: 323–422.
2. Cacace MG, Landau EM, Ramsden JJ. (1997) The Hofmeister series. *Quart Rev Biophys* **30**: 241–277.
3. Israelachvili JN, Wennerström H. (1992) Entropic forces between amphiphilic surfaces in liquids. *J Phys Chem* **96**: 520–531.
4. Bu W, Vaknin D, Travesset A. (2005) Monovalent counterion distributions at highly charged water interfaces: Proton-transfer and Poisson–Boltzmann theory. *Phys Rev E* **72**: 060501(R).
5. Lyklema J. (2000) *Fundamentals of Interface and Colloid Science. Vol. III: Liquid–Fluid Interfaces.* Academic Press, London.
6. Tatulian SA. (1993) Ionization and ion binding. In: Cevc G (ed.), *Phospholipid Handbook*, pp. 511–552. Marcel Dekker, New York.

7. Tocanne JF, Teissié J. (1990) Ionization of phospholipids and phospholipid-supported interfacial lateral diffusion of protons in membrane model systems. *Biochim Biophys Acta* **1031**: 111–142.
8. Parsegian VA, Rand RP, Fuller NL, Rau DC. (1986) Osmotic stress for the direct measurement of intermolecular forces. *Meth Enzymol* **127**: 400–416.
9. Lis LJ, McAlister M, Fuller N, Rand RP, Parsegian VA. (1982) Interactions between neutral phospholipid bilayer membranes. *Biophys J* **37**: 657–665.
10. Rand RP, Parsegian VA. (1989) Hydration forces between phospholipid bilayers. *Biochim Biophys Acta* **988**: 351–376.
11. McIntosh TJ, Simon SA. (1993) Contributions of hydration and steric (entropic) pressures to the interactions between phosphatidylcholine bilayers: Experiments with the subgel phase. *Biochemistry* **32**: 8374–8384.
12. Nagle JF, Tristram-Nagle S. (2000) Structure of lipid bilayers. *Biochim Biophys Acta* **1469**: 159–195.
13. Cevc G, Marsh D. (1987) *Phospholipid Bilayers. Physical Principles and Models.* John Wiley & Sons, New York.
14. Puskin JS. (1977) Divalent cation binding to phospholipids: An EPR study. *J Membr Biol* **35**: 39–55.
15. Hauser H, Hinckley CC, Krebbs J, Levine BA, Phillips MC, Williams RJP. (1977) The interaction of ions with phosphatidylcholine bilayers. *Biochim Biophys Acta* **468**: 364–377.
16. Altenbach C, Seelig J. (1984) Ca^{2+} binding to phosphatidylcholine bilayers as studied by deuterium magnetic resonance. Evidence for the formation of a Ca^{2+} complex with two phospholipid molecules. *Biochemistry* **23**: 3913–3920.
17. Macdonald PM, Seelig J. (1988) Anion binding to neutral and positively charged lipid membranes. *Biochemistry* **27**: 6769–6775.
18. Rydall JR, Macdonald PM. (1992) Investigation of anion binding to neutral lipid membranes using ^{2}H-NMR. *Biochemistry* **31**: 1092–1099.
19. McLaughlin S, Bruder A, Chen S, Moser C. (1975) Chaotropic anions and the surface potential of bilayer membranes. *Biochim Biophys Acta* **394**: 304–313.
20. Tatulian SA. (1983) Effect of lipid phase transition on the binding of anions to dimyristoylphosphatidylcholine liposomes. *Biochim Biophys Acta* **736**: 189–195.
21. Dubois M, Zemb T, Fuller N, Rand RP, Parsegian VA. (1998) Equation of state of a charged bilayer system: Measure of the entropy of the lamellar–lamellar transition in DDABr. *J Chem Phys* **108**: 7855–7869.
22. Ricoul F, Dubois M, Belloni L, Zemb T. (1998) Phase equilibria and equation of state of a mixed cationic surfactant — Glycolipid lamellar system. *Langmuir* **14**: 2645–2655.
23. Ohshima H, Inoko Y, Mitsui T. (1982) Hamaker constant and binding constants of Ca^{2+} and Mg^{2+} in dipalmitoyl phosphatidylcholine/water system. *J Colloid Int Sci* **86**: 57–72.
24. Lis LJ, Parsegian VA, Rand RP. (1981) Binding of divalent cations to dipalmitoylphosphatidylcholine bilayers and its effect on bilayer interaction. *Biochemistry* **20**: 1761–1770.
25. Lis LJ, Lis WT, Parsegian VA, Rand RP. (1981) Adsorption of divalent cations to a variety of phosphatidylcholine bilayers. *Biochemistry* **20**: 1771–1777.

26. Loosley-Millman M, Rand RP, Parsegian VA. (1982) Effects of monovalent ion binding and screening on measured electrostatic forces between charged phospholipid bilayers. *Biophys J* **40**: 221–232.
27. Cunningham BA, Lis LJ. (1986) Thiocyanate and bromide ions influence the bilayer structural parameters of phosphatidylcholine bilayers. *Biochim Biophys Acta* **861**: 237–242.
28. Cunningham BA, Lis LJ. (1989) Interactive forces between phosphatidylcholine bilayers in monovalent salt solutions. *J Colloid Int Sci* **128**: 15–25.
29. Tatulian SA, Gordeliy VI, Sokolova AE, Syrykh AG. (1991) A neutron diffraction study of the influence of ions on phospholipid membrane interactions. *Biochim Biophys Acta* **1070**: 143–151.
30. Petrache HI, Tristram-Nagle S, Harries D, Kucerka N, Nagle JF, Parsegian VA. (2006) Swelling of phospholipids by monovalent salt. *J Lipid Res* **47**: 302–309.
31. Petrache H, Zemb T, Belloni L, Parsegian VA. (2006) Salt screening and specific ion adsorption determine neutral-lipid membrane interactions. *Proc Natl Acad Sci USA* **103**: 7982–7987.
32. Böckmann RA, Grubmüller H. (2004) Multistep binding of divalent cations to phospholipid bilayers: A molecular dynamics study. *Angew Chem Int Ed* **43**: 1021–1024.
33. Parsegian VA. (2005) *Van der Waals forces.* Cambridge University Press, Cambridge, UK.
34. Aroti A, Leontidis E, Dubois M, Zemb T. (2007) Effects of monovalent anions of the Hofmeister series on DPPC lipid bilayers. Part I. Osmotic Stress experiments and in-pane equation of state. *Biophys J* **93**: 1580–1590.
35. Gurtovenko AA, Vattulainen I. (2008) Effect of NaCl and KCl on phosphatidylcholine and phosphatidylethanolamine lipid membranes: Insight from atomic-scale simulations for understanding salt-induced effects in the plasma membrane. *J Phys Chem B* **112**: 1953–1962.
36. Cordomi A, Edholm O, Perez JJ. (2008) Effect of ions on a dipalmitoylcholine bilayer. A molecular dynamics simulation study. *J Phys Chem B* **112**: 1397–1408.
37. Lee SJ, Song Y, Baker NA. (2008) Molecular dynamics simulations of asymmetric NaCl and KCl solutions separated by phosphatidylcholine bilayers: Potential drops and structural changes induced by string Na^+-lipid interactions and finite size effects. *Biophys J* **94**: 3565–3576.
38. Leontidis E, Aroti A, Belloni L, Dubois M, Zemb T. (2007) Effects of monovalent anions of the Hofmeister series on DPPC lipid bilayers. Part II. Modelling the perpendicular and lateral equation of state. *Biophys J* **93**: 1591–1607.
39. Manciu M, Ruckenstein E. (2007) On possible microscopic origins of the swelling of neutral lipid bilayers induced by simple salts. *J Colloid Int Sci* **309**: 56–67.
40. Katz Y. (1988) Surface tension in phospholipid bilayers. *J Colloid Int Sci* **122**: 92–99.
41. Fattal DR, Andelman D, Ben-Shaul A. (1995) The vesicle-micelle transition in mixed lipid-surfactant systems: A molecular model. *Langmuir* **11**: 1154–1161.
42. Aroti A, Leontidis E, Maltseva E, Brezesinski G. (2004) Effect of Hofmeister anions on DPPC Langmuir monolayers at the air-water interface. *J Phys Chem B* **108**: 15238–15245.

43. Aroti A, Leontidis E, Dubois M, Zemb T, Brezesinski G. (2007) Monolayers, bilayers and micelles of zwitterionic lipids as model systems for the study of specific anion effects. *Colloids Surf A* **303**(1–2): 144–158.
44. Leontidis E, Aroti A, Belloni L. (2008) DPPC liquid expanded monolayers as model systems to understand the anionic Hofmeister series: 1. A tale of models. *J Phys Chem B* **113**: 1447–1459.
45. Ninham BW, Yaminsky V. (1997) Ion binding and ion specificity: The Hofmeister effect and Onsager and Lifshitz theories. *Langmuir* **13**: 2097–2108.
46. Boström M, Williams DRM, Ninham BW. (2001) Specific ion effects: Why DLVO theory fails for biology and colloid systems. *Phys Rev Lett* **87**: 168103.
47. Boström M, Deniz V, Ninham BW. (2006) Ion specific surface forces between membrane surfaces. *J Phys Chem B* **110**: 9645–9649.
48. Sachs JN, Nanda H, Petrache HI, Woolf TB. (2004) Changes in phosphatidylcholine headgroup tilt and water order induced by monovalent salts: Molecular dynamics simulations. *Biophys J* **86**: 3772–3782.
49. Pandit SA, Bostick D, Berkowitz ML. (2003) Molecular dynamics simulation of a dipalmitoylphosphatidylcholine bilayer with NaCl. *Biophys J* **84**: 3743–3750.
50. Böckmann RA, Hac A, Heimburg T, Grubmüller H. (2003) Effect of sodium chloride on a lipid bilayer. *Biophys J* **85**: 1647–1655.
51. Leontidis E, Aroti A. (2008) DPPC liquid expanded monolayers as model systems to understand the anionic Hofmeister series: 2. Ion partitioning is mostly a matter of size. *J Phys Chem B* **113**: 1460–1467.
52. Dzubiella J. (2008) Salt-specific stability and denaturation of a short salt-bridge-forming α-Helix. *J Am Chem Soc* (in press, DOI: 10.1021/ja805562g)
53. Jenkins HDB, Marcus Y. (1995) Viscosity B-Coefficients of Ions in Solution. *Chem Rev* **95**: 2695–2724.
54. von Hippel PH, Peticolas V, Schack L, Karlson L. (1973) Model studies of neutral salts on the conformational stability of biological macromolecules. I. Ion binding to polyacrylamide and polystyrene columns. *Biochemistry* **12**: 1256–1271.
55. Voet A. (1937) Quantitative lyotropy. *Chem Rev* **20** : 169–179.
56. Melander W, Horvàth C. (1977) Salt effects on hydrophobic interactions in precipitation and chromatography of proteins: An interpretation of the lyotropic series. *Arch Biochem Biophys* **183**: 200–215.
57. Jorgensen WL, Ulmschneider JP, Tirado-Rives J. (2004) Free energies of hydration from a generalised Born model and an all-atom force field. *J Phys Chem B* **108**: 16264–16270.
58. Zhu J, Alexov E, Honig B. (2005) Comparative study of generalised Born models: Born radii and peptide folding. *J Phys Chem B* **109**: 3008–3022.
59. Collins KD. (1997) Charge density-dependent strength of hydration and biological structure. *Biophys J* **72**: 65–76.
60. Roobottom HK, Jenkins DB, Passmore J, Glasser L. (1999) Thermochemical radii of complex ions. *J Chem Educ* **76**: 1570–1574.
61. Serr A, Netz RR. (2006) Polarisabilities of hydrates and free ions from DFT calculations. *Int J Quant Chem* **106**: 2960–2974.
62. Robertson WH, Diken EG, Johnson MA. (2003) Snapshots of water at work. *Science* **301**: 320–321.

63. Petersen PB, Saykally RJ, Mucha M, Jungwirth P. (2005) Enhanced concentration of polarisable anions at the lipquid water surface: SHG spectroscopy and MD simulations of sodium thiocyanide. *J Phys Chem B* **109**: 10915–10921.
64. Viswanath P, Motschmann H. (2007) Oriented thiocyanate anions at the air-electrolyte interface and its implications on interfacial water — A vibrational sum frequency spectroscopy study. *J Phys Chem C* **111**: 4484–4486.
65. Iso K, Okada T. (2000) Evaluation of electrostatic potential induced by anion-dominated partition into zwitterionic micelles and origin of selectivity in anion uptake. *Langmuir* **16**: 9199–9204.
66. Kabalnov A, Olsson U, Wennerström H. (1995) Salt effects on non-ionic microemulsions are driven by adsoprtion/depletion at the surfactant monolayer. *J Phys Chem* **99**: 6220–6230.
67. Pegram LM, Record MT. (2006) Partitioning of atmospherically relevant ions between bulk water and the water/vapor interface. *Proc Natl Acad Sci USA* **103**: 14278–14281.
68. Pegram LM, Record MT. (2007) Hofmeister salt effects on surface tension arise from partitioning of anions and cations between bulk water and the air-water interface. *J Phys Chem B* **111**: 5411–5417.
69. Pegram LM, Record MT. (2008) Thermodynamic origin of Hofmeister ion effects. *J Phys Chem B* **112**: 9428–9436.
70. Chik J, Mizrahi S, Chi S, Parsegian VA, Rau DC. (2005) Hydration forces underlie the exclusion of salts and of neutral polar solutes from hydroxypropylcellulose. *J Phys Chem B* **109**: 9111–9118.
71. Collins KD, Neilson GW, Enderby JE. (2007) Ions in water: Characterising the forces that control chemical processes and biological structure. *Biophys Chem* **128**: 95–104.
72. Vrbka L, Vondrasek J, Jagoda-Cwiklik B, Vacha R, Jungwirth P. (2006) Quantification and rationalization of the higher affinity of sodium over potassium to protein surfaces. *Proc Natl Acad Sci USA* **103**: 15440–15444.
73. Zhang Y, Furyk S, Bergbreiter DE, Cremer PS. (2005) Specific ion effects on the water solubility of macromolecules. PNIPAM and the Hofmeister series. *J Am Chem Soc* **127**: 14505–14510.
74. Chen, X, Yang T, Kataoka S, Cremer PS. (2007) Specific ion effects on interfacial water structure near macromolecules. *J Am Chem Soc* **129**: 12272–12279.
75. Jungwirth P, Tobias DJ. (2006) Specific ion effects at the air/water interface. *Chem Rev* **106**: 1259–1281.

Chapter 3

Modelling Specific Ion Effects in Engineering Science

Christoph Held* and Gabriele Sadowski*,†

During the last decades, the application of electrolytes in many industrial and research fields has increasingly gained in importance. The modelling of aqueous electrolyte systems is crucial for the design and optimisation of processes in biochemical engineering. It can provide component and solution properties which are substantial for the design of separation units. Liquid densities (pvT data), solubilities as well as vapour pressures (VLE data) are directly obtained by an equation of state (EOS). Water activity coefficients (WAC) and mean ionic activity coefficients (MIAC) of electrolyte solutions can be calculated by either an EOS or by an excess Gibbs Energy model (G^E model).

Here, we describe the application and typical modelling results for a G^E model (MSA-NRTL) as well as for an EOS (ePC-SAFT). In addition to strong electrolytes which are almost fully dissociated, we also consider some weak electrolytes (acids like HF or ion-paired electrolytes) that do only partially dissociate in aqueous solution. Here, ion pairing is accounted for by an association/dissociation equilibrium between the ion pair and the respective free ions in solution.

1. Introduction

The knowledge of real-phase behaviour of electrolyte solutions provides a basis for the design and the simulation of many processes in biological and chemical engineering. Otherwise, salts are systematically used for the recovery of biomolecules and as auxiliary material in separation units. Thus, technical applications of systems containing electrolytes can be found in waste-water and drinking-water treatment, fertilizer production,

*Laboratory of Thermodynamics, Department of Biochemical and Chemical Engineering, Technische Universität Dortmund, Emil-Figge-Str. 70, 44227 Dortmund, Germany.

†E-mail: gabriele.sadowski@bci.tu-dortmund.de

or enhanced oil recovery.[1] Other important processes based on thermodynamic properties of electrolytes are electrolysis,[2] wet flue-gas scrubbing, osmosis and reverse osmosis of aqueous solutions, as well as reactive distillation with an electrolyte serving as entrainer.

Another class of electrolytes important in the engineering domain is the one of acids. The total world production of hydrochloric acid is estimated at 20 Miot per year. Acids like aqueous solutions of HCl, HBr, or HI are used for the production of organic and inorganic compounds or for pH control and neutralisation.

All industrial applications of salts and acids require the development of models allowing for the calculation of thermodynamic component and solution properties. The basis for the consideration of the Coulombic forces among the ions is e.g. provided by the work of Debye and Hückel (DH) in 1923[3] or by the Mean Spherical Approximation (MSA) introduced by Waisman and Lebowitz[4] in 1970. On top of that, state-of-the-art models also account for short-range (SR) interactions between the ions and the solvent, as well as for those among the solvent molecules. Both the Coulombic long-range (LR) and the short-range contributions are usually treated as independent of each other yielding the total excess Gibbs energy of a system:

$$G^{\mathrm{E}} = G^{\mathrm{SR}} + G^{\mathrm{LR}}. \tag{1}$$

The following chapters will introduce the two models we focus on: the G^{E}-model MSA-NRTL and the equation of state ePC-SAFT, both of which composed of different terms for the SR and for the LR contributions in Eq. (1). After that, we will apply the two models to electrolyte solutions in order to describe thermodynamic properties of strong and weak electrolytes.

2. Two Different Approaches: G^{E} Model and Equation of State

2.1. *The MSA-NRTL Model*

Papaiconomou *et al.*[5] combined the NRTL model[6,7] with the semi-restricted version of the MSA[8] model adapted to the Gibbs energy formalism. Simonin *et al.*[9,10] applied the MSA-NRTL to strong electrolytes as well as to weaker acids like HNO_3 or H_2SO_4. Moreover, heat capacities and enthalpies could be described by the model.[11]

2.1.1. *The model*

In analogy to the general approach presented in Eq. (1), the expression for the excess Gibbs energy developed by Papaiconomou *et al.*[5] consists of two parts: the long-range contribution (LR) which corresponds to the electrostatic interactions of ions while all non-Coulombic short-range (SR) interactions between ions and solvent or between two solvent molecules are accounted for in a second part:

$$G^{\mathrm{E}} = G^{\mathrm{SR}} + G^{\mathrm{LR}} = G^{\mathrm{NRTL}} + G^{\mathrm{MSA}}. \tag{2}$$

The long-range contribution is modelled with a semi-restricted version of the MSA model which was adapted to the Gibbs energy formalism. The short-range contribution is calculated with the help of the NRTL model and is described as:

$$G^{\mathrm{NRTL}} = RT \sum_{j=1}^{k} x_j \frac{\sum_{i=1}^{k} x_i P_{ij} \tau_{ij}}{\sum_{i=1}^{k} x_i P_{ij}}, \tag{3}$$

where the τ_{ij} refers to the interaction parameters between two species i and j. P_{ij} is the probability of finding a molecule of species i in the neighbourhood of a central molecule of species j and is assumed to obey a Boltzmann distribution depending on the interaction energy (which is included in the parameters τ_{ij}). Next to the interaction parameters, a 'non-randomness' parameter α_{ij} is also used in the NRTL model, which is adjusted to phase-equilibrium data.

Besides the short-range forces, the long-range interactions are modelled by the MSA contribution to the excess Gibbs energy.[4] The MSA was introduced in 1970 by Waismann and Lebowitz who solved the Ornstein–Zernike equation for a fluid of charged spheres of equal size. Within the MSA, ions are explicitly accounted for and have a defined volume. The MSA-NRTL G^{E} model presented here is formulated in the restricted MSA-framework. It uses a shielding parameter Γ which describes the reciprocal distance within which the long-range forces among ions are shielded by water. This is an 'optimum' parameter minimising the system's energy and consequently has to be determined iteratively. However, Harvey *et al.*[8] showed that the iteration can be circumvented by the following approximation which is applied within the MSA-NRTL model:

$$\Gamma = \frac{1}{2\sigma}\left(-1 + \sqrt{1 + 2\kappa\sigma}\right). \tag{4}$$

Here, κ is the Debye–Hückel shielding parameter and is defined in Eq. (13).

2.1.2. *The parameters*

The resulting MSA-NRTL model uses six model parameters, each of them having a physical meaning. For the NRTL contribution, this is the the non-randomness parameter α as well as four parameters describing the interactions between the different species water (W), anion (A), and cation (C), yielding τ_{WA}, τ_{WC}, $\tau_{WC,AC}(1)$, and $\tau_{WC,AC}(2)$. For the MSA contribution, additionally the ionic diameters σ_+ and σ_- have to be adjusted. Within the restricted MSA hydrated salt, instead of hydrated ion, parameters are applied which are obtained by adding up the adjusted ion diameters and dividing them by ν yielding the hydrated 'mean ionic size' σ required in Eq. (4):

$$\sigma = \frac{(\sigma_+ + \sigma_-)}{(\nu_+ + \nu_-)}. \tag{5}$$

Here, ν_+ and ν_- denote the stoichiometric factor of the anion and the cation, respectively. All six parameters needed within the MSA-NRTL approach are effectively mixture parameters, as they reflect interactions between two species, with $\tau_{WC,AC}(1)$ being even concentration-dependent. All these parameters were obtained in Ref. 5 by adjusting them to experimental MIAC data for the investigated system at 25°C.

2.1.3. *Calculation of thermodynamic properties*

In contrast to equations of state, activity coefficients are directly obtained by G^E models as the excess Gibbs energy is thermodynamically linked to mole-fraction activity coefficients f_i by the general relation:

$$\ln f_i = \frac{1}{k_B T}\frac{\partial G^E}{\partial N_i}. \tag{6}$$

The reference state for this activity coefficient is the pure component i. For solutes j, these activity coefficients are often related to the state of infinite dilution (rational activity coefficient) through:

$$f_j^* = \frac{f_j(T,p,x_j)}{f_j^\infty(T,p,x_j \to 0)}. \tag{7}$$

For electrolytes, the mole-fraction based mean ionic activity coefficient (MIAC) $f_\pm^*$ is defined as the geometrical mean of the mole-fraction based rational activity coefficients of the ions in solution:

$$f_\pm^* = ((f_+^*)^{\nu_+} \cdot (f_-^*)^{\nu_-})^{\frac{1}{(\nu_+ + \nu_-)}}. \tag{8}$$

Here ν is the stoichiometric factor of the anion and cation. For electrolytes in solution, the MSA-NRTL model describes the mole-fraction based MIAC as:

$$\ln f_\pm^* = \ln f_\pm^{*,\mathrm{NRTL}} + \ln f_\pm^{*,\mathrm{MSA}}. \tag{9}$$

The activity coefficient of water (WAC) is calculated by Eq. (6). The determination of other thermodynamic properties (e.g. densities) is not possible using a G^E model.

2.2. *The ePC-SAFT Equation of State*

ePC-SAFT is a combination of the PC-SAFT equation of state by Gross and Sadowski[12] and the Debye–Hückel contribution,[3] which accounts for the Coulomb interactions. It considers the ionic species independent of the salt they are part of. Only two parameters are used to characterise each ion. The first one is the ionic diameter σ_j which is actually the diameter of the hydrated ion. The second ionic parameter is the dispersive-energy parameter u_j/k_B which reflects the strength of ionic hydration.

2.2.1. *The model*

In contrast to a G^E model, ePC-SAFT is formulated in terms of the Helmholtz energy A to also consider density effects of the system. It is again formulated as a sum of LR and SR contributions

$$A = A^{\mathrm{SR}} + A^{\mathrm{LR}}, \tag{10}$$

whereas A^{SR} is described by the PC-SAFT equation of state and the Debye–Hückel theory is used to model the Coulomb interactions among ions (LR contribution).

The PC-SAFT EOS[12] is based on a perturbation theory where the hard-chain system (a chain consisting of repelling hard spheres) is used as the reference system. In order to obtain the residual Helmholtz energy A^{res} (the thermodynamic key quantity of the model), all deviations from the reference system (e.g. the attractive van der Waals forces) are treated as perturbations that can be considered as independent contributions to the Helmholtz energy of the system:

$$\begin{aligned} A^{\text{SR}} &= A^{\text{id}} + A^{\text{res}}, \\ A^{\text{res}} &= A^{\text{hc}} + A^{\text{disp}} + A^{\text{assoc}}. \end{aligned} \tag{11}$$

Here, A^{id} is the ideal-gas contribution; A^{hc} represents the hard-chain repulsion of the reference system. A^{disp} and A^{assoc} account for the Helmholtz energy contributions due to attraction (dispersion) and hydrogen bonding (association), respectively. To account for the fact that at least some of the species carry charges, this model was combined with a Debye–Hückel theory to describe the Helmholtz energy contribution A^{DH} to a system that is caused by charging the species.

$$A^{\text{LR}} = A^{\text{DH}}. \tag{12}$$

One key quantity of the contribution A^{DH} is the ion-specific inverse Debye screening length κ which compares to the MSA Γ [Eq. (4)]; κ describes the reciprocal distance within which the long-range forces among ions are shielded by water, expressed as:

$$\kappa = \sqrt{\frac{\rho_N e^2}{k_{\text{B}} T \cdot \varepsilon} \cdot \sum_j z_j^2 \cdot x_j}. \tag{13}$$

Here, ρ_N, e, and z_j describe the number density of the system, the elementary charge, and the charge number of the ions, respectively. The repulsive interactions of the ions and the attractive interactions with water (hydration) are accounted for in A^{hc} and A^{disp}, respectively.

The fact that both the DH κ as well as the MSA Γ in the Coulombic contributions in Eq. (12) (ePC-SAFT) and in Eq. (2) (MSA-NRTL) are expressed in terms of the dielectric constant ε makes clear that these contributions follow the approach of so-called 'primitive models'. This means that the solvent is not explicitly accounted for in the Coulombic contribution but is rather described as a dielectric continuum shielding electrostatic forces. In contrast, non-primitive models [e.g. Ref. (13)] do also explicitly account for the solvent molecules in the LR contribution.

2.2.2. *The parameters*

Using ePC-SAFT, a molecule is described as a chain of tangent-jointed spheres (segments). Three pure-component parameters are used to describe such a molecule as illustrated in Fig. 1: the segment diameter σ, the number of segments m_{seg}, and the energy parameter u/k_B to describe the attraction between two segments of different molecules.

For associating components like water, two additional parameters are required, namely the association-energy parameter ε_{hb}/k_B and the association volume κ_{hb}. For water, we apply the parameter set presented in Table 1 of Refs. 14 and 15 with the two-site 2B association scheme.[16]

As ions are assumed to be spherical, the segment number of ions is always set to one ($m_{ions} = 1$) yielding finally only to two parameters per ion: the diameter σ_j and the dispersive-energy parameter u_j/k_B of the hydrated ion. The latter gives a direct hint to which extent the ion interacts with water. To describe mixtures (salt solutions), the conventional Berthelot–Lorenz combining rules are used:

$$\sigma_{ij} = \frac{1}{2}(\sigma_i + \sigma_j), \tag{14}$$

$$u_{ij} = \sqrt{u_i u_j}. \tag{15}$$

Equation (15) is applied to interactions between water and ions only. Van der Waals dispersion between two ions is neglected in this work.

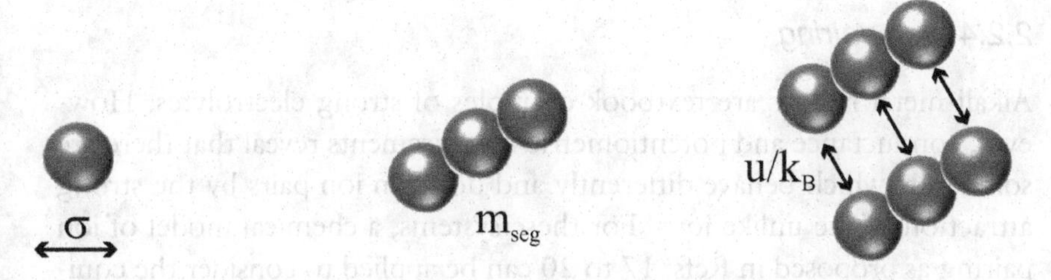

Fig. 1. PC-SAFT adjustable parameters for a non-associating substance.

2.2.3. *Calculation of thermodynamic properties*

The system pressure as well as the fugacity of the components in a mixture can be calculated as derivatives of the residual Helmholtz energy A^{res} of the system:

$$p^{\text{res}} = -\left(\frac{\partial A^{\text{res}}}{\partial V}\right)_{T,n_i},$$
$$RT\ln\varphi_i = \left(\frac{\partial A^{\text{res}}}{\partial n_i}\right)_{T,V,n_{j\neq i}} - RT\ln z; \quad z = p^{\text{res}}\frac{V}{RT}. \tag{16}$$

The water activity coefficient (WAC) directly yields the deviation of the behaviour of water in the mixture from that in the pure-component state ('0W') at the same temperature. It can thus be obtained as the ratio of the two fugacity coefficients:

$$fw = \frac{\varphi_{\text{W}}(T, p, x_{\text{W}})}{\varphi_{0\text{W}}(T, p, x_{\text{W}} = 1)}. \tag{17}$$

By contrast, the rational activity coefficient f_j^* describes the deviation from the infinite-diluted solution. It can be obtained by ePC-SAFT as the ratio of ion fugacity coefficient φ_j at the actual concentration related to the one at infinite dilution φ_j^{∞}:

$$f_j^* = \frac{\varphi_j(T, p, x_j)}{\varphi_j^{\infty}(T, p, x_j \to 0)}. \tag{18}$$

The mole-fraction based mean ionic activity coefficient (MIAC) $f_{\pm}^*$ of the electrolyte is then again the geometrical mean of the mole-fraction based rational activity coefficients of the ions in solution according to Eq. (8).

2.2.4. *Ion pairing*

Alkali-metal halides are textbook examples of strong electrolytes. However, conductance and potentiometric measurements reveal that there are some salts which behave differently and do form ion pairs by the strong attraction of the unlike ions. For these systems, a chemical model of ion pairing as proposed in Refs. 17 to 20 can be applied to consider the equilibrium between the completely dissociated electrolyte and the ion pair formed:

$$\nu_- \mathrm{An}^{z-} + \nu_+ \mathrm{Cat}^{z+} \overset{K_{\mathrm{ip}}}{\longleftrightarrow} [\mathrm{An}_{\nu_-} - \mathrm{Cat}_{\nu_+}]_{\mathrm{ip}}^{z_+\nu_+ + z_-\nu_-}. \tag{19}$$

Here, z and ν denote the valence and the stoichiometric factors of the anion and the cation, respectively. The ion-pairing constant K_{ip} is given by the law of mass action as:

$$K_{\mathrm{ip}} = \prod_j a_j^{*,\nu_j} = \frac{a_{\mathrm{ip}}^*}{a_-^{*,\nu_-} \cdot a_+^{*,\nu_+}}, \tag{20}$$

where the a_j^* are the activities of the ions and the ion pair related to the reference state of infinite dilution. As systems with ion pairing do also contain an ion pair, three more parameters need to be specified when applying ePC-SAFT: segment diameter σ_{ip}, dispersion energy u_{ip}/k_B, and segment number m_{ip} of the ion pair. The latter has to be introduced since an ion pair cannot be considered as a spherical species anymore. In order to reduce the number of adjustable parameters, the following mixture rules are applied:

$$\sigma_{\mathrm{ip}} = \frac{\nu_+\sigma_+ + \nu_-\sigma_-}{\nu}, \qquad m_{\mathrm{ip}} = \nu_+ m_+ + \nu_- m_-, \tag{21}$$

where ν means the stoichiometric factor of the salt (i.e. the number of ions within this salt, e.g. $\nu = 3$ for Na_2SO_4). The dispersive-energy parameter $u_{\mathrm{ip}}/k_{\mathrm{B}}T$ and the ion-pairing constant remain as adjustable parameters of an ion pair. The procedure of parameter estimation is explained in detail in Ref. 21.

3. Modelling Results

Several assumptions have to be made for the calculation of solution densities, vapour–liquid phase equilibria (VLE), and MIAC of electrolyte solutions:

- Within the temperature range considered here, it is reasonable to assume that the vapour phase above the solution consists of pure water only. We therefore restrict the ions to the liquid phase resulting in the fact that only the isofugacity criterion for water must be taken into account for VLE calculations:

$$\varphi_{\mathrm{W}}^{\mathrm{L}}(T, x_{\mathrm{W}}^{\mathrm{L}}) \cdot x_{\mathrm{W}}^{\mathrm{L}} = \varphi_{\mathrm{W}}^{\mathrm{V}}(T, x_{\mathrm{W}}^{\mathrm{V}} = 1) \cdot x_{\mathrm{W}}^{\mathrm{V}}. \tag{22}$$

- Those electrolytes, for which neither an experimental proof nor the probability for an ion-pairing exists, are regarded as strong electrolytes that fully dissociate into their respective cations and anions.
- In case of ePC-SAFT, dispersive (short-range) interactions were only considered for water–water, water–ion, and water–ion pairs but not among ions. Within the ion-specific MSA-NRTL approach, attractive short-range forces are also assumed to occur between anions and cations.

As mentioned before, ion-specific instead of salt-specific parameters are used here for both models. Thus, the ionic parameters determined for an ion are applicable to all electrolytes containing this ion. Obtaining such a universal set of parameters requires a simultaneous regression of several electrolyte solutions, which is described in Refs. 5 and 15 for both models, MSA-NRTL and ePC-SAFT.

Using ePC-SAFT, liquid densities, vapour pressures [directly obtained by Eq. (16); not included in the parameter estimation], and solute activity coefficients (MIAC) are modelled whereas only activity coefficients will be presented for MSA-NRTL as this quantity is the only one that can be obtained by a G^{E}-model. Any deviations from experimental data will be given by absolute relative deviations (ARD):

$$\mathrm{ARD} = 100 \cdot \frac{1}{NP} \sum_{k=1}^{NP} \left| \left(1 - \frac{y_k^{\mathrm{calc}}}{y_k^{\mathrm{exp}}} \right) \right|. \tag{23}$$

In order to compare model results to experimental data, the mole-fraction based MIAC values obtained by the models are converted into the experimentally obtained molality-based ones:

$$\gamma_{\pm}^{*,m} = \frac{f_{\pm}^{*}}{(1 + 0.001 \cdot \nu \cdot M_{\mathrm{W}} \cdot m_{\mathrm{s}})}, \tag{24}$$

where m_{s} is the salt molality and M_{W} is the molar mass of water.

In the following, we will give some examples of modelling results obtained with MSA-NRTL and ePC-SAFT.

3.1. *MSA-NRTL*

Compared to Pitzer-like G^E models, the MSA-NRTL approach has two advantages. Firstly, it seems that a comparably good data description can be achieved using a smaller number of parameters. Some G^E models even use more than 20 adjustable parameters, which are very difficult to determine, making these models unattractive for industrial applications. Secondly, the parameters of the MSA-NRTL model have physical meaning. This is not the case for the Pitzer models, although some attempts have been made to interpret the possible meaning of some Pitzer parameters. Furthermore, MSA-NRTL was shown to be applicable in either an ion-specific or a salt-specific approach.[5] The advantages of an ion-specific modelling are: (a) the number of adjustable parameters can be reduced significantly, and (b) these parameters can directly be used for the modelling of multiple-salt solutions. On the other side, there might also be good reasons for a salt-specific treatment representing the solutions' pseudo-components (water and salt). As mentioned in the introductory chapter of this book, the interactions of ions with water may depend on their counterion which would imply the use of salt-specific parameters. However, in this work, we use the ion-specific approach for NRTL-MSA in order to compare with the ePC-SAFT calculations and model parameters.

As the calculation of solution densities is not possible with a G^E model, for MSA-NRTL we have to focus on the components' activity coefficients. As a typical example, Fig. 2 shows these activity coefficients for LiBr in water. As illustrated, the addition of LiBr leads to highly non-ideal solution behaviour: due to the high charge density of the small lithium ion, the MIAC reaches very high values, indicating a very strong interaction between water and ions. Applying the MSA-NRTL parameter set[5] to Eq. (6), the MIAC values [Fig. 2(a)] and the activity coefficients of water [Fig. 2(b)] can be modelled accurately even at high concentrations where the deviation from an ideal-mixture behaviour is strongest.

Figure 3 illustrates the influence of anions on the activity coefficient of water (WAC). For aqueous alkali bromide solutions [Fig. 3(a)], the following sequence of WAC values can be observed: $WAC_{KBr} > WAC_{NaBr} > WAC_{LiBr}$. If the Br^- anion is replaced by the hydroxide ion, a reversed WAC series compared to alkali halides (chlorides, bromides, or iodides) is experimentally observed. In this case the sequence is: $WAC_{KOH} < WAC_{NaOH} < WAC_{LiOH}$. Thus, the highest WAC value in

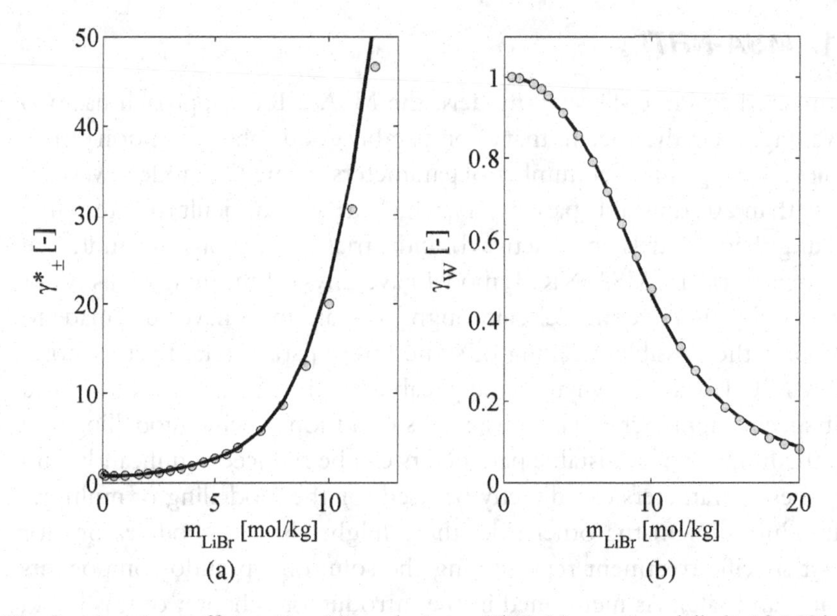

Fig. 2. Activity coefficients in a LiBr solution at 25°C. Lines are modelled with MSA-NRTL; symbols are experimental data from Hamer and Wu.[22] (a) MIAC of LiBr, and (b) activity coefficients of water in an aqueous LiBr solution.

the alkali hydroxide series is found for the LiOH solution whereas in the alkali bromide series the WAC of a LiBr solution is the lowest. Generally, smaller ions have higher surface charge densities (hard ions). Thus, the hydrated water is much stronger bound to smaller ions compared instead to bigger ions of the same charge (soft ions). Keeping this in mind, the thumb-rule 'like seeks like' presented in the introduction of this book helps to qualitatively understand the specific-ion effects occurring here. On the one hand, we consider the case of two unlike ions (hard–soft) in aqueous solutions as for lithium bromide in Fig. 3(a). Here, the hard (Li^+) and the soft (Br^-) ions will certainly not approach each other very closely as the electrostatic attraction of the anions is not strong enough to loose the hydration shell of the strong cation. This results in the fact that both the anion and the cation are still surrounded by water molecules, leading to low WAC values. Replacing Li^+ by the softer K^+ cation, the counterions become more similar so that the reciprocal electrostatic attraction increases and the ions approach each other more closely than in the LiBr solution. Thus, the ions are surrounded by fewer water molecules and the WAC values increase compared to the solution of LiBr.

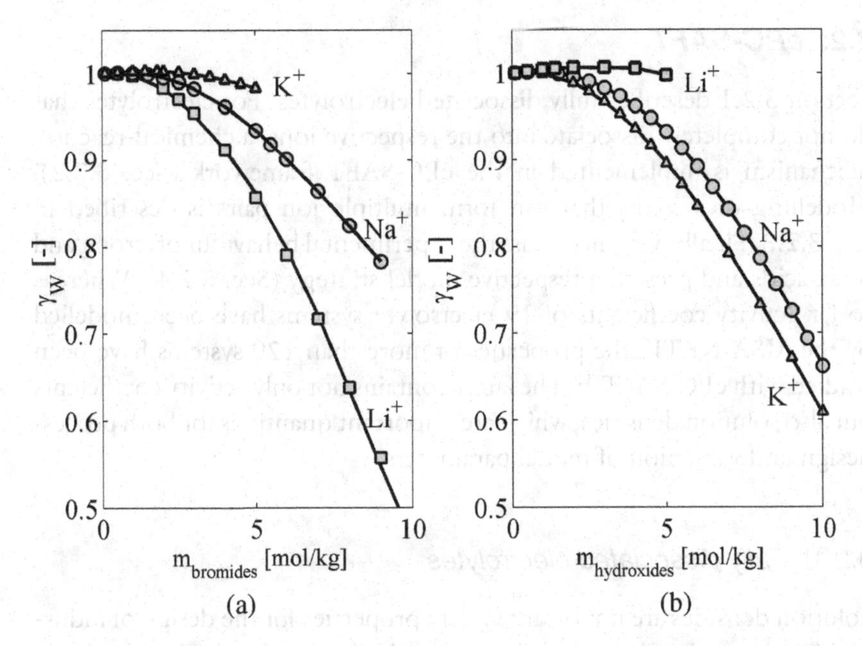

Fig. 3. Water activity coefficients of (a) three bromide and (b) three hydroxide aqueous salt solutions at 25°C as function of salt molality. Experimental data[22]: Li^+ — squares, Na^+ — circles, K^+ — triangles. The lines represent MSA-NRTL calculations. Activity coefficients decrease with decreasing size of the cation $K^+ > Na^+ > Li^+$ for bromide solutions but in the reversed order for hydroxide solutions.

In the case of two like ions (here: Li^+ and OH^-, both of them being hard ions), a very strong reciprocal attraction is expected. These counterions do approach very closely. They are surrounded by only a small number of water molecules, which lead to higher WAC values and even to a reversal in the WAC series.

This behaviour can quantitatively be captured with MSA-NRTL using a single parameter set per ion (Fig. 3). In Sec. 3.2.2 we will see that the acetate anion also causes a reverse in the WAC (or MIAC) series compared to alkali halides. However, no acetate parameters exist for MSA-NRTL so that we cannot provide direct comparisons for the acetates.

Although accurate results are obtained with the MSA-NRTL, it should be mentioned that applying an ion-specific treatment gives rise to an ARD value which is threefold compared to a salt-specific treatment for modelling the MIAC which was shown for 19 salts in water.[5]

3.2. ePC-SAFT

Section 3.2.1 describes fully dissociated electrolytes. For electrolytes that do not completely dissociate into the respective ions, a chemical-reaction mechanism is implemented in the ePC-SAFT framework (Sec. 3.2.2). Modelling of systems that can form multiple ion pairs is described in Sec. 3.2.3. Finally, we will discuss the experimental behaviour of strong and weak acids and present a respective model strategy (Sec. 3.2.4). Whereas so far activity coefficients of 19 electrolyte systems have been modelled by the MSA-NRTL, the properties of more than 120 systems have been studied with ePC-SAFT.[15] The latter contains not only activity coefficients but also solution densities, which are important quantities for both process design and validation of model parameters.

3.2.1. *Fully dissociated electrolytes*

Solution densities are important system properties for the design of industrial apparatus. Furthermore, they are typically used for model-parameter estimations and validation of model consistencies. As an example, the liquid densities of six caesium-salt solutions are shown in Fig. 4. All solution densities are presented as density differences $\Delta\rho$ between the densities of the aqueous salt solution and pure water at the same temperature. As to be seen, the experimental data can be described with high accuracy even at high salt concentrations of up to 6 mol/kg. Obviously, ePC-SAFT is a powerful model for the description of electrolyte solution densities (which in general cannot be described by a G^E model). This holds true as well for the prediction of these data at different temperatures.

Figure 5 shows the influence of alkali cations [Fig. 5(a)] and halide anions [Fig. 5(b)] on the water activity coefficients in aqueous solutions of alkali halides.

As already seen in Fig. 3, for all salt solutions, the experimental water activity coefficient first slightly increases with increasing salt concentration. After passing a maximum, the values decrease continuously. This behaviour holds true for every strong electrolyte in aqueous solution. Moreover, the experimental data show decreasing water activity coefficients for increasing sizes of the anion but for decreasing sizes of the cations. This means that the ion–water interactions increase in the order $Cl^- < Br^- < I^-$ for the anions, but in the order $K^+ < Na^+ < Li^+$ for the cations. Again — as

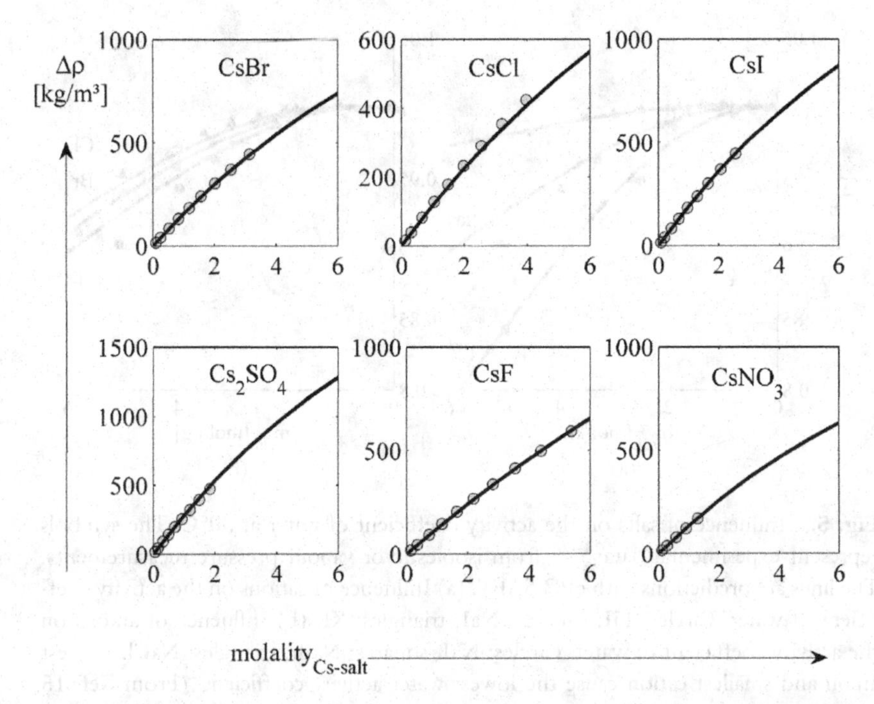

Fig. 4. Liquid densities of aqueous solutions of six caesium salts related to the density of pure water at 20°C as function of salt molaity. The lines represent calculations with ePC-SAFT. The circles represent experimental data.[23] (From Ref. 15, © Elsevier, reprinted with permission.)

explained before — the thumb of rule 'like seeks like' explains this specific behaviour: compared to the iodide anion, Li^+ differs most (→ low WAC value) whereas K^+ is most equal to I^- (→ high WAC value). Within the iodide series in Fig. 5(b), Li^+ is most equal to Cl^- and differs most from I^-. Thus, WAC values are high for LiCl and low for LiI aqueous solutions. As shown in Fig. 5, ePC-SAFT is able to predict the water activity coefficients in good agreement with experimental data.

An even more sensitive property characterising electrolyte solutions is the mean ionic activity coefficient (MIAC) which deviates much more from unity than the activity coefficient of water. Due to the Coulomb forces, the MIAC first decreases with increasing concentration (Fig. 6). After reaching a minimum, it often increases with increasing salt concentration, reaching in some cases very high values. Since a high value of this activity coefficient means a favourable interaction between the ions and water,

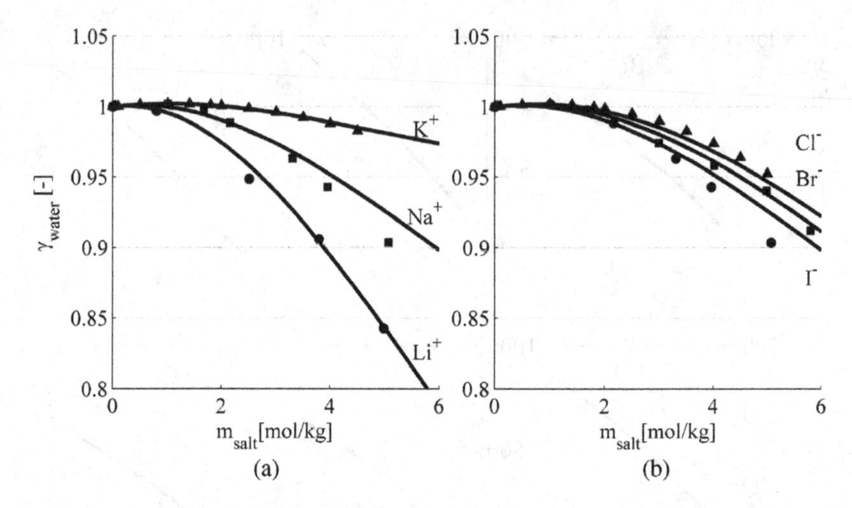

Fig. 5. Influence of salts on the activity coefficient of water at 30°C. The symbols represent experimental data[24–26] from isopiestic or vapour pressure measurements. The lines are predictions with ePC-SAFT. (a) Influence of cations on the activity coefficient of water. Circles: LiI, squares: NaI, triangles: KI. (b) Influence of anions on the activity coefficient of water. Circles: NaI, squares: NaBr, triangles: NaCl. Largest anion and smallest cation cause the lowest water activity coefficient. (From Ref. 15 © Elsevier, reprinted with permission.)

the latter phenomenon provides evidence for extensive hydration of some salts.[29]

As it can be seen form Fig. 6, ePC-SAFT performs well in modelling the MIAC. As a typical example, the modelling results for the MIAC of some alkali bromides at 25°C are shown. In analogy to WAC values, analysing experimental MIAC data (e.g. alkali metal bromides in Fig. 6) again allows for an interpretation of ionic hydration. The following MIAC sequence is observed: $\gamma^*_{LiBr} > \gamma^*_{NaBr} > \gamma^*_{KBr} > \gamma^*_{RbBr} > \gamma^*_{CsBr}$. As already concluded from Fig. 5, the smallest alkali cation (highest surface charge density) is most strongly hydrated and strong hydration leads to high MIAC values (low water activity coefficients). The MIAC calculations with ePC-SAFT (Fig. 6) compare to the MSA-NRTL results for LiBr (already shown in Fig. 2) as well as for the alkali bromide solutions (Fig. 3). Next to the activity coefficients in the alkali bromide solutions (Fig. 6), the reversed series for the hydroxides (Fig. 3) can also be described with ePC-SAFT.[15]

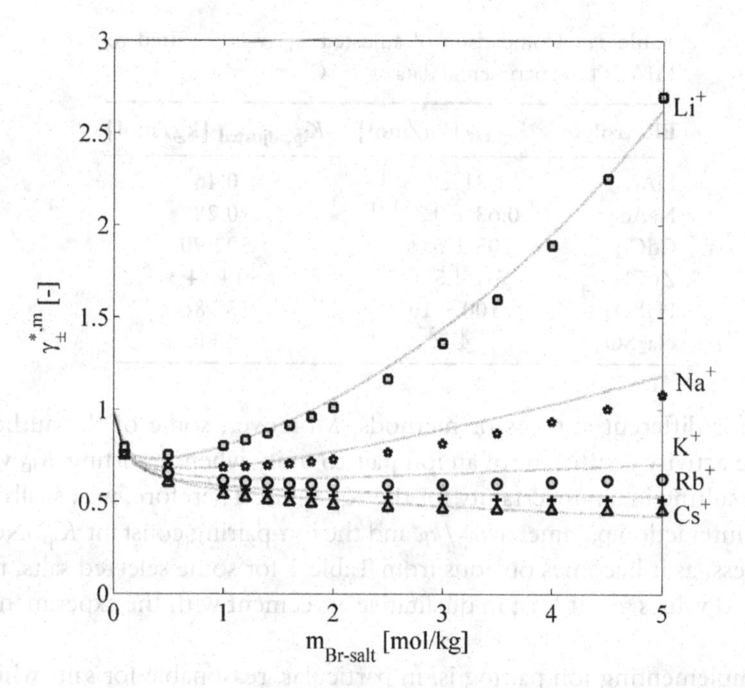

Fig. 6. Mean ionic activity coefficients of five bromide salts in aqueous solution at 25°C as function of salt molality. Experimental data[26]: LiBr — squares, NaBr — stars, KBr — circles, RbBr — crosses, CsBr — triangles. The dotted lines represent ePC-SAFT calculations. Activity coefficients decrease with increasing size of the cation $Li^+ > Na^+ > K^+ > Rb^+ > Cs^+$. (From Ref. 15, © Elsevier, reprinted with permission.)

3.2.2. *Salts with ion pairing*

Until now, we have discussed thermodynamic properties of strong electrolytes in water. In contrast to these systems, weak electrolytes do not fully dissociate in solution. Since in these cases fewer ions are available in solution, the MIAC of weak electrolytes can reach very low values (weak salt–water interactions). Implementing a chemical-reaction approach according to Sec. 2.2 provokes a reduced number of free hydrated ions in the modelling, and consequently the calculated MIAC also decreases.

In general, two additional adjustable parameters are needed to describe the properties of an ion pair: the dispersive interaction parameter $u_{\text{ip}}/k_{\text{B}}$ and the ion-pairing constant K_{ip}. The latter could in general be obtained from independent measurements. However, literature data scatter quite

Table 1. Comparison of adjusted K_{ip} values (fitted to MIAC) to experimental data at 25°C.

Electrolyte	$K_{ip,Lit}$ [kg/mol]	$K_{ip,adjusted}$ [kg/mol]
LiAc	1.81 ± 0.5[30]	0.46
NaAc	0.63 ± 0.3[30,31]	0.28
$CdCl_2$	105 ± 65[32,33]	502.90
$ZnCl_2$	5 ± 4.5[34–36]	14.84
H_2SO_4	100 ± 10[34]	137.86
Na_2SO_4	4.5[37]	4.50

a lot for different authors or methods. Moreover, some of the authors set the activity coefficient of an ion pair to unity when reporting K_{ip} values, resulting in an uncertainty for the K_{ip} value. Therefore, we usually fit both, interaction parameter u_{ip}/k_B and the ion-pairing constant K_{ip}. Nevertheless, as it becomes obvious from Table 1 for some selected salts, the adjusted values are at least in qualitative agreement with the experimental ones.

Implementing ion pairing is, in particular, reasonable for salts which are derivatives of weak acids, like e.g. acetates. As already mentioned above, electrolytes containing the acetate anion show a reversed sequence of experimental MIAC values compared to alkali salts as shown in Figs. 4 and 7. Here the sequence is $\gamma^*_{KAc} > \gamma^*_{NaAc} > \gamma^*_{LiAc}$ (see experimental data in Fig. 7).

The implementation of the ion-pairing equilibrium allows for a precise modelling of the MIAC of weak electrolytes, such as alkali acetates or cadmium halides in aqueous solution. As a typical example, Fig. 7 illustrates the improvement of the MIAC description for some alkali acetates when ion pairing is considered in the modelling. Neither the classical ePC-SAFT nor PC-SAFT-MSA (results not shown here) could describe this reversed MIAC series of the acetate salts as illustrated in Fig. 7(a). Accounting for ion pairing in these systems within ePC-SAFT allows for a correct description of this effect [Fig. 7(b)].

3.2.3. *Complex formation*

Cadmium and zinc halides are asymmetrical electrolytes where different types of ion pairs can be formed due to a complex regime of ion-pair

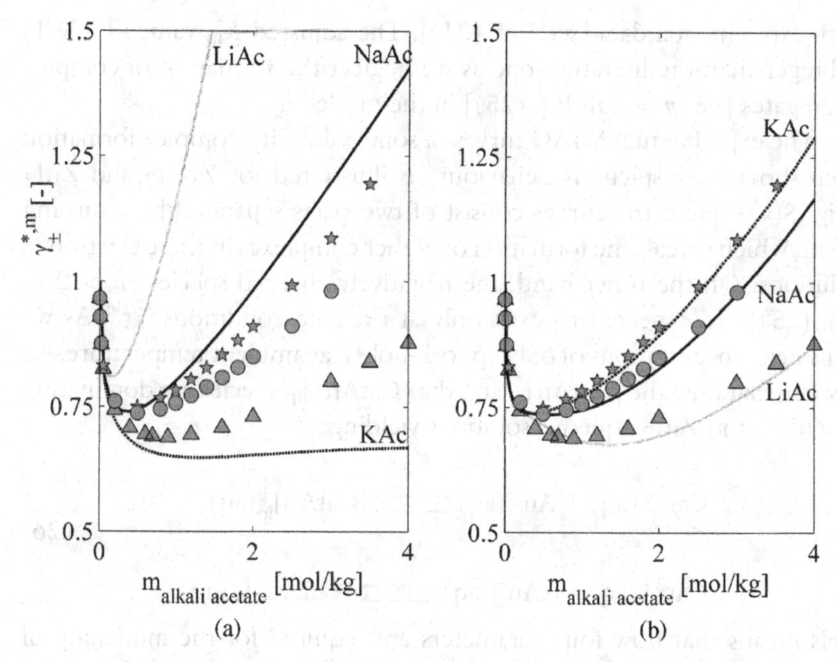

Fig. 7. Mean ionic activity coefficients of LiAc, NaAc, and KAc. The symbols represent experimental data from Lobo *et al.*[26] at 25°C, the lines are modelling results: KAc — stars and dashed lines; NaAc — circles and full lines; LiAc — triangles and thin lines. Experimental MIAC values increase in the order LiAc < NaAc < KAc. (a) Calculations were performed with the classical ePC-SAFT, neglecting ion pairing. Modelled MIAC values increase in the order LiAc > NaAc > KAc. (b) Calculations were performed with the ePC-SAFT, accounting for ion pairing. Modelled MIAC values increase in the experimentally validated order LiAc < NaAc < KAc. (From Ref. 21, © Elsevier, reprinted with permission.)

formation.[33,38] In the case of a 2:1 electrolyte, the following generally formulated ion-pairing steps might be possible[29,36,39]:

$$Cat^{2+}(aq) + nAn^{-}(aq) \overset{K_{ip,n}}{\rightleftharpoons} [CatAn_n]_{ip}^{2-n}(aq), \quad n = 1, \ldots, 4. \quad (25)$$

Barthel and Buchner[33] showed that for $CdCl_2$, the positively charged ion pair $[CdCl]_{ip}^{+}$ dominates in aqueous solution. For the modelling we therefore propose to regard this type of ion pair (e.g. $[CdCl]_{ip}^{+}$) as the only one being present in solution. Following this approach [i.e. $n = 1$ in Eq. (25)], the modelled MIAC of $CdCl_2$ is in excellent agreement with experimental data [see Ref. (21)]. The adjusted K_{ip} value (Table 1) is bigger than the literature one as we neglect the formation of complex aggregates [i.e. $n > 1$ in Eq. (25)] in the modelling.

The experimental MIAC curves of some salts with complex formation even show a conspicuous behaviour, as illustrated for $ZnBr_2$ and ZnI_2 [Fig. 8(a)]. Here the curves consist of two parts separated by a turning point, which reveals the formation of higher complexes in these electrolyte solutions. On the other hand, the negatively charged species [$n > 2$ in Eq. (25)] are expected to exist only at extreme conditions.[35,36] As we consider concentrations of only up to 4 mol/L at ambient temperature, we assume that only the $[CatAn]^+_{ip}$ and the $[CatAn_2]_{ip}$ species are dominating in $ZnBr_2$ and ZnI_2 aqueous solutions yielding:

$$\begin{aligned} &Cat^{2+}(aq) + An^-(aq) \xrightleftharpoons{K_{ip,1}} [CatAn]^+_{ip}(aq), \\ &Cat^{2+}(aq) + 2An^-(aq) \xrightleftharpoons{K_{ip,2}} [CatAn_2]_{ip}(aq). \end{aligned} \tag{26}$$

This means that now four parameters are required for the modelling of ion pairing: constants $K_{ip,1}$ and $u_{ip,1}/k_B$ (according to the formation of $[ZnAn]^+_{ip}$) as well as $K_{ip,2}$ and $u_{ip,2}/k_B$ (according to the formation of $[ZnAn_2]_{ip}$). Figure 8(a) illustrates that applying the two-step reaction mechanism allows for a very good correlation of the experimental MIACs of $ZnBr_2$ and ZnI_2. Simultaneously, the liquid densities [Fig. 8(b)] can be modelled accurately even at elevated temperatures, showing the consistency of the thermodynamic modelling.

3.2.4. *Acids*

Acids that fully dissociate into their respective ions in aqueous solution can be treated in the same manner as strong electrolytes.

The acid constants for aqueous solutions of HCl, HBr, and HI at 25°C are very high (see Table 2).

As experimental MIAC data show a similar characteristic compared to the strong electrolytes treated in Sec. 3.2.1, the classical ePC-SAFT approach can be applied to model liquid densities, WAC, and MIAC of those systems at 25°C. Because of the fact that the parameters reflect the properties of the hydrated ions, the H^+ parameters directly represent the H_3O^+ ion.

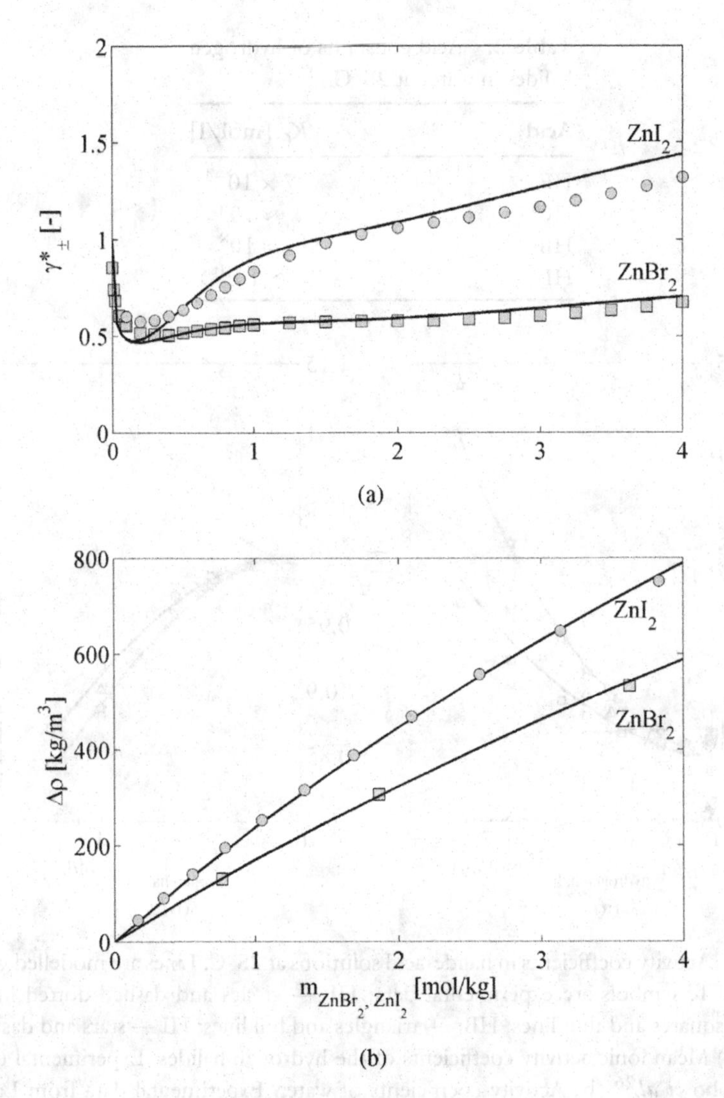

Fig. 8. Solution densities and MIAC of $ZnBr_2$ and ZnI_2 in water. The symbols represent experimental data, the lines show the calculations performed with ePC-SAFT: ZnI_2 — circles; $ZnBr_2$ — squares. (a) MIAC at 25°C. Experimental data from Lobo *et al.*[26] (b) Solution densities at 67.46°C ($ZnBr_2$) and at 19.50°C (ZnI_2) presented as density differences $\Delta\rho$ between salt solution and pure water at the same temperature. Experimental data from Wimby and Berntsson[40] for $ZnBr_2$ and from D'Ans *et al.*[23] for ZnI_2. (From Ref. 21, © Elsevier, reprinted with permission.)

Table 2. Acid constants of hydrogen halides in water at 25°C.

Acid	K_a [mol/l]
HF	7×10^{-4}
HCl	$\approx 10^6$
HBr	$\approx 10^8$
HI	$\approx 10^{10}$

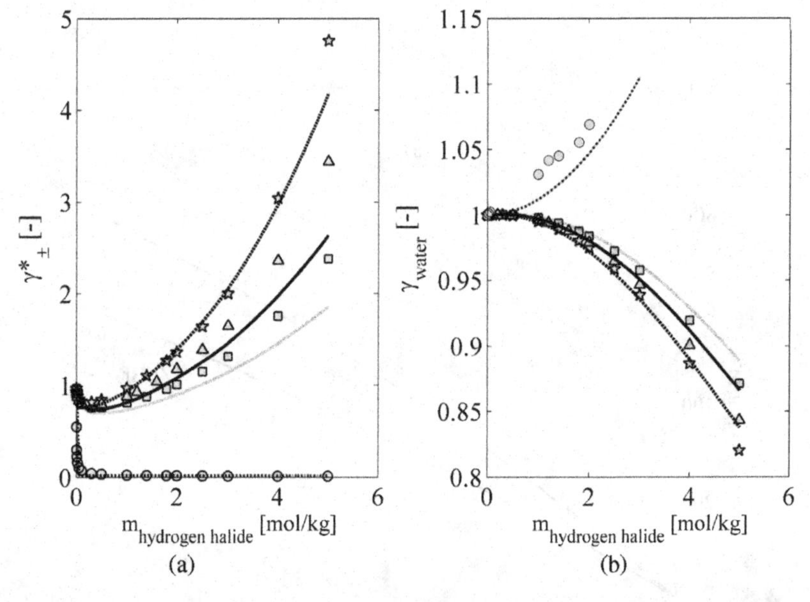

Fig. 9. Activity coefficients in halide-acid solutions at 25°C. Lines are modelled with ePC-SAFT, symbols are experimental data: HF — circles and dashed-dotted lines; HCl — squares and thin lines; HBr — triangles and full lines; HI — stars and dashed lines. (a) Mean ionic activity coefficients of the hydrogen halides. Experimental data from Lobo *et al.*[26] (b) Activity coefficients of water. Experimental data from Lobo *et al.*[26] (From Ref. 21, © Elsevier, reprinted with permission.)

Figure 9 shows experimental data and the ePC-SAFT calculations of the components' activity coefficients in aqueous systems containing hydrogen halides. The calculations for HCl, HBr, and HI in water were performed with the universal parameter set.[15] The modelled MIAC of the acids are illustrated in Fig. 9(a). The influence of the considered hydrogen halides on the water activity coefficients (WAC) can be seen in Fig. 9(b).

As for every strong electrolyte, a slight increase of the experimental WAC data is observed for aqueous HCl, HBr, and HI solutions at very low hydrogen-halide concentrations. After passing a maximum, the values decrease continuously. This behaviour is correctly predicted by ePC-SAFT. Figure 9(a) illustrates that the experimental MIAC data of strong acids show the same characteristics as those of salts in water. Starting from unity, the MIAC values first decrease with increasing acid concentration. After reaching a minimum, they again increase with rising hydrogen-halide concentration, reaching the highest value for the acid containing the anion which differs most from the cation (in this case I^-).[29]

Compared to the other hydrogen-halide acids, the acid constant of aqueous HF is very small. Thus, HF in water is a very weak acid which also has consequences for the experimentally observed activity coefficients. Whereas the WAC in the other acid systems is decreasing after the initial increase, it continuously increases in an aqueous HF solution. This indicates very weak water–ion interactions due to the low degree of dissociation (low K_a value) in these solutions. For this reason, we do not apply the universal ePC-SAFT ion parameters for the HF solution, but consider the association/dissociation equilibrium of HF following the approach described in Sec. 3.1.2 using the acid constant K_a (which is the reverse of K_{ip}) of fluoric acid. Following this approach, the MIAC can accurately be modelled [Fig. 9(a)]. Furthermore, the water activity coefficient is predicted reasonably well [Fig. 9(b)]. Although the data are not shown here, the solution densities at 283.15 K (experimental data from Ref. 35) are simultaneously (with the same parameter set) modelled with an absolute relative deviation ARD of $< 1\%$ up to 6 mol/kg.

3.3. *Parameter Discussion*

Until now the calculations showed that both MSA-NRTL and ePC-SAFT are powerful models for the calculation of thermodynamic properties that are required for designing technical processes. In this section, we finally focus on the meaning of the model parameters and how they can be used to interpret experimentally observed phenomena.

The first parameter which shall be investigated is the hydrated-ion diameter. In case of ePC-SAFT, the size of this parameter is determined by the volume of the bare ion and its surface charge density, both of which influencing the degree of hydration. Thus, it gives an indirect hint to the strength of the interaction between ion and water. Table 3 presents the hydrated diameters of some selected ions as determined for the two models as well as the Pauling diameters of the bare ions.

For both models, all cation parameters are considerably larger than their respective Pauling radii. This is physically correct as the hydrated-size parameter represents the diameter of the hydrated ions rather than the diameter of the bare ions. In contrast, the anion diameters are comparable to the Pauling radii as anions are weakly hydrated compared to cations. Besides their lower charge density, this can also be explained by the structure of the water molecules: the hydrogen atoms of water can approach the anions about 0.85 Å more closely than the oxygen atoms approaching the cations.[41] The anionic MSA-NRTL parameter sizes were directly taken from the Pauling diameters.

However, the cation-size series of ePC-SAFT and MSA-NRTL differ in their sequence. Obviously, the adjusted MSA-NRTL parameters do not reflect the series of the Pauling diameters. This is due to the fact that they directly reflect the strength of hydration and that no volumetric properties were used for the respective parameter estimation. Thus, the most strongly hydrated cation (Li^+) within the alkali halide series considered here seems to have the largest diameter. In contrast, the ePC-SAFT diameters for the alkali cations were obtained by adjusting them to MIAC *and* volumetric data (solution densities). They follow the same trend and are in the same order of magnitude as the Pauling diameters are.

For the cations Li^+, Na^+, and K^+, Collins *et al.*[41] reported X-ray and neutron diffraction measurements of aqueous salt solutions, providing the distance between a central cation and the nearest water oxygen. This directly compares to the ePC-SAFT parameter σ, since this is in fact the diameter of the hydrated cation. Comparing these experimental values σ_{exp} to the σ parameters given in Table 3 shows an excellent agreement of experimental data and adjusted parameters.

The second parameter which is used in the electrolyte models is a parameter representing the short-range interactions between water and an ion. In Table 4, anion–water and cation–water parameters are compared for the two models. Both the ePC-SAFT u/k_B as well as the MSA-NRTL $\tau_{W-\text{ion}}$ interaction parameters directly reflect the strength of ionic hydration. The higher u/k_B and the more negative $\tau_{W-\text{ion}}$ are, the more strongly hydrated the ion is (strong interaction with water) within the considered alkali halides.

Table 3. Comparison of hydrated-ion sizes: Experimental values from X-ray and neutron diffraction measurements[41] versus ePC-SAFT and MSA-NRTL parameters.

Ion	σ_{Pauling} [Å]	σ_{exp} [Å]	$\sigma_{\text{ePC-SAFT}}$ [Å]	$\sigma_{\text{MSA-NRTL}}$ [Å]
Li^+	1.20	1.86	1.81	5.43
Na^+	1.90	2.40	2.41	3.87
K^+	2.66	3.10	2.97	3.45
Cl^-	3.62	—	3.06	3.62
Br^-	3.90	—	3.46	3.90
I^-	4.32	—	3.93	4.32

For the sake of completeness, the Pauling diameters of the bare ions[42] are given, too.

Table 4. Comparison of water/ion interaction parameters for ePC-SAFT and MSA-NRTL.

Ion	u/k_B [K]	$\tau_{\text{W-ion}}$ [−]
Li^+	2697.27	−7.5
Na^+	646.05	−7.0
K^+	271.05	−6.8
Cl^-	47.29	−1.9
Br^-	60.22	−2.2
I^-	80.44	−2.4

In both models, the following series can be observed for anions and cations:

$$\begin{array}{ll} \text{interaction water-cation} & Li^+ > Na^+ > K^+ \\ \text{interaction water-anion} & Cl^- < Br^- < I^- \\ & \text{small ions} \longrightarrow \text{large ions} \end{array}$$

Obviously, the smallest cation but the largest anion interact most strongly with water within the considered systems. Analysing the experimental activity coefficients of water, these parameter series can be confirmed (Fig. 5 and accompanying text). Additionally, it is obvious that the cations are much stronger hydrated than the anions, which is also reflected by the ePC-SAFT and MSA-NRTL parameters u/k_B and τ where the general relation is valid:

$$u_{\text{cation}}/k_B \gg u_{\text{anion}}/k_B \quad \text{and} \quad |\tau_{\text{cw}}| \gg |\tau_{\text{aw}}|.$$

Table 5. Comparison of relative bare Pauling ionic diameters.[42] The smallest cation as well as the smallest anion correspond to the 100% value.

Cation	σ/σ_{Li^+}	Anion	σ/σ_{Cl^-}
Li^+	100%	Cl^-	100%
Na^+	160%	Br^-	110%
K^+	220%	I^-	120%

It can be further observed from Fig. 5 that the influence of the halide anions [Fig. 5(b)] on the water activity coefficient is much weaker than that of the alkali cations [Fig. 5(a)]. This can be explained by the fact that the relative size difference in Pauling diameters is much bigger for the alkali cations than it is for the halide anions (Table 5).

From the above discussion, it becomes obvious that the parameters of both models have physical meanings and are reasonably sequenced. Moreover, by the illustrated differences between anion and cation as well as between small and large ions, it seems possible to qualitatively estimate parameters without having any experimental data.

4. Summary

In this study, the ePC-SAFT EOS as well as the MSA-NRTL G^E model were applied to describe thermodynamic properties of numerous aqueous electrolyte solutions. Whereas only activity coefficients are obtained by the G^E model, volumetric properties can be calculated with an EOS. Ion-specific parameters were used independent of the electrolyte which the ions are part of. The model parameters possess a physical meaning and show reasonable trends within the ion series. Two ion parameters are needed in ePC-SAFT, whereas six parameters are necessary for applying MSA-NRTL. Next to the standard alkali halide electrolyte systems, both models even capture the non-ideal behaviour of solutions containing acetate or hydroxide anions where a reversed MIAC series is experimentally observed. Until now, thermodynamic properties of more than 120 aqueous systems could be successfully modelled with ePC-SAFT. The MSA-NRTL parameter set has also been applied to a couple of systems (so far 19 solutions). Implementing an ion-pairing reaction in ePC-SAFT, it is possible to correlate MIACs of weak electrolytes where up to five components are theoretically present in solution.

List of symbols

a_i	[–]	activity of component i
A	[J]	Helmholtz free energy
e	[C]	elementary charge, 1.6022×10^{-19}C
f_i	[–]	symmetrical activity coefficient of component i (related to pure component) based on mole-fraction scale
f_i^*	[–]	asymmetrical activity coefficient of component i (related to infinite dilution) based on mole-fraction scale
G	[J]	Gibbs energy
k_B	[J/K]	Boltzmann constant, 1.38065×10^{-23} J/K
K	[kg/mol]	equilibrium constant
m	[mol/kg]	molality (moles of solute i per kg solvent)
M	[g/mol]	molecular weight
m_{seg}	[–]	number of segments
N	[–]	total number of particles
NP	[–]	number of data points
p	[bar]	pressure
P_{ij}	[–]	probability of finding a molecule of species i in the neighbourhood of a central molecule of species j
R	[J/molK]	ideal gas constant
T	[K]	temperature
u/k_B	[K]	dispersion-energy parameter
V	[m^3]	volume
x	[–]	mole fraction
z	[–]	charge number
Z	[–]	compressibility factor

Greek letters

α	[–]	NRTL non-randomness parameter
γ_i^*	[–]	asymmetrical activity coefficient of component i (related to infinite dilution) based on molality scale

Γ	[1/Å]	MSA screening length
ε	[C/Vm]	dielectric constant of a medium
φ_i	[–]	fugacity coefficient of component i
κ	[1/Å]	Debye length
μ_i	[J]	chemical potential
ρ	[kg/m^3]	density
ρ_N	[1/Å^3]	number density (number of particles per volume)
ν	[–]	stoichiometric factor
σ_i	[Å]	temperature-independent segment diameter of molecule i
τ_{ij}	[–]	NRTL interaction parameters between species i and j

Subscripts

a	acid
a, An, –	anion
c, Cat, +	cation
i, j, k	component indexes
ip	ion pair
seg	segment
w	water
±	mean ionic
0	pure substance

Superscripts

assoc	association
calc	calculated
disp	dispersion
E	excess
exp	experimental
hc	hard chain
hs	hard sphere
ion	ionic
LR	long range
m	based on molality
res	residual
SR	short range

sym	symmetrical (related to pure component)
x	based on mole fraction
+, −	positive or negative charge
∞	infinitely diluted

Abbreviations

ARD	absolute average relative deviation, defined in Eq. (23)
DH	Debye–Hückel model
EOS	equation of state
G^E	excess Gibbs energy
MIAC	mean ionic activity coefficient
MSA	Mean Spherical Approximation
pvT	pressure–volume–temperature behaviour (density data)
VLE	vapour–liquid equilibrium

References

1. Enick RM, Klara SM. (1992) Effects of CO_2 solubility in brine on the compositional simulation of CO_2 floods. *SPE Reserv Eng* 7: 253–258.
2. Luckas M, Krissmann J. (2001) *Thermodynamik der Elektrolytlösungen: Eine einheitliche Darstellung der Berechnung komplexer Gleichgewichte*. Springer, Berlin.
3. Debye P, Hückel E. (1923) Zur theorie der elektrolyte. I. Gefrierpunktserniedrigung und verwandte erscheinungen. *Phys Z* 9: 185–206.
4. Waisman E, Lebowitz JL. (1970) Exact solution of an integral equation for structure of a primitive model of electrolytes. *J Chem Phys* **52**: 4307–4311.
5. Papaiconomou N, Simonin JP, Bernard O, Kunz W. (2002) MSA-NRTL model for the description of the thermodynamic properties of electrolyte solutions. *Phys Chem Chem Phys* **4**: 4435–4443.
6. Chen CC, Britt HI, Boston JF, Evans LB. (1982) Local composition model for excess Gibbs energy of electrolyte systems. Part I: Single solvent, single completely dissociated electrolyte systems. *AIChE J* **28**: 588–596.
7. Renon H, Prausnitz JM. (1968) Local compositions in thermodynamic excess functions for liquid mixtures. *AIChE I* **14**: 135–144.
8. Harvey AH, Copeman TW, Prausnitz JM. (1988) Explicit approximations to the mean spherical approximation for electrolyte systems with unequal ion sizes. *J Phys Chem* **92**: 6432–6436.
9. Simonin JP, Krebs S, Kunz W. (2006) Inclusion of ionic hydration and association in the MSA-NRTL model for a description of the thermodynamic properties of aqueous ionic solutions: Application to solutions of associating acids. *Ind Eng Chem Res* **45**: 4345–4354.

10. Simonin JP, Bernard O, Krebs S, Kunz W. (2006) Modelling of the thermodynamic properties of ionic solutions using a stepwise solvation-equilibrium model. *Fluid Phase Equilibr* **242**: 176–188.
11. Simonin JP, Bernard O, Papaiconomou N, Kunz W. (2008) Description of dilution enthalpies and heat capacities for aqueous solutions within the MSA-NRTL model with ion solvation. *Fluid Phase Equilibr* **264**: 211–219.
12. Gross J, Sadowski G. (2001) Perturbed-chain SAFT: An equation of state based on a perturbation theory for chain molecules. *Ind Eng Chem Res* **40**: 1244–1260.
13. Seyfkar N, Ghotbi C, Taghikhani V, Azimi G. (2004) Application of the nonprimitive MSA-based models in predicting the activity and the osmotic coefficients of aqueous electrolyte solutions. *Fluid Phase Equilibr* **221**: 189–196.
14. Cameretti LF, Sadowski G, Mollerup JM. (2005) Modeling of aqueous electrolyte solutions with perturbed-chain statistical associated fluid theory. *Ind Eng Chem Res* **44**: 3355–3362; ibid., 8944.
15. Held C, Cameretti LF, Sadowski G. (2008) Modeling aqueous electrolyte solutions. Part 1: Fully dissociated electrolytes. *Fluid Phase Equlilib* **270**: 87–96.
16. Huang SH, Radosz M. (1990) Equation of state for small, large, polydisperse, and associating molecules. *Ind Eng Chem Res* **29**: 2284–2294.
17. Fawcett WR, Tikanen AC. (1996) Role of solvent permittivity in estimation of electrolyte activity coefficients on the basis of the mean spherical approximation. *J Phys Chem* **100**: 4251–4255.
18. Fawcett WR, Tikanen AC. (1997) Application of the mean spherical approximation to the estimation of electrolyte activity coefficients in methanol solutions. *J Mol Liq* **73–4**: 373–384.
19. Tikanen AC, Fawcett WR. (1996) The role of solvent permittivity in estimation of electrolyte activity coefficients for systems with ion pairing on the basis of the mean spherical approximation. *Ber Buns Phys Chem* **100**: 634–640.
20. Tikanen AC, Fawcett WR. (1997) Application of the mean spherical approximation and ion association to describe the activity coefficients of aqueous 1 : 1 electrolytes. *J Electroanal Chem* **439**: 107–113.
21. Held C, Sadowski G. (2009) Modeling aqueous electrolyte solutions. Part 2 Weak electrolytes. *Fluid Phase Equilibria* **279**: 141–148.
22. Hamer WJ, Wu Y-C. (1972) Osmotic coefficients and mean activity coefficients of uni-univalent electrolytes in water at 25°C. *J Phys Chem Ref Data* **1**: 1047.
23. D'Ans J, Surawski H, Synowietz C. (1977) *Landolt-Börnstein, Group IV Volume 1b: Densities of Binary Aqueous Systems and Heat Capacities of Liquid Systems*, Springer, Berlin.
24. Patil KR, Tripathi AD, Pathak G, Katti SS. (1991) Thermodynamic properties of aqueous-electrolyte solutions. 2. Vapor-pressure of aqueous-solutions of NaBr, NaI, KI, KBr, KI, RbCl, CsCl, CsBr, CsI, $MgCl_2$, $CaCl_2$, $CaBr_2$, CaI_2, $SrCl_2$, $SrBr_2$, SrI_2, $BaCl_2$, and $BaBr_2$. *J Chem Eng Data* **36**: 225–230.
25. Patil KR, Tripathi AD, Pathak G, Katti SS. (1990) Thermodynamic properties of aqueous-electrolyte solutions. 1. Vapor-pressure of aqueous-solutions of LiCl, LiBr, and LiI. *J Chem Eng Data* **35**: 166–168.
26. Lobo VMM, Quaresma JL. (1989) *Handbook of Electrolyte Solutions.* Parts A and B. Elsevier, Amsterdam.

27. Steudel R. (1998) *Chemie der Nichtmetalle.* 2. Auflage. W. de Gruyter, Berlin.
28. Hamann CH, Hamnett A, Vielstich W. (2007) *Electrochemistry.* 2nd edn., Wiley-VCH, Weinheim.
29. Robinson RA, Stokes RH. (1970) *Electrolyte Solutions*, 2nd edn., Butterworth, London.
30. Archer DW, Monk CB. (1964) Ion-association constants of some acetates by pH (glass electrode) measurements. *J Chem Soc* : 3117–3121.
31. Fournier P, Oelkers EH, Gout R, Pokrovski G. (1998) Experimental determination of aqueous sodium-acetate dissociation constants at temperatures from 20 to 240 degrees C. *Chem Geol* **151**: 69–84.
32. Reilly PJ, Stokes RH. (1970) Activity coefficients of cadmium chloride in water and sodium chloride solution at 25 degrees. *Aust J Chem* **23**: 1397–1402.
33. Barthel J, Buchner R. (1991) High-frequency permittivity and its use in the investigation of solution properties. *Pure Appl Chem* **63**: 1473–1482.
34. Simonin JP, Bernard O, Blum L. (1998) Real ionic solutions in the mean spherical approximation. 3. Osmotic and activity coefficients for associating electrolytes in the primitive model. *J Phys Chem B* **102**: 4411–4417.
35. Lutfullah, Dunsmore HS, Paterson R. (1976) Re-determination of the standard electrode potential of zinc and mean molal activity coefficients for aqueous zinc chloride at 298.15 K. *J Chem Soc Faraday Transact 1* **72**: 495–503.
36. Ninkovic R, Miladinovic J, Todorovic M, Grujic S, Rard JA. (2007) Osmotic and activity coefficients of the {xZnCl(2)+(1−x)ZnSO4}(aq) system at 298.15 K. *J Solution Chem* **36**: 405–435.
37. Izatt RM, Eatough D, Christensen JJ, Bartholomew CH. (1969) Calorimetrically determined log K delta H degree and delta S degree values for interaction of sulphate ion with several Bi- and Ter-valent metal ions. *J Chem Soc A*: 47–53.
38. Mortazavi-Manesh S, Taghikhani V, Ghotbi C. (2007) A new model in correlating the activity coefficients of aqueous electrolyte solutions with ion pair formation. *Fluid Phase Equilib* **261**: 313–319.
39. Sillén LG. (1964) *Stability Constants of Metal Ion Complexes.* 2nd ed., Pergamon Press, Oxford.
40. Wimby JM, Berntsson TS. (1994) Viscosity and density of aqueous-solutions of Libr, Licl, Znbr2, Cacl2, and Lino3 .1. Single salt-solutions. *J Chem Eng Data* **39**: 68–72.
41. Collins KD, Neilson GW, Enderby JE. (2007) Ions in water: Characterizing the forces that control chemical processes and biological structure. *Biophys Chem* **128**: 95–104.
42. Horvath AL. (1985) *Handbook of Aqueous Electrolyte Solutions: Physical Properties, Estimation and Correlation Methods.* Ellis Horwood, Chichester.

Part B

PROMISING EXPERIMENTAL TECHNIQUES

Chapter 4

Linear and Non-linear Optical Techniques to Probe Ion Profiles at the Air–Water Interface

Hubert Motschmann* and Patrick Koelsch†

Aqueous ions at interfaces play a substantial role in numerous processes in our environment. The diversity of ion-specific phenomena has been extensively investigated both theoretically and experimentally, and has expanded our knowledge about ion-specific effects. In particular, the *in situ* application of novel experimental tools such as non-linear optical spectroscopic techniques has led to a more detailed picture of interfacial ion profiles. Among these, second-harmonic-generation (SHG) spectroscopy and sum-frequency-generation (SFG) spectroscopy have emerged in recent years as suitable methods for characterizing aqueous interfaces. The selection rules for SHG and SFG dictate that a signal can only be generated in a non-centrosymmetric environment, effectively resulting in the suppression of isotropic bulk signal. This feature makes non-linear spectroscopy inherently surface-specific with submonolayer resolution, allowing for the tracking of subtle modifications in the interfacial layer. Being all optical techniques, SHG and SFG are also non-invasive and applicable under aqueous conditions, making them suitable for *in situ* characterization of ions at the air–water interface. In this chapter, an overview is given about linear and especially non-linear optical techniques used to investigate ions at charged and uncharged air–water interfaces. The theoretical background of non-linear optics is briefly reviewed and some recent experimental results are critically examined.

1. Introduction

Various manifestations of specific ion effects in physicochemical and biological systems have been discussed. For instance, the solubility of small

*Institute of Physical and Theoretical Chemistry, University of Regensburg, 93040 Regensburg, Germany.
†Institute of Toxicology and Genetics, Karlsruhe Institute of Technology (KIT), Hermann-von-Helmholtz-Platz 1, D-76344 Eggenstein-Leopoldshafen, Germany.

molecules such as benzene in aqueous solution depends strongly on the nature of the added salts as demonstrated in the work of McDevit *et al.*[1] All salting-in and salting-out effects are strongly influenced by the presence and the nature of the dissolved electrolytes. The solubility of proteins in aqueous solutions is governed by electrolytes in an ion specific fashion.[2] The catalytic activity of certain enzymes depends strongly on the nature of the salts. Pinna *et al.*[3] studied the hydrolytic activity of Aspergillus Niger Lipase and found that the activity is determined by the concentration and the nature of the dissolved electrolytes. The analysis of all these phenomena leads to the Hofmeister sequence of ions and its existence and widespread applicability suggests an underlying common principle.

Obviously, ions are not simple charged hard spheres. If so, ions with comparable diameter and valence should behave in a similar fashion, but that is not the case. However, why do ions act in a specific way? Are we able to identify a single parameter which can predict how a salt solution behaves in a certain experiment?

The answers which are discussed in this book are based on the following three concepts. The first one introduces ion specificity through collective dispersion type interactions; an ion specificity is thereby obtained by the explicit consideration of the size and the polarisability of the ions. Based on molecular dynamics (MD) simulation with polarisable force fields, Jungwirth and Tobias[4] state that induction interactions close to the free surface may be responsible for the preference of heavier ions at interfacial solvation sites. The asymmetric, incomplete solvation shell induces a sizable dipole on the anion at the interface, which is assumed to be the driving force for the interfacial propensity of the ions. MD simulation provides a very detailed picture of the interfacial architecture; however, the results depend strongly on the interaction potentials which are not exactly known. Hence, experiments are needed to verify the predictions. Indeed, this task is challenging and many sophisticated surface analytical techniques, even when pushed to the limits, may still yield only inconclusive results.

The second concept used for the interpretation of ion specificity focuses on the impact of the ion on the water structure. It is assumed that the ions distort the hydrogen bonding network leading to a particularly apt description of structure-breaking and structure-making ions. This should reflect the impact of the ions on the long range structuring of water. However, convincing experimental and theoretical evidence is missing. In contrary, there are strong experimental arguments against such a picture.

Smith *et al.*[5] analysed Raman spectra of water OH vibrations in potassium halide solutions. The spectral signature of the OH vibrational band is sensitive to details of the hydrogen-bonding network. Spectra of fluoride solutions are blue-shifted compared with neat water, whereas solutions of the heavier halides are red-shifted. These effects were previously explained in the framework of the structure-making and structure-breaking abilities. However, Monte Carlo simulations revealed that these perturbations are largely confined to the first solvation shell. Mancinelli *et al.*[6] identified further contradictions of the structure-maker/structure-breaker concept based on neutron diffraction study of NaCl and KCl solutions. Both cations are commonly classified as water structure-breakers, however, the analysis of the neutron diffraction experiment does not support this finding, instead pronounced differences in the solvation shell of both ions have been found. The sodium ions are more tightly solvated and more disruptive to water–water correlation as compared to potassium. So again, a suggestive picture is not really supported by experiments.

Last but not least, the idea of a specific chemical interaction has been proposed that leads to the formation of contact ion pairs, which is assumed to control the ion binding to proteins and their solubility. Kim Collins[7] introduces the law of matching water affinities that interprets the majority of ion specific effects, such as the formation of inner contact ion pairs. Here the driving force is not a specific chemical interaction between the oppositely charged ions; instead it is the consequence of the attractive force which is put in relation to the strength of the prevailing ion–water interaction. Small (hard) ions possess a tightly bound hydration shell, but the force between the ions is sufficiently strong to squeeze out the hydration shell. Big (soft) ions possess a loosely bound hydration shell, the electrostatic interaction is much weaker but still sufficiently strong to expel the hydration shell. The combination of hard and soft ions leads to a mismatch and both ions remain separated by a hydration layer. This simple concept has an amazingly predictive power.[7]

In this chapter, we focus on the simplest model system, namely the air–water interface, and its modification by the presence of ions. The major conceptual advantage is the simplification due to the absence of interfacial binding and the absence of ions in the gas phase. The reduced number of possibilities makes the air–water interface a crucial model system to study fundamental interactions. In a second step, complexity can be added by modifying the surface with a soluble or insoluble charged amphiphile, allowing it to control the surface charge to a certain extent and to analyse the consequences experimentally.

Linear and non-linear optical techniques contribute to our understanding of the architecture of aqueous electrolytes–air interfaces. We discuss recent developments and accomplishment as well as the limitations and the underlying assumptions of these techniques.

2. Surface Tension Measurements

The conventional picture of the interface of simple aqueous salt solutions is based on thermodynamic analysis of the equilibrium surface tension isotherm. Valuable sources for the equilibrium surface tension isotherm of a simple aqueous electrolyte solution are the papers of Jarvis and Scheiman,[8] and P. Weissenborn and Robert J. Pugh (See Chap. 1, Fig. 8).[9]

In general, ions increase the surface tension in a specific manner. However, it is worth mentioning that certain combinations of ions decrease the surface tension or have a negligible effect on it. The thermodynamic analysis of the surface isotherm leads to the picture that the interfacial zone is depleted of ions. The surface deficiency is calculated using Gibbs equation as the derivative of the surface tension isotherm with a dividing plane chosen at a location that the surface excess of water vanishes.

$$\Gamma_{\mathrm{el}}^{\mathrm{H_2O}} = -\frac{1}{RT}\left(\frac{\mathrm{d}\gamma}{\mathrm{d}\ln a_{\mathrm{el}}}\right).$$

Here, γ is the equilibrium surface tension, a_{el} is the electrolyte activity and R is the gas constant and T denotes the temperature. Because of the fact that the slope of the isotherm is positive, it is concluded that the interfacial region is depleted of ions. This conclusion was further supported by a continuum dielectric model introduced by Onsager and Samaras,[10] suggesting that an ion is repelled from the interface between two media of different dielectric constant by an image charge of the same size and polarity located on the other side of the interface.

These observations and interpretations defined the textbook picture. This point of view has been challenged by the progress in understanding atmospheric reactions which in turn motivated molecular dynamics (MD) simulations.[4,11,12] MD simulations using polarisable force fields predict that soft ions such as halides are enriched at the interface with non-monotonic ion profiles. The book chapter of Pavel Jungwirth covers this in greater detail.

This is only an apparent contradiction to the conclusion drawn from the analysis of the equilibrium surface tension isotherm. Thermodynamics can accommodate several conflicting interfacial models provided that the integral excess or depletion is in accordance to Gibbs equation. Therefore, experiments are needed that yield direct insights in the interfacial architecture. These data can be obtained by optical techniques.

3. Challenges in the Investigation of Liquid–Air Interface

The investigation of aqueous electrolyte–air interface encounters a couple of intrinsic challenges. Firstly, the majority of material is dissolved in the bulk and the interfacial region comprises only a tiny fraction of the total material of the system. Consequently, spectroscopic investigations with classical techniques such as Infrared, Raman or UV-spectroscopy are often hampered by the lack of surface specificity and the signals are dominated by bulk contributions. Secondly, the processes at the air–water interface are highly dynamic. On a molecular scale there is a tremendous traffic towards both adjacent bulk phases. Molecules evaporate and condense at the interface and diffuse towards the bulk phase. There is no defined static molecular arrangement and as a consequence, fairly broad spectral features are expected. Moreover, many powerful surface specific techniques such as electron loss spectroscopy have special requirements to the sample and the environment (e.g. UHV-conditions) and cannot be applied to the liquid–air interface.

An elegant alternative which provides an intrinsic surface specificity is provided by non-linear optical reflection techniques based on a second-order effect (Fig. 1). Second-harmonic-generation (SHG) and sum-frequency-generation (SFG) spectroscopy contributed significantly to our current understanding of liquid–air, solid–liquid and liquid–liquid interfaces. Many fundamental insights on the structure of liquid–air interfaces are based on SHG and SFG experiments which are discussed in the next section.

3.1. *Basis of SHG and SFG Spectroscopy*

Second harmonic generation and sum-frequency generation have the same theoretical grounding outlined by Bloembergen and Pershan[13]. Shen

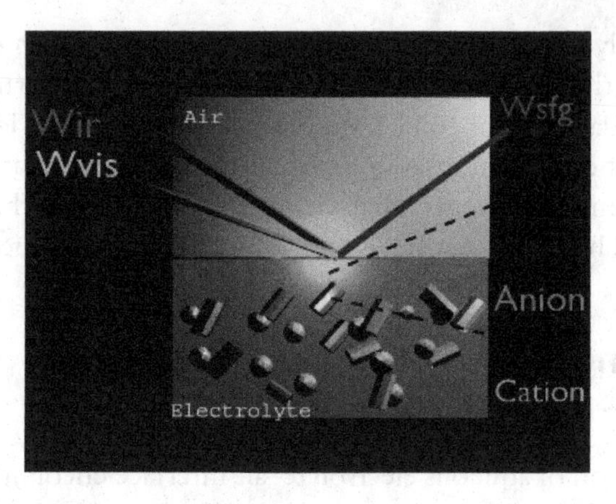

Fig. 1. Principle of an IR-VIS SF spectrum. The spatial and temporal overlap of an infrared and visible laser pulse at the interface generates light at the sum frequency which is emitted in a defined direction. A vibrational spectrum is obtained by scanning the infrared frequency. All second-order non-linear optical effects probe specifically the molecules at interfaces, and contributions from the bulk phase are to a large extent suppressed. *Note.* In an electrolyte solution, the majority of material is dissolved in the bulk and the interfacial region comprises only a tiny fraction of total material of the system. (Reprinted with Permission from Ref. 63. Copyright 2007 American Chemical Society.)

et al.[14] were the first who recognised the potential as a surface specific probe in the mid-1990s.

SHG and SFG are the result of a non-linear polarisation wave that is generated by interacting ultrashort laser pulses at the interface of the two isotropic media: air and water. Both differ in the resonances probed. SHG results from the combination of two visible or near-infrared photons and probes the electronic states of the surface molecules. In an SFG experiment, two laser pulses are coincident in time and space at the interface of a sample. One visible laser pulse is fixed in frequency, while the other is tunable through the infrared region of 1000–3800 cm^{-1}. If the intensities of the incident laser pulses are sufficiently strong to induce a polarisation of the second order in the medium, an SFG signal oscillating at the sum-frequency of the two incident beams may be generated. If the frequency of the IR beam meets a frequency of a vibrational mode of the interfacial molecules, a resonance enhancement of the SFG signal may occur. To put it in simple terms, SFG spectroscopy provides a surface-specific vibrational spectrum while SHG probes the electronic signature of the molecules at the interface. Where does this interface specificity come from and what are the underlying assumptions?

The electric field propagating through a medium exerts a force on the valence electrons. The binding features of an electron can be described by a Morse type function. For small fields, the displacement of the electrons scales linearly with the applied field strength, while for stronger fields, deviations from the linear behaviour occur. Hence, the borderline between linear and non-linear optics is given by the binding potential of the electrons. Linear optics holds as long as the displacement of the electron occurs in a region where the binding potential can be approximated by a parabola. The induced dipole moment $\boldsymbol{\mu}$ can be represented as a power series of the electric field $\mathbf{E}$.

$$\boldsymbol{\mu} = \mu_0 + \alpha\mathbf{E} + \beta\mathbf{E}^2 + \gamma\mathbf{E}^3$$

The static dipole moment is denoted by μ_0, α is the linear polarisability of the electrons, β is the second order hyperpolarisability and so on. Non-linear optical effects are only noticeable once the electric field of the laser pulse is comparable to the inner-molecular fields.

The dominant term in the surface spectroscopy is the one of the second order, and the theory commonly used for the interpretation of SHG and SFG surface spectra is based on the electric dipole approximation. Within this approximation, the effect of optical magnetic fields and multipoles are neglected. This assumption can be at stake in some scenarios, especially for the interpretation of non-resonant SHG phenomena. Quadrupolar contributions are much weaker than the dipolar ones, but they scale with the number density and can amount to noticeable levels which may interfere with the surface term.[15–17] There is no general solution to this problem, and it is often not known *a priori* in SHG and SFG studies whether interfacial contribution is dominant over that of the bulk in an interface system.[18–20] Furthermore, it is commonly assumed that the dipoles induced in the molecules are solely due to the external laser field and that contribution from neighbouring molecules can be neglected. The local fields are hard to capture at a molecular level and require further theoretical work.[21] In this framework, the simplest description of SFG is given by

$$\mathbf{P}_{\mathrm{SF}}^{(2)} = \varepsilon_0 \chi^{(2)} \mathbf{E}_{\mathrm{VIS}} \mathbf{E}_{\mathrm{IR}},$$

where $\mathbf{P}_{\mathrm{SF}}^{(2)}$ is the second-order non-linear polarisation, ε_0 is the vacuum permittivity, $\chi^{(2)}$ is the second-order non-linear susceptibility, a third rank tensor describing the relationship between the incident electric fields vectors $\mathbf{E}_{\mathrm{VIS}}$ and $\mathbf{E}_{\mathrm{IR}}$ and the resultant polarisation. A simple symmetry consideration reveals that within the dipolar approximation, SHG or SFG cannot occur within media with inversion symmetry. Inversion symmetry means **P** goes to −**P** if the direction of **E** is inverted, leading to vanishing polarisation response $\mathbf{P}^{(2)}$.

$$\mathbf{P} \stackrel{\text{i}}{=} -\mathbf{P} \quad \text{if } \mathbf{E} \stackrel{\text{i}}{=} -\mathbf{E}, \quad \text{hence } \mathbf{P}^{(2)} = \mathbf{0}.$$

In the framework of the dipole approximation, SHG or SFG cannot occur in isotropic bulk media such as fluids or gases. At the interface of two isotropic media, the symmetry is broken and the molecules possess a net orientational order reflecting the asymmetry of the environment. These molecules are probed by SHG and SFG leading to an interface specific response. It is important to realise that SHG or SFG *per se* does not provide information of the interfacial width without further assumptions. The signal is generated within the interfacial region and determined by the integral of the molecular orientational distribution function in between both bulk media.

The laboratory coordinate system is defined by the plane of incidence given by the beam direction and the normal of the reflecting surface. This is a distinct coordinate system since *p*- and *s*-polarisations are eigen-polarisations of an isotropic interface and furthermore, *p*- and *s*-lights do not interfere because of the mutual perpendicular orientation of the fields. The electric field vector is resolved in components parallel and perpendicular to the surface. Nevertheless, a variety of different axis systems and beam geometries are used in the literature and care must always be taken to relate equations derived for specific field directions and axis systems. A good overview is presented by Hirose *et al.*[22]

The reflectivity is given by the Fresnel factors that relate the reflected light E^r and transmitted light to the incoming field E^i as described in Ref. 23:

$$r_s = \left(\frac{E_s^r}{E_s^i}\right) = \frac{n_1 \cos\theta_1 - n_2 \cos\theta_2}{n_1 \cos\theta_1 + n_2 \cos\theta_2},$$

$$r_p = \left(\frac{E_p^r}{E_p^i}\right) = \frac{n_1 \cos\theta_1 - n_1 \cos\theta_2}{n_1 \cos\theta_2 + n_2 \cos\theta_2},$$

where n is the refractive index of media 1 or 2, and θ is the angle of incidence or refraction. The total electric field at the interface is given by the summation of the incident and reflected beam. The local electric fields at the surface drive a non-linear polarisation wave that is oscillating at the sum frequency in the case of SFG, or at twice the frequency of the incoming light in the case of SHG. The resulting electromagnetic wave is coherently emitted in a direction given by momentum transfer. The SHG/SFG light is reflected and transmitted and the more accessible one is detected.

The non-linear susceptibility $\chi^{(2)}$ is a third-rank tensor comprising in total 27 elements; each one relates different combinations of the interacting field components to the induced non-linear polarisation. The generated polarisation wave is obtained by the summation of all different combinations and reads in the laboratory frame of reference:

$$P_{\mathrm{SF}}^{(2)} = \sum_{i}^{x,y,z} P_{i,\mathrm{SF}}^{(2)} = \varepsilon_0 \sum_{i}^{x,y,z} \sum_{j}^{x,y,z} \sum_{k}^{x,y,z} \chi_{ijk}^{(2)} E_{j,\mathrm{VIS}} E_{k,\mathrm{IR}}.$$

E is the surface field acting on the molecules which can be calculated by an evaluation of the Fresnel equations. It is common practice to transform the free propagating laser fields to the surface fields by the so-called K-factors:

$$K_x = \mp \frac{2n_{\mathrm{I}} \cos\theta_{\mathrm{I}} \cos\theta_{\mathrm{T}}}{n_{\mathrm{I}} \cos\theta_{\mathrm{T}} + n_{\mathrm{T}} \cos\theta_{\mathrm{I}}},$$

$$K_y = (1 + r_s) = \frac{2n_{\mathrm{I}} \cos\theta_{\mathrm{I}}}{n_{\mathrm{I}} \cos\theta_{\mathrm{I}} + n_{\mathrm{T}} \cos\theta_{\mathrm{T}}},$$

$$K_z = \sin\theta_{\mathrm{I}}(1 + r_p) = \frac{2n_{\mathrm{I}} \sin\theta_{\mathrm{I}} \cos\theta_{\mathrm{I}}}{n_{\mathrm{I}} \cos\theta_{\mathrm{T}} + n_{\mathrm{T}} \cos\theta_{\mathrm{I}}}.$$

The non-linear surface polarisation wave can be expressed in terms of the K-factors as

$$\mathbf{P}_{i,\mathrm{SF}}^{(2)} = \varepsilon_0 \chi_{ijk}^{(2)} K_j E_{p/s,\mathrm{VIS}}^{\mathrm{I}} K_k E_{p/s,\mathrm{IR}}^{\mathrm{I}}.$$

The relation between the polarisation wave and the emitted SFG radiation in the far field is provided by the so-called L-factors which are obtained by an evaluation of the continuity condition of the fields across the interface and the phase matching condition.

$$\mathbf{E}_{i,\mathrm{SF}} = L_i \mathbf{P}_{i,\mathrm{SF}}^{(2)},$$

$$L_x^{\mathrm{R}} = -\frac{i\omega_{\mathrm{SF}}}{c\varepsilon_0} \frac{\cos\theta_{\mathrm{SF}}^{\mathrm{T}}}{n_{\mathrm{T}} \cos\theta_{\mathrm{SF}}^{\mathrm{I}} + n_{\mathrm{I}} \cos\theta_{\mathrm{SF}}^{\mathrm{T}}},$$

$$L_y^{\mathrm{R}} = \frac{i\omega_{\mathrm{SF}}}{c\varepsilon_0} \frac{1}{n_{\mathrm{I}} \cos\theta_{\mathrm{SF}}^{\mathrm{I}} + n_{\mathrm{T}} \cos\theta_{\mathrm{SF}}^{\mathrm{T}}},$$

$$L_z^{\mathrm{R}} = \frac{i\omega_{\mathrm{SF}}}{c\varepsilon_0} \frac{(n_{\mathrm{T}}/n_{\mathrm{layer}})^2 \sin\theta_{\mathrm{SF}}^{\mathrm{T}}}{n_{\mathrm{I}} \cos\theta_{\mathrm{SF}}^{\mathrm{T}} + n_{\mathrm{T}} \cos\theta_{\mathrm{SF}}^{\mathrm{I}}}.$$

Experimentally, the intensity of the s- or p-polarised SF light is detected. For p-polarisation, it reads:

$$\begin{aligned} I_{p,\mathrm{SF}} &\propto |\mathbf{E}_{x,\mathrm{SF}}|^2 + |\mathbf{E}_{z,\mathrm{SF}}|^2 \\ &\propto |L_x \mathbf{P}^{(2)}_{x,\mathrm{SF}}|^2 + |L_z \mathbf{P}^{(2)}_{z,\mathrm{SF}}|^2 \\ &\propto \left| L_x \sum_j^{x,y,z} \sum_k^{x,y,z} \varepsilon_0 \chi^{(2)}_{xjk} K_j E^{\mathrm{I}}_{\mathrm{VIS}} K_k E^{\mathrm{I}}_{\mathrm{IR}} \right|^2 \\ &\quad + \left| L_z \sum_j^{x,y,z} \sum_k^{x,y,z} \varepsilon_0 \chi^{(2)}_{zjk} K_j E^{\mathrm{I}}_{\mathrm{VIS}} K_k E^{\mathrm{I}}_{\mathrm{IR}} \right|^2 . \end{aligned}$$

The decisive information is contained in the second-order non-linear susceptibility. The oriented gas model relates the macroscopic quantity susceptibility $\chi^{(2)}$ to the microscopic hyperpolarisability β. Subject to certain simplifying assumptions, it states that the susceptibility is given by the summation of all hyperpolarisabilities of all molecules. The underlying assumptions have been assessed in Ref. 24 and found to be valid.

Both tensors are described in a different frame of reference: the $\chi^{(2)}$ tensor is given in the laboratory frame of reference based on the plane of incidence, whereas the hyperpolarisability tensor in the molecular frame of reference defined by the symmetry of the molecule. An Euler transform relates both, leading to rather lengthy expressions between the corresponding tensor components, which are best handled and manipulated with the aid of a symbolic programming language such as Maple or Mathematica.

$$\begin{aligned} \beta_{IJK} &= \mathbf{U}_{Ii}(\phi,\theta,\psi)\beta_{ijk}\mathbf{U}^{-1}_{Jj}(\phi,\theta,\psi)\mathbf{U}^{-1}_{Kk}(\phi,\theta,\psi), \\ \mathbf{U} &= \mathbf{R}_c\mathbf{R}_b\mathbf{R}_a, \end{aligned}$$

$$R_c(\psi) = \begin{pmatrix} \cos\psi & \sin\psi & 0 \\ -\sin\psi & \cos\psi & 0 \\ 0 & 0 & 1 \end{pmatrix},$$

$$R_b(\phi) = \begin{pmatrix} \cos\phi & \sin\phi & 0 \\ -\sin\phi & \cos\phi & 0 \\ 0 & 0 & 1 \end{pmatrix},$$

$$R_a(\theta) = \begin{pmatrix} \cos\theta & 0 & -\sin\theta \\ 0 & 1 & 0 \\ \sin\theta & 0 & \cos\theta \end{pmatrix}.$$

The individual tensor components are given by an orientational average of the adsorbed molecules within a volume defined by the coherence length of the incident light. The non-linear susceptibility $\chi^{(2)}$ contains information about the number density N and the orientational average of the molecules.

$$\chi^{(2)}_{ijk} = \frac{N \sum\limits_{\alpha\beta\gamma} \langle R(\psi)R(\theta)R(\phi)\beta_{\alpha\beta\gamma}\rangle}{\varepsilon_0(\omega_\upsilon - \omega_{\mathrm{IR}} - i\Gamma)},$$

where ω_{IR} is the frequency of the incident infrared light, ω_υ is the eigenfrequency of a vibrational mode and Γ^{-1} is the relaxation time of the vibrationally excited states in the resonance. A further analysis requires assumptions about the prevailing orientational distribution function. In the majority of studies, a narrow Gaussian or δ-shaped orientational distribution is assumed for the sake of simplicity.

Vibrational modes can only be excited in an SFG experiment if they are simultaneously infrared and Raman active. The hyperpolarisability reads

$$\beta_{\alpha\beta\gamma} = \frac{1}{2\hbar}\frac{M_{\alpha\beta}A_\gamma}{\omega_\upsilon - \omega_{\mathrm{IR}} - i\Gamma},$$

where $\hbar$ is the reduced Planck constant, $M_{\alpha\beta}$ and A_γ are the Raman and infrared transition moments.

$$M_{\alpha\beta} = \sum_s \left[\frac{\langle g|\mu_\alpha|s\rangle\langle s|\mu_\beta|\upsilon\rangle}{\hbar(\omega_{\mathrm{SF}} - \omega_{sg})} - \frac{\langle g|\mu_\beta|s\rangle\langle s|\mu_\alpha|\upsilon\rangle}{\hbar(\omega_{\mathrm{VIS}} + \omega_{sg})}\right],$$

$$A_\gamma = \langle \upsilon|\mu_\gamma|g\rangle.$$

μ is the electric dipole operator, g denotes the ground state and υ the excited state, and s refers to any intermediate state.

In an SFG experiment, the frequency of the incoming infrared light is changed. A resonance enhancement occurs if the infrared frequency matches the frequency of the vibrational mode leading to a resonant SFG signal. The frequency dependence of the non-linear susceptibility component can be split into a real and an imaginary part leading to

$$\chi_{\mathrm{R}} \propto \frac{1}{\omega_\upsilon - \omega_{\mathrm{IR}} - i\Gamma} = \frac{\omega_\upsilon - \omega_{\mathrm{IR}}}{(\omega_\upsilon - \omega_{\mathrm{IR}})^2 + \Gamma^2} + i\frac{\Gamma}{(\omega_\upsilon - \omega_{\mathrm{IR}})^2 + \Gamma^2}.$$

Beside the resonant part χ_R, there is also a non-resonant part χ_{NR} to the non-linear susceptibility which further complicates the analysis. The measured SFG signal is given by the interference between both:

$$\chi^{(2)} = \chi_R^{(2)} + \chi_{NR}^{(2)},$$

$$I_{SF} \propto ||\chi_{R,ijk}^{(2)}|e^{i\delta} + |\chi_{NR,ijk}^{(2)}|e^{i\varepsilon}|^2,$$

with the phases δ and ε. At the air–water interface, the non-resonant part is usually fitted to a single value and phase, and the resonant part is described by assuming discrete resonances. In the case of metallic or semi-conductor supports, the line shape and the signal strength is determined by the interference between resonant and non-resonant parts. In any case, SFG analysis must occur after a deconvolution in component peaks. A direct comparison of the raw spectra to Raman- or IR-spectra may be misleading because of interference effects. This theoretical framework outlines the modelling strategy for fitting the spectra.

SFG spectroscopy is the only spectroscopy capable of producing vibrational water spectra of the air–water interface. The analysis is simplified by the symmetry of the interface and the symmetry of the molecule which reduces the number of independent tensor elements of the hyperpolarisability. The water molecule belongs to the point group $C2v$ and the number of independent tensor elements is reduced to five: β_{caa}, β_{aca}, β_{aac}, β_{bbc}, and β_{ccc}. Of these, only β_{bbc}, β_{aac}, and β_{ccc} are non-zero for the symmetric stretch and β_{caa} and β_{aca} are non-zero for the anti-symmetric stretch. We denote the tilt angle between the molecular $C2$ axis and the surface normal Θ and ψ as the twist angle of the molecular plane with respect to the surface plane. Then, the relationship between the molecular hyperpolarisability, and the macroscopic hyperpolarisability, for the symmetric stretch after Euler angle transformations yields:

$$\chi_{xxz} = \chi_{yyz} = \frac{N_s}{2}\beta_{ccc}[(r+1)\langle\cos\theta\rangle - (1-r)\langle\cos^3\theta\rangle]$$

$$\chi_{xzx} = \chi_{yzy} = \chi_{zxx} = \chi_{zyy} = \frac{N_s}{2}\beta_{ccc}(\langle\cos\theta\rangle - \langle\cos^3\theta\rangle)(1-r)$$

$$\chi_{zzz} N_s\beta_{ccc}[r\langle\cos\theta\rangle + (1-r)\langle\cos^3\theta\rangle]$$

This framework outlines the underlying complexity in analysing SFG spectra.

3.2. *Experimental Setups*

Picosecond (ps) and femtosecond (fs) laser systems are most often used for SFG spectroscopy. Ps SFG spectrometers cover a wide spectral scanning range (about 1000 cm^{-1} without realignment). The narrow band ps light pulses are tuned through the spectral region, while the SFG signal is generally collected for a few minutes for each wave number to gain reasonable S/N ratios (the repetition rate for ps light pulses is usually in the order of 10–25 Hz). Detecting spectral regions of about 1000 cm^{-1} usually takes several tens of minutes. In this setup, the advantage of a wide scanning range is offset by the difficulty of performing time-dependent experiments.

Broadband SFG spectroscopy is based on fs laser pulses (usually around 120 fs) covering a spectral range of about 200 cm^{-1}.[25–29] The repetition rate of these systems is two orders of magnitude higher compared to ps systems (of the order of 1 kHz). This allows around 1000 acquisitions per second in a spectral range of about 200 cm^{-1}, with a reasonable S/N ratio in about 2–5 minutes. In a fs SFG spectrometer, a narrow band ps visible pulse is mixed at the interface with a broadband fs infrared pulse. The resulting broadband SFG signal is dispersed on a grating within a spectrometer and subsequently imaged by an intensified or backlight illuminated CCD camera.

3.3. *SFG and SHG Measurements at the Air–Water Interface*

The SFG spectra of the air–water interface measured in different laboratories with many different instrumentations agree reasonably well.[30–34] A typical spectrum of water in *ssp* polarisation is presented in Fig. 2. The first letter *s* denotes the polarisation of the SFG light, the second letter *s* refers to the one of the visible light and the last letter refers to the polarisation of the IR radiation which is set to *p*.

The spectrum shows a broad band in the 3000–3600 cm^{-1} region which is attributed to hydrogen bonded water and a narrow peak at 3700 cm^{-1}. Water is stabilised by the formation of a tetrahedral hydrogen bonding network. At the surface, this network is necessarily distorted leading to water molecules that cannot fully participate in the hydrogen bonding network and possess a non-hydrogen bonded OH. There is broad consensus within the scientific community about the nature of

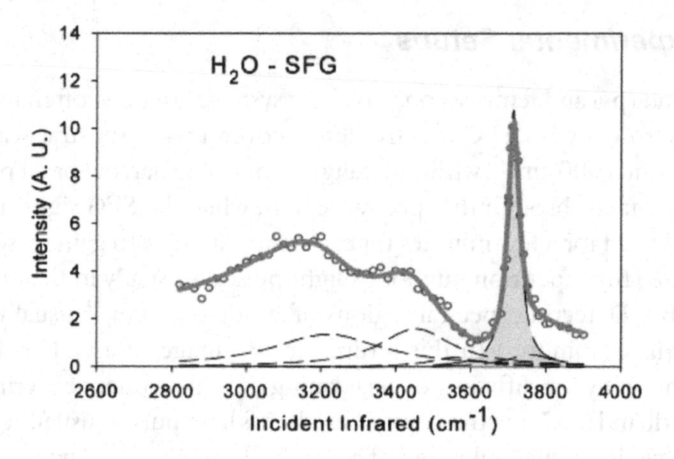

Fig. 2. SSP–polarised spectrum of neat water in the O–H stretching region. The deconvoluted peaks shaded in grey have a phase opposite to the non-shaded peaks. (Reprinted with permission from: Gopalakrishnan S, Jungwirth P, Tobias DJ, Allen HC. (2005) *J Phys Chem B* **109**(18): 8861–8872. Copyright 2005 American Chemical Society.)

the narrow peak at 3700 cm^{-1}: this is unambiguously attributed to dangling OH pointing towards the air–water interface. However, there is still a big controversy about the appropriate interpretation of the hydrogen bounded region of the spectra at 3000–3600 cm^{-1}. As a matter of fact, very different conclusions have been drawn based on the very same set of experimental data. The interpretation may even differ in qualitative aspects of the interfacial water structure.

First insights are gained by comparing the SFG surface spectra with the IR and Raman spectra of bulk water and hexagonal ice (Fig. 3). The bulk absorbance of hexagonal ice at 100 K shows a broad peak at 3200 cm^{-1} which is attributed to vibrational modes from the four oscillating dipoles associated with four-coordinate, hydrogen-bonded water molecules.[35] The absorbance of bulk water contains a broad peak centred at 3400 cm^{-1}.[36] Hence, the broad band observed in SFG spectrum resembles features of hexagonal ice and water.

Shen *et al.* introduced the term 'ice like band' to characterise the peak at 3200 cm^{-1}. This term is in a way misleading and needs some clarification. Scattering experiments reveal the complete absence of any long-range ice-like structural ordering within the surface layer of water. Instead, the term 'ice like' is meant to denote a highly dynamic surface of water forming a heavily distorted tetrahedral hydrogen bonding network

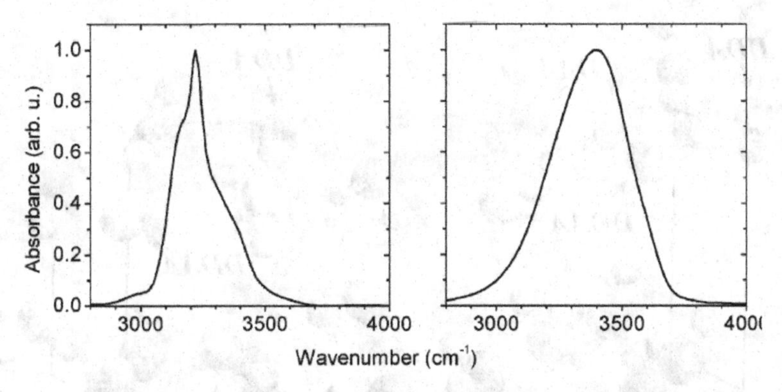

Fig. 3. Bulk absorbance of hexagonal ice at 100 K (left) and the bulk absorbance of liquid water (right). (Reprinted with permission from: Bettie JE, Labbe HJ, Whally E. (1969) *J Chem Phys* **50**: 4501; Querry MR, Wieliczka DM, Segelstein D. In *Handbook of Optical Constants of Solids II*, Academic Press: Boston, MA, 1991.)

of the hexagonal ice surface. The interfacial molecules are H-bonded to neighbours with dynamically varying strengths and geometries leading to fairly broad spectral feature (Fig. 4).

The previous section outlined the established fitting strategy for analysing the spectra and the underlying complexity. The non-linear resonant susceptibility is given by a summation of all individual resonances, which is then fitted to the spectral signature. However, the obtained fitting is not unique and there are different sets of resonances assigned to certain molecular arrangements that fully describe the experiment. As a consequence, the interfacial water architecture is interpreted in a different fashion and there is a controversial debate within the scientific community.[30,31,33,34,37–39]

Allen and Co-workers[40] attributed the 3200 cm^{-1} band to the vibrations of OH oscillators from surface water molecules that have one completely free OH (the 3700 cm^{-1} OH oscillator) and one OH that is hydrogen bonded to other water molecules (i.e. the other end of the free OH), the so-called DAA water molecules. This interpretation is supported by cluster studies.[41,42] Richmond's group[43,44] used molecular dynamics simulations to determine the population densities of different species of water molecules as functions of interfacial depth and orientation. The different configurations have been assigned different individual resonances and are used to reproduce the experiment. It is found that surface water molecules that possess one proton donor bond and one proton acceptor bond make the dominant contribution to both the *ssp*- and *sps*-polarised

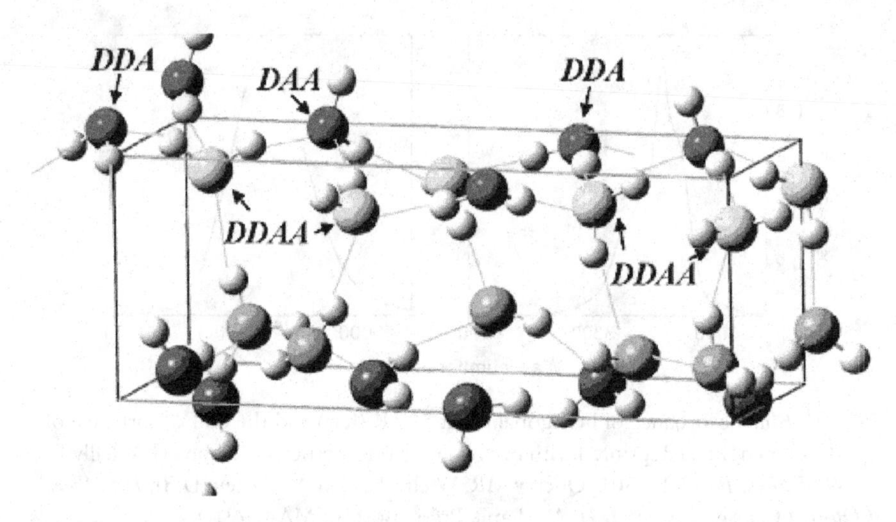

Fig. 4. Cartoons describing the structures of water vapour interfaces of neat water. The interfacial water structure is modelled after a distorted hexagonal ice. The top layer of the topmost layer (oxygen marked red) is occupied by DAA and DDA molecules, and the second layer (oxygen marked yellow) is occupied by DDAA molecules. Some DDAA molecules connecting to the molecules in the top layer with two donor bonds have an overall ice-like tetrahedral bonding structure. They contribute to the ice-like part of the vibrational spectrum. Other DDAA molecules are less symmetrically bonded to neighbours and contribute to the spectrum in the higher frequency region. The third and fourth layers (marked orange and red, respectively) have molecules with increasing randomness in position and orientation, closely resembling that of liquid water. (Reprinted with permission from Ref. 57. Copyright 2008 American Physical Society.)

spectral responses and are located within an Angstroem of the Gibbs dividing surface.

As a personal statement, we regard the assumption of discrete resonances in the water–air spectrum as inappropriate. The interface is highly dynamic. The continuous variations of hydrogen-bonding geometry and strength around interfacial water molecules shift and spread the OH stretching frequency into a highly inhomogeneously broadened band. The famous mathematician Neuman stated once, 'with four parameters I can fit an elephant, and with five I can make him wiggle his trunk.'[45] The assumption of too many individual resonances assigned to certain prevailing water configurations is a dead end for the description of a broad band spectrum with so few features.

Theoretical efforts to simulate the SFG spectra of water are able to reproduce the experimental features of the experiments. However, the imaginary part of thc susceptibility deduced from such studies differs with the recent phase-sensitive SFG experiments by Shen and co-workers. The aim of phase-sensitive measurements is the separate determination of the real and imaginary part of the non-linear susceptibility, instead of fitting the data to the square of the susceptibility. The measurement of amplitude and phase of the susceptibility tensor reduces the ambiguity in the interpretation of the SFG spectra. This is achieved by generating interference between a SFG signal generated in a material of known susceptibility, such as a quartz plate, and the one reflected from the water surface. The potential of phase-sensitive measurement has been recognised early but it remains a time-consuming and challenging experiment.[46]

Shen and Ostroverkhov performed phase-sensitive measurements of the air–water interface in a further developed experimental setup using a quartz reference plate and a phase modulator (Fig. 5).[47,48] In particular, by directly measuring the imaginary component of the non-linear

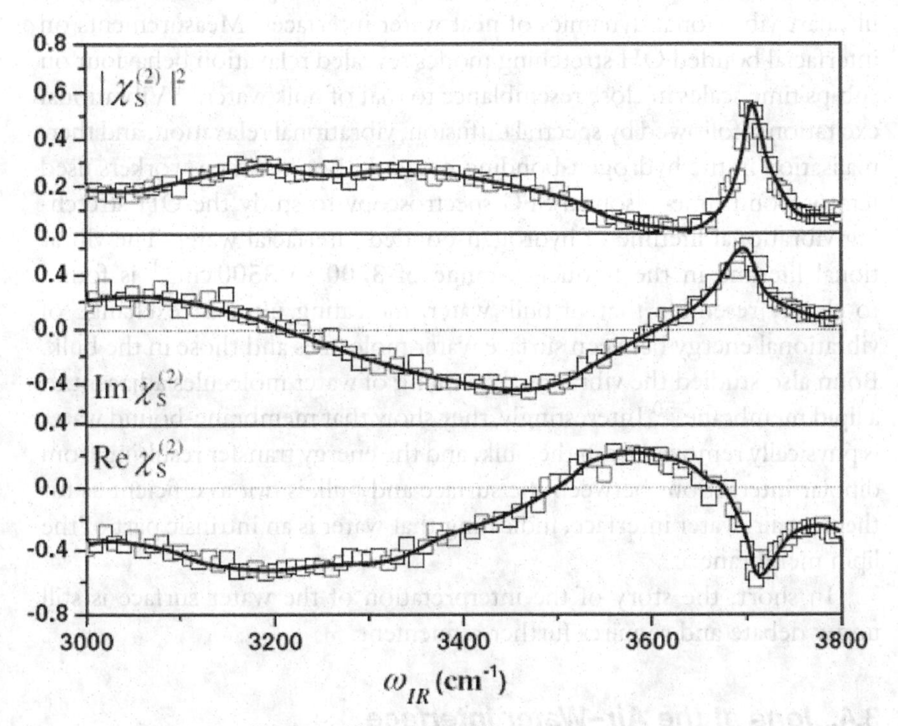

Fig. 5. Spectra of $|\chi_S^{(2)}|^2$, $\mathrm{Im}\chi_S^{(2)}$, and $\mathrm{Re}\chi_S^{(2)}$ for the neat water vapour interface in the OH stretching range obtained with phase sensitive sum frequency spectroscopy. (Reprinted with permission from Ref. 57. Copyright 2008 American Physical Society.)

susceptibility in the water region, they found that the lower frequency component ($< 3200\,cm^{-1}$) is in phase with that of free OH, while the mid-frequency component (3200–$3600\,cm^{-1}$) is out of phase. In their model, the loosely coined ice-like and liquid-like features are further refined and their contribution to the interfacial spectrum of water has been reported. This is in contradiction to the above mentioned theoretical modelling which assumes that both the ice-like and liquid-like bands are opposite in sign with respect to the dangling OH peak.

Recently, an alternative assignment of the modes has been put forward for the surface vibrational spectrum of water. The focus is again on the band which occurs between 3000–$3600\,cm^{-1}$ where the vibrational coupling (intramolecular) between the stretch and the bending overtone contribution to the spectrum has been suggested instead of structural effects.[49]

SFG spectroscopy is usually used to study the static properties of aqueous interfaces. The use of a pump-probe geometry can additionally allow detection of time-resolved SFG spectra.[50–54] This permits the study of ultrafast vibrational dynamics of neat water interfaces. Measurements on interfacial bonded OH stretching modes revealed relaxation behaviour on sub-ps time scales in close resemblance to that of bulk water.[54] Vibrational excitation is followed by spectral diffusion, vibrational relaxation, and thermalisation in the hydrogen-bonding network. Bonn and co-workers used femtosecond time–resolved SFG spectroscopy to study the OH stretching vibrational lifetime of hydrogen-bonded interfacial water. The vibrational lifetime in the frequency range of 3200 to $3500\,cm^{-1}$ is found to closely resemble that of bulk water, indicating ultrafast exchange of vibrational energy between surface water molecules and those in the bulk. Bonn also studied the vibrational coupling of water molecules adjacent to a lipid membrane.[50] Interestingly, they show that membrane-bound water is physically removed from the bulk, and the energy transfer resulting from dipolar interactions between the surface and bulk is not as efficient as for the neat air–water interface, indicating that water is an intrinsic part of the lipid membrane.

In short, the story of the interpretation of the water surface is still under debate and requires further refinement.

3.4. *Ions at the Air–Water Interface*

As outlined in the introduction, MD simulations predict a non-monotonous concentration profile for large and polarisable anions at the air–water interface in contrast to the established textbook picture which suggests a surface depleted of ions. However, up till now the direct measurement of this ion profile remains an unsolved experimental challenge. Most of the evidence is based on non-linear optical techniques covered in this section and on X-ray scattering techniques as well as on X-ray fluorescence which are discussed in Chap. 5.

SFG spectroscopy has been used to study the influence of the ions on the interfacial water structure. These measurements are indirect; it is not the salt that is studied but the impact of the ions on the local water environment. Two groups investigated a series of aqueous sodium halogenide salt solutions in a systematic fashion to analyse the impact of the anion on the interfacial water structure. The spectra reveal profound changes with respect to the neat water surface and are shown in Fig. 6. The SFG spectra of 2-M NaI solution showed a significant enhancement of the liquid-like band and the appearance of a broad shoulder above the dangling OH peak as compared to the neat air–water SFG spectrum.

Based on this data set, Allen and co-workers[33] concluded a confirmation of the MD simulation results on the enhanced adsorption of the more polarisable anions within the interfacial region. The observed spectral changes were attributed to I^- ions disrupting the interfacial H-bonding network and creating more asymmetrically H-bonded molecules. Furthermore, a comparison of the intensity changes in the SFG spectra with the corresponding bulk IR and Raman spectra in the 3000–3600 cm^{-1} region was interpreted as an increase of the effective interfacial layer thickness, providing corroborative evidence for the MD predictions. Richmond and co-workers challenged this point of view. According to their interpretation, the observed changes in the SFG spectra do not necessarily indicate the increase of the thickness of the surface water layers at the electrolyte solution surfaces. Instead, the observed changes of the spectrum can also be attributed to a blue shift and a narrowing of the mode at $\sim$ 3400 cm^{-1} with no significant changes in the mode strengths of bonded OH.[39]

As expected, NaI showed the most profound effect on the interfacial water structure; the impact of the other salts followed the order given in the periodic table. The intensity of the hydrogen-bonded water in the broad 3400 cm^{-1} band increased in the order of NaCl, NaBr, and NaI; while the intensity of the broad band around 3250 cm^{-1} decreased with the same order. The SFG spectra of the NaF aqueous solution surface measured in Allen and Richmond laboratory did not agree in the 3000–3600 cm^{-1}

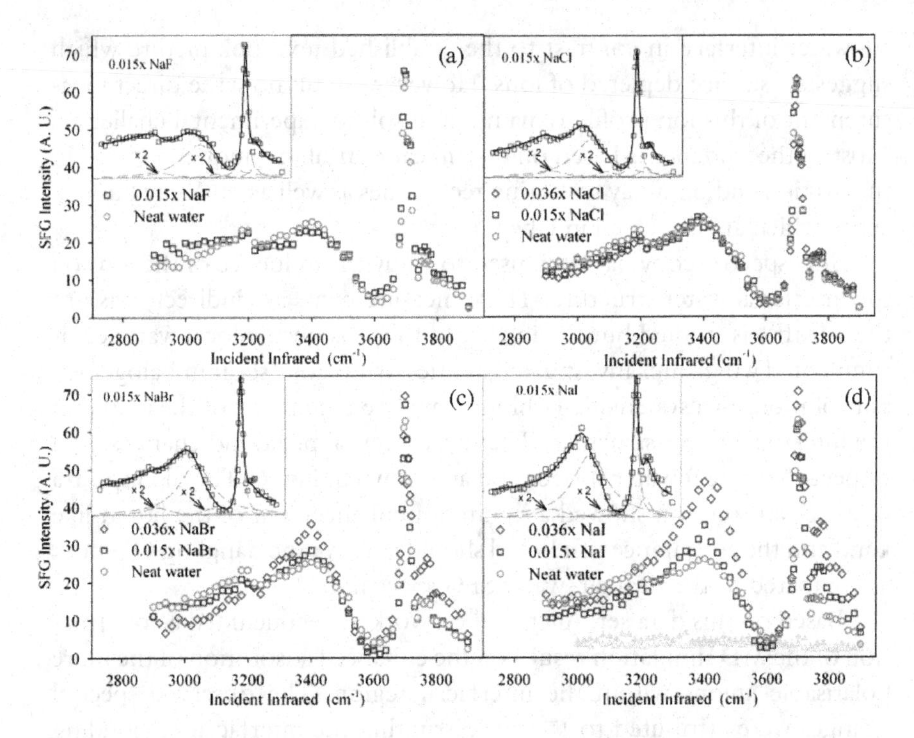

Fig. 6. SSP polarised SFG spectra of (a) 0.015× NaF, (b) 0.015× and 0.036× NaCl, (c) 0.015× and 0.036× NaBr, and (d) 0.015× and 0.036× NaI. The neat water SFG spectrum is plotted in each figure for comparison. The open yellow and closed yellow triangles within (d) show the sum frequency intensity of the 0.015× NaI in D2O and pure D2O, respectively. Insets: SFG sodium halide aqueous solutions spectral fits. Component Lorentzian peaks are shown as dashed pink lines, and the calculated fits from the component peaks are shown as black lines that go through most of the data points. (Reprinted with permission from: Liu D, Ma G, Levering LM, Allen HC. (2004) *J Phys Chem B* **108**: 2252. Copyright 2004 American Chemical Society.)

spectral region, while the $3700\,\mathrm{cm}^{-1}$ remained almost unchanged from that of the neat air–water interface.

The differences may be due to impurities within the sample. Surface active impurities are a major concern in these studies. It turns out that salts purchased with a claimed purity of 99.999% may contain traces of organic surface active components. These impurities do not play any role for investigations of bulk properties; however, they are extremely disturbing for the investigation of interfacial properties. To underline the importance of the purification of salts, our group[55] compared the SFG spectra of the as purchased salt from Sigma–Aldrich claimed to be of purity in the third decimal places with that of the specially purified salt solution. The organic contaminants show up directly in the SFG spectra around 2800–3000 cm^{-1} region. Further the free OH intensity around 3700 cm^{-1} is considerably lowered and the broad region around 3000–3600 cm^{-1} is significantly modified. Hence, as good code of conduct we suggest to extend the presentation of the interfacial water SFG spectra to the spectral range of 2800–3000 cm^{-1} for instance as an inset. Surface active impurities would leave the spectral signature of the hydrocarbon chains. Hence, the absence of any peaks in this region is absolutely convincing for monitoring the desired purity and it is more sensitive than a discussion based on the 3700 cm^{-1} OH peak.

Shen and co-workers[56,57] employed the phase-sensitive SFG-VS measurement and studied the *ssp* SFG-VS spectra of the NaI aqueous solution surfaces in comparison with the neat air–water interface. The authors concluded that the presence of I^- anion near the interface region may not disturb the water molecular structure at the topmost surface layer, while it can reorientate the water molecules in the subphase. Their data confirmed the influence of the interfacial water hydrogen bonding structure by the I^- anions of the previous SFG studies.

In a recent study, the influence of indifferent electrolytes on the adsorption behaviour of cationic soluble surfactant solutions has been investigated by surface tension measurements, ellipsometry and surface second harmonic generation (SHG).[58–60] Each technique addresses different structural aspects and the combined data provide a detailed picture of the interfacial architecture. The analysis gives an indirect proof of the existence of a phase transition between the free and condensed state of the counterions caused by a small increase of surface charge close to the critical micelle concentration (cmc).

In addition, SFG spectroscopy can be used to indirectly detect ion distributions at charged interfaces using water vibrational signatures.[58–61] The strength of the SFG response depends on the number of oriented water molecules. At a charged aqueous interface, the electric field at the surface aligns the polar water molecules, which in turn increases the SFG response. This enhancement of the water vibrational signal can be used to indirectly detect the depth of the electric field in the solution, which consequently depends on the ion distribution in the vicinity of the interface.

In Ref. 60, the surface charge itself was tuned by the amount of adsorbed ionic surfactants (1-dodecyl-4-dimethylaminopyridinium bromide — DMPB[58]). Charging up the interface by a small amount of DMP leads to an increase in SF intensity by a factor of ten. A further increase of the surface charge leads to stronger electric fields at the interface, however, the charge will also be more effectively screened by the counterions. This leads to an overall decreased depth of the electric field, which goes along with a decreased amount of oriented water molecules at the interface. Less oriented water molecules reduce the SFG response which was observed in the experiments.

The enhancement of the electric field by about a factor of ten is not specifically related to the surfactant used, as it is also observed in the context of charged lipid monolayers at the air–water interface. Wurpel and co-workers used λ-phage DNA bound to a cationic lipid monolayer at the air–water interface to study screening effects of counterions by detecting water signals in the OD stretching region at different concentrations.[61]

Further studies in this direction may also involve solid surfaces of different surface chemistry which can be used as electrodes to change the surface potentials in a fast and controlled manner. Diamond for example has a large electrochemical potential window, making this material particularly suitable for studying ion distributions.[62] The comparably flat surface of diamond (roughness in the nm regime) can also be functionalised to obey a hydrophilic or hydrophobic surface (hydrogen or oxygen terminated).

The SFG measurements applied in all these studies are indirect. Not the ion is monitored but instead the impact of the presence of ions in the interface on the water structure. In this way the studies of Viswanath and Motschmann[63,64] are unique. The extreme candidate of the Hofmeister series of ions, thiocyanate, was studied at the air–water interface using SFG spectroscopy. The stretching vibration of the ion is monitored and polarisation-dependent measurements have been used to retrieve the orientation of the ion. Secondly, the impact of the ion on the interfacial water is studied, allowing a correlation of the perturbation of solvent features due to the presence of ions at the interface. Figure 7 shows the CN vibrational stretch of the thiocyanate anion in all polarisation combinations around 2064 cm^{-1}, confirming its enrichment at the interface. This rod-like anion adopts a preferential orientation in the order of 45 degrees with respect to the normal to the surface. The presence of the SCN^- ion at the interface leads to a decrease of 3200 cm^{-1} and a slight increase of 3400 cm^{-1} bands.

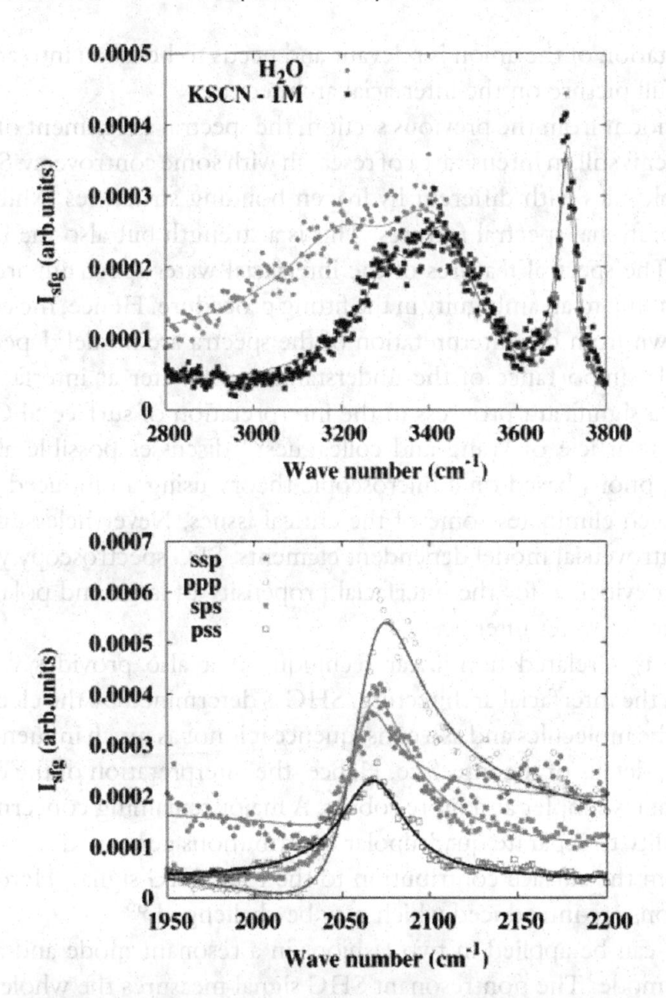

Fig. 7. Vibrational sum frequency spectra of water and ~ 1 M potassium thiocyanate solution. The presence of the ions at the interface decreases the 3200 cm^{-1} band. The presence of the ion has also been directly monitored by observing the CN stretch of thiocyanate anion for ~ 1 M potassium thiocyanate solution. The points and continuous lines represent the experimental data and fits, respectively. (Reprinted with permission from Refs. 63 and 64. Copyright 2007 and 2008 American Chemical Society.)

The combined data are a convincing piece of evidence for the interfacial propensity of the ion to adsorb at the interface. This study demonstrates the applicability of the SFG technique on air–electrolyte solutions to probe directly the presence of an oriented anion in addition to its implications on interfacial water signature. Furthermore, this work shows that the orientation of the anion is relevant and needs to be taken into account to get a full picture on the interfacial architecture.

As evident from the previous section, the spectral assignment of interfacial water is still an intense area of research with some controversy. Surface water molecules with different hydrogen bonding structures exhibit different vibrational spectral features. This is a strength but also the burden of SFG. The spectral features of the interfacial water spectrum are fairly broad, leading to an ambiguity in the fitting procedure. Hence, the conclusions drawn from the interpretation of the spectra are model dependent. Due to the importance of the understanding of water at interfaces, we expected a significant progress in the interpretation of surface SFG data. The review article of Wang and colleagues[65] discusses possible alternative descriptions based on a microscopic theory using an induced dipole lattice which eliminates some of the critical issues. Nevertheless, despite some controversial model dependent elements, SFG spectroscopy yields a mounting evidence for the interfacial propensity of large and polarisable ions at the air–water interface.

SHG is a related non-linear technique that also provides valuable insight in the interfacial architecture. SHG is determined by the electronic states of the molecules and as a consequence it is not as much influenced by the subtle details of the interface. Hence, the interpretation of the experimental data is simpler and more robust. A major remaining concern is the impossibility to separate quadrupolar contributions generated in the bulk phase from the surface contribution to the total SHG signal. Here some assumptions are introduced which may be challenged.[66]

SHG can be applied in two fashions in a resonant mode and a non-resonant mode. The non-resonant SHG signal measures the whole water surface layer without any discrimination on the surface water molecules in different hydrogen-bonding configurations. Therefore, non-resonant SHG is an appropriate tool to estimate the number of oriented molecules contributing to the SHG signal in a semi-quantitative fashion. The SHG response is governed by the number density and the orientational distribution of the molecules.[67] A careful polarisation-dependent measurement allows separation of both contributions.[68] The SHG intensity allows then an estimation of the relative increase of the average thickness of the total surface water.

Wang and co-workers[69,70] performed polarisation-dependent SHG measurements from the interfacial water molecules of the NaF, NaCl, and

NaBr aqueous electrolyte solutions. The quantitative polarisation analysis of the measured SHG data showed that the average orientation of the interfacial water molecules changed slightly with the bulk concentration of the NaF, NaCl and NaBr salts as compared to the neat air–water interface. The increase in the SHG signal with the bulk salt concentration was attributed to an increase of the thickness of the interfacial water molecular layer. The average thickness of the surface water layers increased with the bulk electrolyte concentration in an almost linear fashion in the following order: KBr > NaBr > KCl > NaCl ≅ NaF. The change of the average thickness of the surface water layers was less than 35% as compared to the neat air–water interface, even at very high salt concentration up to five molar solutions for the NaBr and four molar solution for the KBr. The absence of the electric-field–induced SHG (EFISHG) effect indicated further that the electric double layer at the salt aqueous solution surface is weaker as expected from molecular dynamics (MD) simulations.[71]

A second alternative to apply SHG is to use it in a resonant mode which enables direct monitoring of selected ions. Saykally *et al.* investigated the surfaces of electrolyte solution using resonant SHG. These experiments utilise the charge-transfer-to-solvent (CTTS) transitions in the UV and enabled direct monitoring of the anion. CTTS transitions exhibit very large non-linear cross sections as the direct consequence of the charge separation that is associated with the transition. By measuring the SHG intensity on and off resonance with the anions, the total SHG intensity can be separated into the contributions from the anions and the water background, and the surface concentration of the anions can be extracted. Several ions fulfil these prerequisites — e.g. iodide (I^-),[72] ferrocyanide [$Fe(CN)_6^{4-}$],[73] azide (N^{3-}),[74] and thiocyanate (SCN^-)[75] — and have been studied by resonant SHG (see Fig. 8). In the same spectral region, the water hyperpolarisability is non-resonant and contributes to the signal via its concentration-dependent non-resonant susceptibility term.

These data also provide a strong proof for the presence of the polarisable ions at the interface. Due to the relatively small hyperpolarisability of iodide, it is still not possible to determine whether the iodide anions are enhanced at the interface or whether the surface mole fraction increases linearly with the bulk.

In the last decade, tremendous progress has been made theoretically and experimentally in understanding the distribution of aqueous ions at the charged and uncharged interface. Among other experimental techniques,

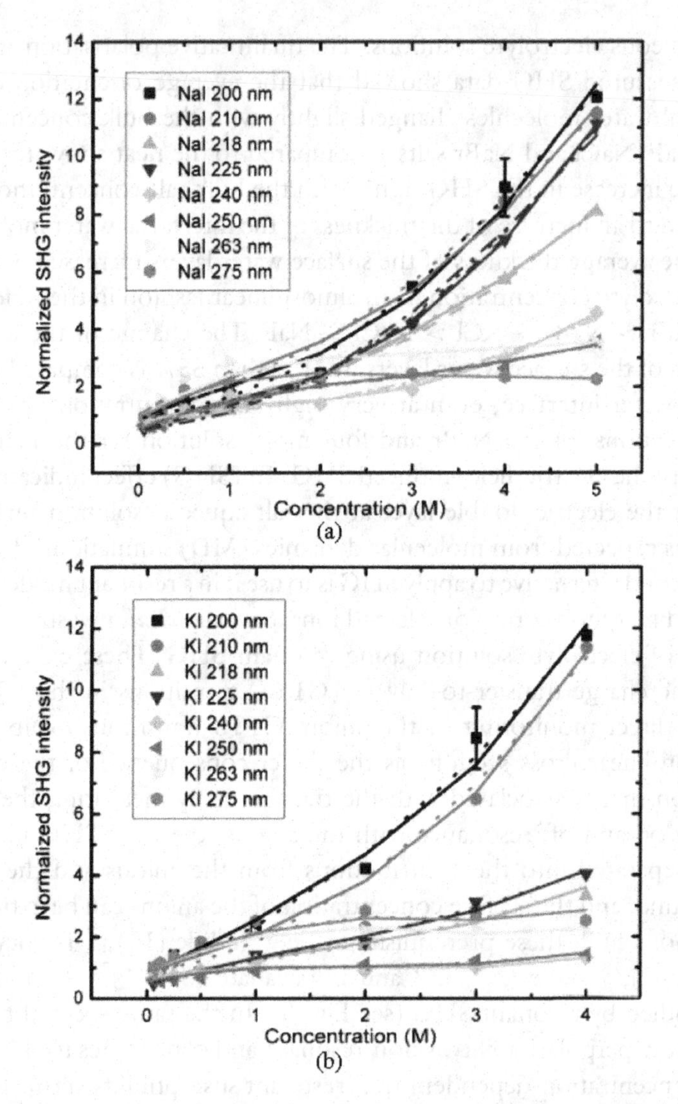

Fig. 8. Resonant SHG of NaJ and KJ. The high concentration increase in the resonant SHG intensity is due to iodide anions at the interface forming a dense ionic double layer with the cations. Due to the relatively small hyperpolarisability of iodide, it is not possible to determine whether the iodide anions are enhanced at the interface or whether the surface mole fraction increases linearly with the bulk. (Reprinted with permission from Ref. 76. Copyright 2006 American Physical Society.)

linear and non-linear optical methods can be applied under normal conditions and the only requirement is that the sample can be reached by light. In addition, spectroscopy methods can distinguish between several molecular species present at the interface. Water can be used as a probe to study ion distributions at charged solid or liquid interfaces. More complex ions with vibrational signatures can directly be detected and orientational distributions can be determined with the aid of polarised laser light sources. In a further approach, these interfaces may be designed to be more complex. In a systematically manner, for example, simple biomolecules under changing surface chemistry in different environmental conditions can be studied in various spectral regions related to the biomolecule, water and/or complex ions. Furthermore, with the aid of time-resolved SFG spectroscopy, the dynamics of water molecules and the energy exchange to bulk water molecules at charged or uncharged interfaces can be studied. Overall, optical techniques provide the experimental tools to get molecular level-based information needed to further our understanding of ions at interfaces.

References

1. McDevit *et al.* (1952) *J Am Chem Soc* **74**: 1773.
2. Wyman JT, Edsall J. (1951) *Biophysical Chemistry*. Academic Press.
3. Pinna MC, Salis A, Monduzzi M, Ninham BW. (2005) *J Phys Chem B* **109**: 5406–5408.
4. Jungwirth P, Tobias DJ. (2006) *Chem Rev* **106**: 1259–1281.
5. Smith D, Saykally RJ, Geissler PL. (2007) *J Am Chem Soc* **129**: 13847.
6. Mancinelli R, Botti A, Bruni F, Ricci MA, Soper AK. (2007) *J Phys Chem B* **109**: 13570.
7. Collins KD. (2004) *Methods* **34**: 300–311.
8. Jarvis and Scheiman. (1968) *J Phys Chem* **72**: 74.
9. Pugh R. (1996) *Journal of Colloid and Interface Science* **184**: 550–563.
10. Onsager L, Samaras NNT. (1934) *J Chem Phys* **2**: 528.
11. Gopalakrishnan S, Liu DF, Allen HC, Kuo M, Shultz MJ. (2006) *Chem Rev* **106**: 1115.
12. Tobias DJ, Hemminger JC. (2008) *Science* **319**: 1197.
13. Bloembergen N, Pershan PS. (1962) *Phys Rev* **128**: 606.
14. Du Q, Superfine R, Freysz E, Shen YR. (1993) *Phys Rev Lett* **70**: 2313.
15. Heinz TF, Divincenzo DP. (1990) *Phys Rev A* **42**: 6249.
16. Zhu XD, Shen YR. (1990) *Phys Rev A* **41**: 4549.
17. Andrews DL, Blake NP. (1990) *Phys Rev A* **41**: 4550.
18. Shen YR. (1999) *Appl Phys B* **68**: 295.
19. Wei X, Hong SC, Lvovsky AI, Held H, Shen YR. (2000) *J Phys Chem B* **104**: 3349.

20. Held H, Lvovsky IA, Wei X, Shen YR, (2002) *Phys Rev B* **66**: 205110.
21. Boyd RW. (1992) *Non-linear Optics.* Academic Press, San Diego.
22. Hirose C, Akamatsu N, Domen K. (1992) *Appl Spectr* **46**: 1051–1072.
23. Born M, Wolf E. (1999) *Principles of Optics: Electromagnetic Theory of Propagation, Interference and Diffraction of Light (Paperback).* Cambridge University Press.
24. Motschmann H, Penner T, Armstrong N, Enzenyilimba M. (1993) *J Phys Chem* **97**: 3933.
25. Hommel EL, Ma G, Allen HC. (2001) *Analytical Sciences* **17**(11): 1325–1329.
26. Hommel EL, Allen HC. (2001) *Analytical Sciences* **17**(1): 137–139.
27. Richter LJ, Petralli-Mallow TP, Stephenson JC. (1998) *Opt Lett* **23**(20): 1594–1596.
28. Ishibashi T, Onishi H. (2001) *Chemical Physics Letters* **346**(5–6): 413–418.
29. vanderHam EWM, Vrehen QHF, Eliel ER (1996). *Optics Letters* **21**(18): 1448–1450.
30. Wei X, Shen YR. (2001) *Phys Rev Lett* **86**: 4799.
31. Gragson DE, McCarty BM, Richmond GL. (1996) *J Phys Chem* **100**: 14272.
32. Baldelli S, Schnitzer C, Shultz MJ, Campbel DJ. (1997) *J Phys Chem B* **101**: 10435.
33. Liu D, Ma G, Levering LM, Allen HC. (2004) *J Phys Chem B* **108**: 2252.
34. Gan W, Wu D, Zhang Z, Feng R, Wang H. (2006) *J Chem Phys* **124**: 114705.
35. Bertie JE, Labbe HJ, Whally E. (1969) *J Chem Phys* **50**: 4501.
36. Querry MR, Wieliczka DM, Segelstein D. (1991) In *Handbook of Optical Constants of Solids II.* Academic Press, Boston, MA.
37. Brown MG, Raymond EA, Allen HC, Scatena LF, Richmond GL. (2000) *J Phys Chem A* **104**: 10220.
38. Raymond EA, Tarbuck TL, Richmond GL. (2002) *J Phys Chem B* **106**: 2817.
39. Raymond EA, Richmond GL. (2004) *J Phys Chem B* **108**: 5051.
40. Gopalakrishnan S, Jungwirth P, Tobias DJ, Allen HC. (2005) *J Phys Chem* **109**: 8861.
41. Devlin JP, Sadlej J, Buch V. (2001) *J Phys Chem A* **105**: 974.
42. Steinbach C, Andersson P, Kazimirski JK, Buck U. (2004) *J Phys Chem A* **108**: 6165.
43. Walker DS, Hore DK, Richmond GL. (2006) *J Phys Chem B* **110**: 20451–20459.
44. Walker DS, Richmond GL. (2008) *J Phys Chem C* **112**: 201–209.
45. Dyson F. (2004) *Nature* **427**: 297.
46. Ong TH, Davies PB, Bain CD. (1993) *J Phys Chem* **97**: 12047–12050.
47. Ji N, Ostroverkhov V, Tian CS, Shen YR. (2008) *Phys Rev Lett* **100**: 096102.
48. Tian CS, Shen YR. (2009) *J Am Chem Soc* **131**: 2790–2791.
49. Sovago M, Campen RK, Wurpel GWH, Muller M, Bakker HJ, Bonn M. (2008) *Phys Rev Lett* **100**: 173901.
50. Bredenbeck J, Ghosh A, Smits M, Bonn M. (2008) *J Am Chem Soc* **130**: 2152–2153.
51. Smits M, Ghosh A, Sterrer M, Muller M, Bonn M. (2007) *Phys Rev Lett* **98**: 098302.
52. Hayashi M, Shiu YJ, Liang KK, Lin SH, Shen YR. (2007) *J Phys Chem A* **111**: 9062–9069.

53. Backus EH, Grecea ML, Kleyn AW, Bonn M. (2007) *J Phys Chem B* **111**: 6141–6145.
54. McGuire JA, Shen YR. (2006) *Science* **313**: 1945–1948.
55. Koelsch P, Viswanath P, Motschmann H, Shapovalov VL, Brezesinski G, Moehwald H *et al.* (2007) *Colloids & Surfaces A-Physicochemical & Engineering Aspects* **303**: 110.
56. Tian CS, Ji N, Waychunas GA, Shen YR. (2008) *J Am Chem Soc* **130**: 13033.
57. Ji N, Ostroverkhov V, Tian CS, Shen YR. (2008) *Phys Rev Lett* **100**: 096102.
58. Koelsch P, Motschmann H. (2005) *Langmuir* **21**(8): 3436–3442.
59. Koelsch P, Motschmann H. (2004) *Current Opinion in Colloid & Interface Science* **9**(1–2): 87–91.
60. Koelsch P, Motschmann H. (2004) *J Phys Chem B* **108**(48): 18659–18664.
61. Wurpel GWH, Sovago M, Bonn M. (2007) *J Am Chem Soc* **129**(27): 8420.
62. Hartl A, Schmich E, Garrido JA, Hernando J, Catharino SCR, Walter S, Feulner P, Kromka A, Steinmuller D, Stutzmann M. (2004) *Nat Mater* **3**(10): 736–742.
63. Viswanath P, Motschmann H. (2007) *J Phys Chem C* **111**(12): 4484.
64. Viswanath P, Motschmann H. (2008) *J Phys Chem C* **112**(6): 2099.
65. Zheng DS, Wang Y, Liu AA, Wang HF. (2008) *International Reviews in Physical Chemistry* **27**: 629–664.
66. Andrews DL. (1993) *J Mod Opt* **40**: 939.
67. Simpson GJ, Rowlen KL. (2000) *Anal Chem* **72**: 3399.
68. Simpson GJ, Rowlen KL. (2000) *Anal Chem* **72**: 3407.
69. Bian HT, Feng RR, Xu YY, Guo Y, Wang HF. (2008) *Phys Chem Chem Phys* **10**: 4920.
70. Zhang WK, Wang HF, Zheng DS. (2006) *Phys Chem Chem Phys* **8**: 4041.
71. Brown MA, D'Auria R, Kuo I-FW, Krisch MJ, Starr DE, Bluhm H, Tobias DJ, Hemminger JC. (2008) *Phys Chem Chem Phys* **10**: 4778.
72. Petersen PB, Johnson JC, Knutsen KP, Saykally RJ. (2004) *Chem Phys Lett* **397**: 46–50.
73. Petersen PB, Saykally RJ. (2005) *J Am Chem Soc* **127**: 15446–15452.
74. Petersen PB, Saykally RJ. (2004) *Chem Phys Lett* **397**: 51–55.
75. Petersen PB, Saykally RJ, Mucha M, Jungwirth P. (2005) *J Phys Chem B* **109**: 10915–10921.
76. Petersen PB, Saykally RJ. (2006) *J Phys Chem B* **110**: 14060–14073.

[illegible] Viet Khuong [illegible] Bonn M (2007) J [illegible] 111, 6111–6114.

[illegible] (1996) Nature [illegible] 1948.

[illegible]

[illegible] Mole[illegible] [illegible] 110 [illegible]

[illegible] Chem [illegible] 130, 170 [illegible]

[illegible] Shen YR (2008) [illegible] 100, 096102

58. [illegible] (2008) [illegible] 21(6): 5 [illegible]

59. [illegible] (2004) [illegible] 5741–5749.

[illegible] 48: 12659 [illegible]

[illegible] (2003) [illegible] 2[illegible] 4420

[illegible]

[illegible] 112, 17 [illegible]

[illegible] (2006) [illegible] 11(2): 2064–2069

[illegible] Information [illegible] Physics Communications [illegible]

[illegible]

[illegible] (1986) [illegible] Chem 72 [illegible]

[illegible] Rowlen KL (1990) [illegible] 72, 1407

[illegible] Wang H [illegible] (2008) [illegible] Chem [illegible]

[illegible] 4920

[illegible] (2009) [illegible]

[illegible] 108 [illegible]

[illegible] (2003) [illegible]

[illegible] Petersen PB, Saykally RJ (2004) Chem Phys Lett [illegible] 35 [illegible]

[illegible] Petersen PB, Saykally RJ [illegible] (2005) [illegible] 109 [illegible]

[illegible] Saykally RJ [illegible] 110, 14060 [illegible]

Chapter 5

X-Ray Studies of Ion Specific Effects

Padmanabhan Viswanath*, Luc Girard*[,a], Jean Daillant*[,‡], Luc Belloni*, Olivier Spalla* and Dmitri Novikov[†]

There is a broad range of phenomena in biological, environmental, and physical sciences where ions of the same valency have a dramatically different effect. Since the seminal work of Hofmeister in 1888 on protein stability, this has been illustrated in many examples ranging from enzymatic activity and amyloidosis to humics stability and halide heterogeneous chemistry. These 'ion specific' effects are not completely understood yet and unravelling the short-range solvent-mediated couplings dominant at interfaces might provide a clue towards understanding them. This points out the need for quantitative and direct measurements on aqueous salt solutions to access the surface composition at the relevant nm or sub-nm lengthscale. In this chapter, we discuss how this can be achieved using X-rays, mainly X-ray reflectivity, anomalous X-ray reflectivity, grazing incidence X-ray fluorescence and X-ray standing waves.

1. Introduction

Ion specific effects generally follow direct or reverse order of the so-called Hofmeister series, which for monovalent anions is, $SCN^- > ClO_4^- \approx I^- > Br^- > Cl^- > F^-$. In many cases, interfacial effects appear to play a key role. In the simplest case of ions at the air–water interface for example, the specificity is directly reflected in the stability of foams[1] or in the surface tension and surface potential of aqueous salt solutions.[2] In biology, beyond protein crystallisation,[3,4] the enzymatic activity can be controlled by addition of alkali halides with large differences between fluoride, chloride, bromide, or iodide.[5] Specific types of cations are also identified together with

*CEA, IRAMIS, LIONS, CEA Saclay F-91191 Gif-sur-Yvette Cedex, France.
[†]HASYLAB, Notkestr. 85, 22607 Hamburg, Germany.
[‡]E-mail: jean.daillant@cea.fr
[a]Present address: ICSM UMR 5257, CEA Marcoule, BP17171 30207 Bagnols sur Cèze Cedex and ENSCM, 8 Rue de l'Ecole Normale 34296 Montpellier Cedex 5.

pathological aggregation of proteins relevant to amyloidosis.[6] In environmental sciences, the interactions between ions, humics and minerals influence the mobility, availability of pollutants and micro-nutrients as well as the storage or loss of carbon to the atmosphere.[7] In atmospheric science, ion specificity has been shown to play an important role in heterogeneous processes that take place in atmospheric aerosols.[8] For example, the segregation of bromide with respect to chloride at the surface of sea-spray covered snowpack is crucial for tropospheric ozone destruction during polar sunrise.[9]

However, there is presently no first-principle theory or general agreement on the mechanisms involved even in the apparently simple case of ions at the air–water interface, where the surface tension and the surface potential follow the Hofmeister series in reverse order. Contrary to acids, inorganic salts and bases are found to increase the surface tension of pure water[2] which, based on Gibbs equation, implies depletion. The original description of the surface tension of dilute (Debye–Hückel) electrolytes by Onsager and Samaras[10] took into account the repulsive image force experienced by any ion at the surface and predicted an increase in surface tension ($d\gamma/dc \sim 1.2$ (mN/m)/(mol/L)), smaller than generally observed. Since then, different ideas have been used for refining the theory and introducing ion specificity. The simplest phenomenological approach consists in including an ion-free layer,[11] but ion hydration[12] and solvation[13] have also been incorporated in modified Poisson–Boltzmann equations. In 1997, Ninham and Yaminsky suggested that dispersion forces should play an important role and would introduce specificity.[14] Further extension on this idea was carried out using hypernetted chain (HNC) integral equation approximation at the primitive model level of description (ionic spheres immersed in a continuous dielectric solvent). This proved to be extremely successful in describing the osmotic coefficients[15] but failed to predict the surface tension, pointing out the need for molecular level description of the solvent structure and the ions at the interface. Recently, molecular dynamics (MD) simulation incorporating polarisable potentials predict a non-monotonic distribution of ions with surface enhancement of anions and depletion of cations.[16] The simplified picture emerging from the simulations is that the polarisation of ion and solvent in the asymmetric surface environment can compensate for the partial loss of solvation in sufficiently large and polarisable ions, leading in the end to a surface affinity.[17]

Stimulated by these advancements, sophisticated interface sensitive techniques were employed to investigate the surface of aqueous salt solutions. Using sum frequency generation (SFG) and second harmonic generation (SHG) which are described in another chapter of this book, the presence of ions in the surface layer could be demonstrated and distortion of the hydrogen-bonding network was evidenced.[18–20] Though these techniques are sensitive to non-centrosymmetric environment, they are limited by the knowledge of the interfacial depth. Electron spectroscopies have in principle the required depth (1 nm or less) and chemical sensitivity but cannot be used under ambient conditions. Using aqueous NaI solution in liquid jet, it was concluded that the near-surface ion density is smaller than in the bulk,[21] and using crystals at deliquescence, the anion to cation ratio was found to be much higher than what is predicted in MD simulations.[22]

2. X-ray Reflectivity

The refractive index of matter for X-rays $n = n' + in'' \approx 1 - \delta - i\beta$ far from absorption edges of the atoms in the medium with $\delta = 1 - \lambda^2/(2\pi)r_e\rho$ and $\beta = \lambda\mu/4\pi$ where $\lambda = 12400/E(\text{eV})$ is the wavelength of X-rays of energy E, $r_e = 2.815 \times 10^{-15}$ m refers to the classical electron radius, ρ is the electron density of the medium and μ is the linear absorption coefficient of the material. As n is slightly less than 1, total external reflection occurs for grazing angles of incidence smaller than a critical angle $\theta_c = \sqrt{(2\delta)}$. The reflectivity technique consists in recording the reflected intensity from an interface as a function of the grazing angle of incidence θ.[23] The adsorption of ions will in general lead to only small changes in the refractive index profile, and it will be possible to use the approximate formula:

$$R = R_F \frac{1}{\rho_{\text{sub}} - \rho_{\text{sol}}} \int \left(\frac{\partial \rho(z)}{\partial z} \right) dz, \qquad (1)$$

where R_F is the Fresnel reflectivity of the solution/substrate interface, $\rho(z)$ is the electron density at z, ρ_{sol} the bulk solution electron density and ρ_{sub} the substrate electron density.

When working close to an absorption edge, n' and n'' will undergo variations which can be used in order to change the contrast of an ion.[23] The method is called anomalous reflectivity, or resonant reflectivity when working right at the edge.

Reflectivity was used in Ref. 24 in order to probe ion distributions at a liquid–liquid interface which was shown to substantially differ from the Gouy–Chapman prediction. The surface–normal electron density profile $\rho(z)$ of concentrated aqueous salt solutions of RbBr, CsCl, LiBr, RbCl, and $SrCl_2$ was determined by X-ray reflectivity in Ref. 25. For all salts but RbBr and $SrCl_2$ $\rho(z)$ increases monotonically with depth z. Anomalous (resonant) XR of RbBr revealed a depletion at the surface of Br^- ions to a depth of 10 Å. However, as the deviations from the other solutions were small, the evidence for a different ion composition in the surface and the bulk was not strongly conclusive.

Another example of resonant reflectivity will be discussed in Sec. 4.2.

3. Grazing Incidence X-Ray Fluorescence at the Air–Solution Interface

3.1. *Theoretical Background*

Grazing incidence X-ray fluorescence (GIXF) allows the determination of surface composition with a nm resolution. Figure 1 shows schematically the GIXF setup in which the fluorescence from the ions (chemical sensitivity) is recorded as a function of the grazing angle of incidence (depth sensitivity). This is a direct, element specific, surface sensitive and non-destructive technique which can be used to determine quantitatively the interfacial composition. It was mentioned in the previous section that, as n is slightly less than 1, total external reflection occurs for grazing angles of

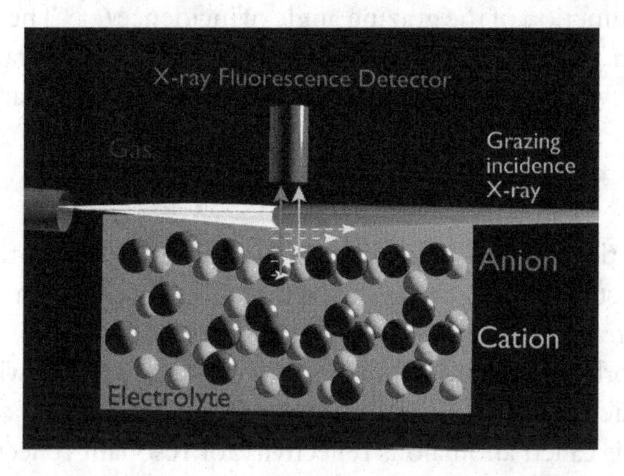

Fig. 1. Schematic sketch of the grazing incidence X-ray fluorescence experiment.

incidence smaller than a critical angle $\theta_c = \sqrt{(2\delta)}$. For example, for 8 keV incident energy, $\theta_c \approx 2.5$ mrad for water and below θ_c, an evanescent wave is propagated with a penetration depth of about 4.6 nm. More generally, the penetration depth for a grazing angle of incidence θ is equal to $1/2\mathcal{I}m(k_z)$, where k_z is the vertical component of the wave-vector in the liquid: $\mathcal{I}m(k_z) = 1/\sqrt{2}k_0\sqrt{[(\theta^2 - 2\delta)^2 + 4\beta^2]^{1/2} - (\theta^2 - 2\delta)}$. Beyond θ_c, the penetration depth increases to microns [Fig. 2(a)]. It is now limited by absorption and depends on the ions in solution and their concentration. The fluorescence for grazing angles much larger than θ_c also provides a convenient reference for bulk concentration. In an experiment, a full spectrum including fluorescence and elastic scattering from the bulk is recorded for each value of the grazing angle of incidence θ. K and L fluorescence lines in between P K_α (2014 eV) and Br K_α (11 924 eV) can easily be detected using standard detectors in a He atmosphere (better than air for absorption) giving access to Cl^-, ClO_4^-, Br^-, I^-, K^+, and Cs^+ (see Table 1).

After energy calibration, the first step consists in analyzing the fluorescence spectrum. If the spectrum only consists of a few well separated lines, numerical integration can be performed. In other cases, a model must be used in order to describe the fluorescence lines. The procedure given in Ref. 26 gives an accurate description of the peak lineshapes. This includes for each line a Gaussian peak with an energy-dependent width, a shelf and a tail [Fig. 2(b)]. The shelf accounts for detector related phenomena, in particular transport of charge carriers in the detector, whereas the tail describes radiative Auger transitions and satellite lines. Elastic and Compton intensities can be analysed using similar methods.[27]

The fluorescence of each element is then determined by fitting. The fluorescence amplitude can be determined by linear fitting, whereas the parameters of the width, tail and shelf fraction are determined using, for example, a non-linear Levenberg–Marquardt algorithm. Reabsorption corrections (i.e. the absorption of the fluorescence intensity by the solution) can be introduced at this point and the result of the fitting procedure is the fluorescence intensity of each element as a function of θ:

$$I_i \propto I_0|t(\theta)|^2(2\mathcal{I}m(k_z)+\mu_i)\int_0^\infty dz c_i(z)\exp[-(2\mathcal{I}m(k_z)+\mu_i)z]. \quad (2)$$

$t(\theta)$ is the transmission coefficient of the air solution interface for the grazing angle of incidence θ. The fluorescence intensities are then normalised

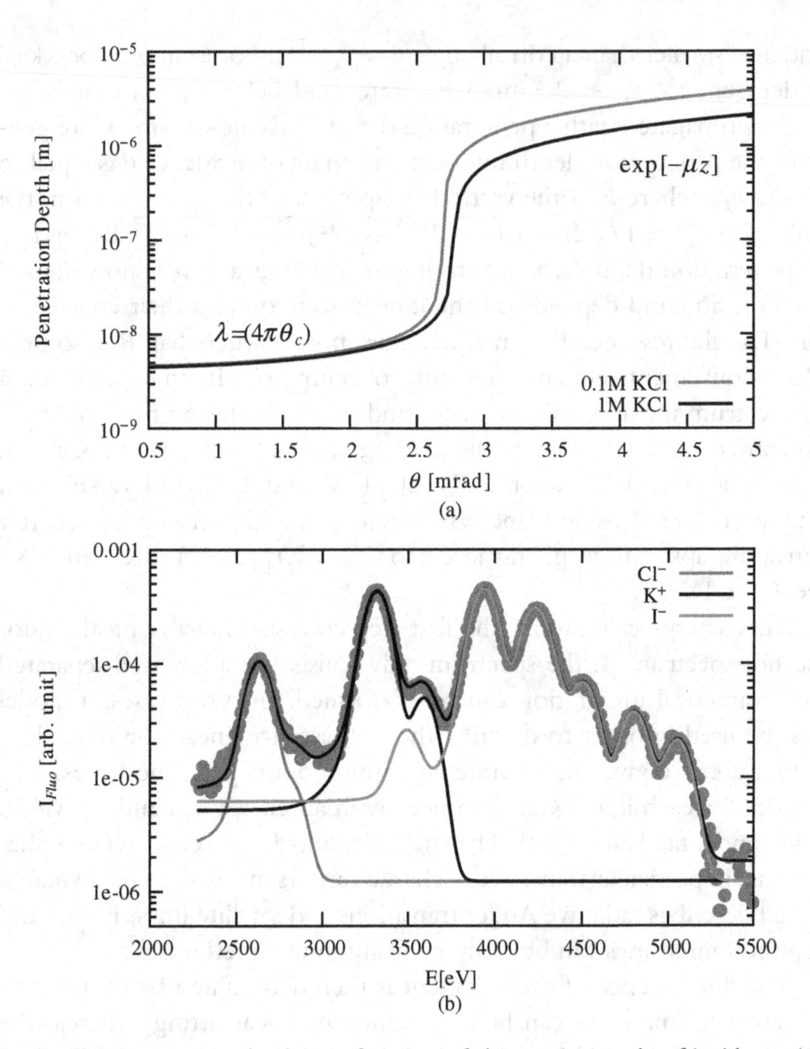

Fig. 2. (a) Penetration depth as a function of the grazing angle of incidence (θ) for 0.1-M and 1-M KCl solutions at 8026 eV. The penetration depth tends to 46 Å for water for low θ and is limited by absorption for high θ. (b) Typical fluorescence spectrum and its fit for a mixture of 0.1-M solutions of KCl and KI.

to the elastic scattering intensity I_{elas} in order to get rid of any dependency of the geometrical factor related to the experimental configuration. For the dilute solutions, we simply have: $I_{\text{elas}} \times (2\mathcal{I}m(k_z) + \mu_{\text{inc}}) \propto k_B T \kappa_T$, with μ_{inc} being the linear absorption coefficient for the incident energy as the photons are scattered with the same energy E, and κ_T the isothermal

Table 1. Fluorescence energies after the X-ray data booklet, Lawrence Berkeley National Laboratory. The * indicates lines which are easily recorded using standard detectors in a He atmosphere.

Ions	Emission (eV)
F	676 (K_{α_1})
P*	2014 (K_{α_1}), 2013 (K_{α_2}), 2139 (K_{β_1})
S*	2308 (K_{α_1}), 2307 (K_{α_2}), 2464 (K_{β_1})
Cl*	2622 (K_{α_1}), 2620 (K_{α_2}), 2815 (K_{β_1})
Br*	1480 (L_{α_1}), 1525 (L_{β_1})
I*	3937 (L_{α_1}), 3926 (L_{α_2}), 4220 (L_{β_1}), 4507 (L_{β_2}), 4800 (L_{γ_1})
Li	54 (K_{α_1})
Na	1040 (K_{α_1}), 1071 (K_{β_1})
K*	3313 (K_{α_1}), 3311 (K_{α_2}), 3589 (K_{β_1})
Rb*	13 395 (K_{α_1}), 13 335 (K_{α_2}), 14 961 (K_{β_1}), 1752 (L_{β_1})
Cs*	4286 (L_{α_1}), 4272 (L_{α_2}), 4619 (L_{β_1}), 4935 (L_{β_2}), 5280 (L_{γ_1})

compressibility. Finally noting that the penetration length is larger than the expected characteristic lengths in the distribution, one can expand:

$$\left(\frac{I_i}{I_{\text{elas}}}\right) \Big/ \left(\frac{I_i}{I_{\text{elas}}}\right)_{\text{bulk}} = 1 + \frac{\alpha_i \Gamma_i}{c_{i\infty}} - \frac{\alpha_i^2}{c_{i\infty}} \int_0^\infty dzz[c_i(z) - c_{i\infty}] + \cdots \tag{3}$$

with $\alpha_i = 2\mathcal{I}m(k_z) + \mu_i$ and k_z the normal component of the wavevector in the solution and μ_i the linear absorption coefficient at the fluoresence energy of the ion i. Moreover, $\Gamma_i = \int_0^\infty dz(c_i(z) - c_{i\infty})$ is the Gibbs surface excess, which is negative for depletion, with $c_i(z)$ and $c_{i\infty}$ the concentration of ion i, at depth z and in bulk, respectively. Note that the second term in the expansion is related to the surface potential $\psi = (1/\epsilon) \int_0^\infty zdz \sum_i q_i[c_i(z) - c_{i\infty}]$, where q_i is the charge of ion i and ϵ the permittivity of water. The linear fits for alkali-halides are consistent with their very small surface potentials. Equation (3) provides direct access to the surface composition through Γ_i.

3.2. Example: Determination of the Surface Composition of Alkali-Halide Mixtures

The surface composition of alkali-halide mixtures has been determined using grazing incidence X-ray fluorescence in Ref. 28. Salts of high purity (>99.9–99.999%) purchased from Sigma–Aldrich were baked overnight (except iodide) in accordance to their stability at 0.5 atm. Fresh stock solutions of 1-M concentration were prepared in Millipore® water (>18 MΩ cm, < 10 ppb of organic contents) which is further diluted in steps of 0.1 M. Surface tension measurements for each concentration were carried out in a cleaned Teflon® dish using du-Nouy ring method (Krüss tensiometer) at 25°C. For GIXF experiments, Nitrogen was bubbled before preparing the solution to reduce the content of dissolved oxygen in water, and for the photosensitive iodide salts, surface tension measurements were carried out in a light protected chamber. For hygroscopic salts, the concentration was determined using capillary ion analysis. The experiments were performed at the ESRF beamline ID10B at energies of 8026 eV or 13 800 eV depending on the nature of the ion which needed to be excited.

The incident beam of dimension 1 mm × 0.15 mm ($H \times V$) was used to precisely define the grazing angle of incidence, and the solution was filled in a large Teflon trough ≈ 400 mm in diameter in order to avoid meniscus-related problems. A particular value of the grazing angle of incidence fixes the penetration depth and a full spectrum including fluorescence and elastic scattering from the bulk was recorded using a Röntec XFlash 1000 detector (Peltier cooled drift diode). K and L fluorescence lines in between Cl K_α (2622 eV) and Br K_α (11 924 eV) could be detected giving access in these experiments particularly to Cl^-, ClO_4^-, Br^-, I^-, K^+, Cs^+, Ni^{2+} and La^{3+}. All the measurements were carried out in Helium-filled enclosure to minimise absorption and avoid interference of Argon from the atmosphere.

Directly comparing the fluorescence intensities from chloride (in HCl) and perchlorate (in $HClO_4$) anions where all the correction factors are similar since the fluorescence from chlorine only is considered, one can clearly see [Fig. 3(a)] that, below θ_c, the fluorescence from ClO_4^- is indeed more intense than that of Cl^-, agreeing well with the expectation that $HClO_4$ adsorbs more strongly at the interface than HCl. Quantitative analysis using Eq. (3) for 0.1 M potassium chloride (KCl) solution shows that the normalised fluorescence intensity is smaller below θ_c than above [Fig. 3(b)], implying a depletion of ions at the interface. Comparison to a model calculation using only the long-range Coulombic interaction shows that the depletion is stronger than what is predicted, whereas a better result is found by incorporating an additional

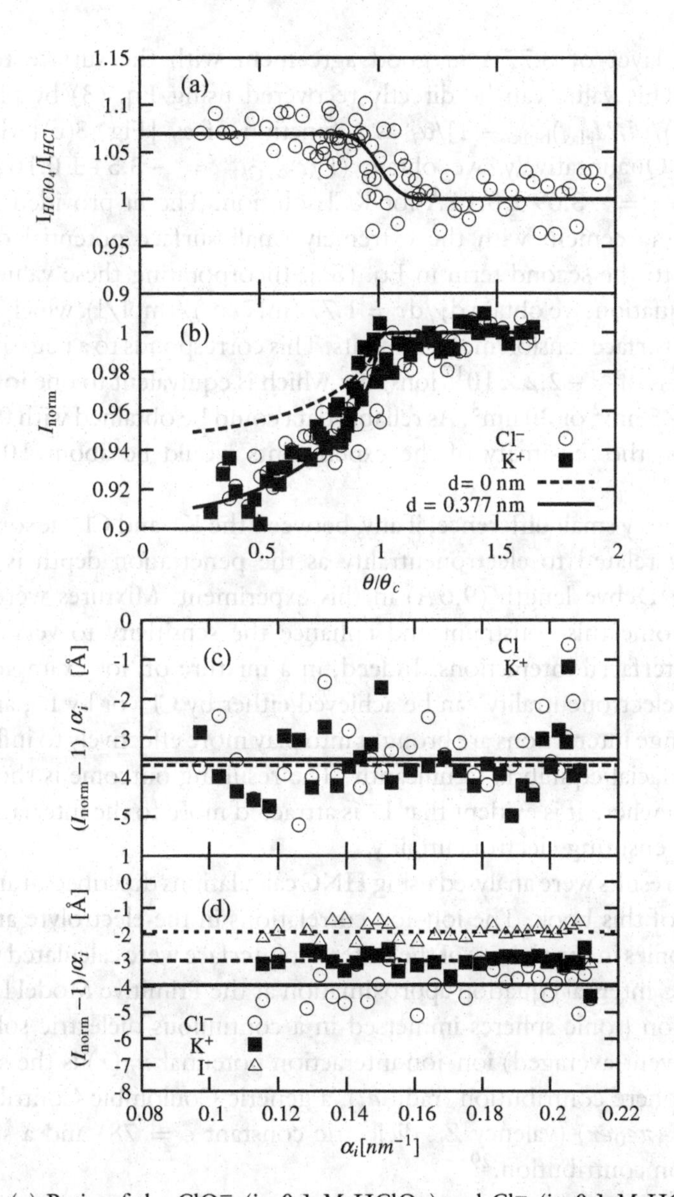

Fig. 3. (a) Ratio of the ClO_4^- (in 0.1-M $HClO_4$) and Cl^- (in 0.1-M HCl) fluorescence intensities. (b) Normalised K^+ and Cl^- intensities for 0.1-M KCl solution. Calculation considering the Coulombic effect interactions only (dashed line) and an additional ion free layer of 3.77 Å thickness as expected from surface tension measurements (continuous line).[11] $[(I_i/I_{\text{elas}})/(I_i/I_{\text{elas}})_{\text{bulk}} - 1]/\alpha_i$ as a function of α_i yielding $\Gamma_i/c_{i\infty}$ (c) for 0.1-M KCl solution and (d) for a mixture of 0.1-M KCl and 0.1-M KI solution.

ion free layer of 3.77 Å in good agreement with the surface tension data.[11] This value can be directly recovered using Eq. (3) by plotting $[(I_i/I_{\text{elas}})/(I_i/I_{\text{elas}})_{\text{bulk}} - 1]/\alpha_i$ as a function of α_i [Fig. 3(c)] yielding $\Gamma_i/c_{i\infty}$. Quantitatively, we obtain $(\Gamma/c_\infty)_{Cl^-} = -3.54 \pm 0.16$ Å and $(\Gamma/c_\infty)_{K^+} = -3.69 \pm 0.14$ Å for KCl solution. The fit provided a zero slope in agreement with the extremely small surface potential of KCl [related to the second term in Eq. (3)]. Incorporating these values into Gibbs equation, we obtain $d\gamma/dc = 1.77$ (mN/m)/(mol/l), which agrees with the surface tension measurements. This corresponds to a negative surface excess of $\approx -2.2 \times 10^{17}$ ions/m^2 which is equivalent to one ion lacking per 4.5 nm^2 or 30 nm^3. As reliable data could be obtained with 0.01 M solutions, the sensitivity of the experiments would be about 10 times better.

The very small difference, if any, between the K^+ and Cl^- adsorptions is clearly related to electroneutrality as the penetration depth is larger than the Debye length (9.6 Å) in this experiment. Mixtures were used to overcome this constraint and enhance the sensitivity to very short-range interfacial interactions. Indeed, in a mixture of, for example, KCl and KI, electroneutrality can be achieved either by Cl^- or by I^-, and the short-range interactions are brought into play more effectively to influence the interfacial equilibria significantly. The resulting outcome is shown in Fig. 3(d) where it is evident that I^- is attracted more to the interface than Cl^-, K^+ ensuring electroneutrality.

The results were analysed using HNC calculations described in another chapter of this book. The ion–ion correlations in the electrolyte and the ionic profiles in the vicinity of the water–air interface were calculated within the HNC integral equation approximation at the Primitive Model level of description (ionic spheres immersed in a continuous dielectric solvent). The (solvent-averaged) ion–ion interaction potential $u_{ij}(r)$ is the sum of a hard-sphere contribution (radii a_i), a generic Coulombic Contribution $Z_i Z_j e^2/(4\pi\epsilon_0\epsilon r)$ (valency Z_i, dielectric constant $\epsilon = 78$) and a specific dispersion contribution.[29]

$$u_{ij}^{\text{disp}}(r) = -\frac{3}{2}\alpha_i^*\alpha_j^*\frac{I_i I_j}{I_i + I_j} \times \left[1 + \frac{1 - n_w^2}{n_w} I_w \frac{n_w I_w + I_i + I_j}{(n_w I_w + I_i)(n_w I_w + I_j)}\right]\frac{F(r)}{r^6}.$$

Table 2. Parameters used in the calculations: radius r, excess polarisability α^*, ionisation potential I and K_i, the short-range exponential potential at contact.

	r (Å)	α^* (Å^3)	I (kJ/mol)	K_i (k_BT)
Cl	1.81	4.7	1251.1	−4.0
Br	1.96	6.1	1139.9	−4.7
I	2.20	6.1	1008.4	−4.4
H	2.50	0.0	1312	−3.15
Li	2.35	−1.7	520.2	0.0
Na	2.05	1.0	495.8	0.0
K	1.89	0.5	418.8	0.0
Cs	1.81	2.0	375.7	0.0

α_i^* is the excess polarisability of ion i (compared to water medium), I_i its ionization potential (I_w is that of water, with refractive index $n_w = 1.34$). $F(r)$ differs from the asymptotic value 1 only at short distance and accounts for the finite size of the ions.

In the first step, the ion–ion correlations in the bulk electrolyte are calculated using the Ornstein–Zernike equation with HNC closure condition at a temperature, $T = 298\,\text{K}$ with the required ionic densities ρ_i. The radii and polarisabilities were adjusted[15] to reproduce the thermodynamical properties of the electrolyte up to a few molar (see Table 2).

In the second step, the electrolyte is put in contact with a single big sphere (extra component at infinite dilution). As long as the radius R of this sphere is large compared to the characteristic distances in the electrolyte, its precise value is irrelevant and the sphere-electrolyte interface mimics a flat air–water interface. In practice, $R = 50\,\text{Å}$ is sufficient. The HNC equation is solved for the pair distribution functions $g_i(z)$ between the big sphere and the ions and gives the local ionic profiles $\rho_i(z) = \rho_i g_i(z)$ where $z = r - R$ is the normal distance to the interface. The interface-ion pair potential $u_i(z)$ contains the hard-sphere contribution, the dispersion contribution

$$u_i^{\text{disp}}(z) = \frac{n_w^2 - 1}{16 n_w} \frac{I_i I_w}{n_w I_w + I_i} \left[1 + \frac{1 - n_w^2}{2} \frac{I_w^2}{I_3} \frac{n_w I_w + I_i + I_3}{(I_3 + I_i)(n_w I_w + I_3)} \right] \frac{\alpha_i^*}{z^3}$$

(with $I_3 = \sqrt{(1+n_w^2)/2I_w}$), the *generic* screened image force contribution

$$u_i^{\text{image}}(z) = Z_i^2 \frac{e^2}{16\pi\epsilon_0\epsilon} \frac{\epsilon - 1}{\epsilon + 1} \frac{\exp(-2\kappa z)}{z}$$

(screening constant κ). In the absence of a first-principle theory to predict the surface properties and to capture the specific interactions, we extend the approach given in Ref. 15 by including an effective short-range interaction potential, u_i^{extra}, (exact shape is not important) which is of the form,

$$u_i^{\text{extra}}(z) = K_i \exp(-z/d).$$

Provided that the range d is shorter than the other characteristic distances, the same profiles can be obtained with different couples K_i, d. We have fixed here $d = 1$ Å and play with the strength K_i (potential depth when $\sigma_i < 0$). Note that the present HNC model differs from the more approximated mean-field Poisson–Boltzmann approach due to the proper account of the ion–ion correlations, which play a non-negligible role in the molar range. Finally, the amount of adsorbed ions is obtained by integrating the excess local densities:

$$\Gamma_i = \rho_i \int_0^{\infty} (g_i(z) - 1)dz.$$

The accuracy of these data proved to be good enough to allow for a detailed analysis in terms of ion-surface potential. Except H^+, as cations are known to have a smaller effect on surface tension, we decided to use the effective potential for anions and H^+ only. The strengths of u_i^{extra} are taken to be, -4.0 k_BT at contact for Cl^-, -4.7 k_BT for Br^-, -4.4 k_BT for I^-, and -3.15 k_BT for H^+. The surface compositions of different alkali-halide solutions and mixtures are given in Table 3.

The values of Γ_i calculated using this method are in excellent agreement with the measured values except for lithium salts, possibly indicating the need for a slightly repulsive short-range potential for this cation. It is quite remarkable that it was possible to reproduce about 20 independent measurements with only four adjustable parameters, supporting the idea that specific effects are indeed due to short-range couplings. Also, it is interesting to note that this simple model provides a quantitative description of the concentration dependence of the surface excess. In the case of HCl, we were able to reproduce even the sign reversal of the surface excess which is negative for 0.1-M concentration and positive (with higher

Table 3. Measured and calculated surface excess (Γ/c) for different alkali-halide solutions.

	NaCl	NaCl	HCl	HCl	KCl		LiCl	NaCl + NaI		KCl + KI			CsCl + CsI			LiCl + LiBr	
c (mol/L)	0.01	0.1	0.1	1.0	0.1		1.0	0.1		0.1			0.1			0.1	
	Cl^-	Cl^-	Cl^-	Cl^-	Cl^-	K^+	Cl^-	Cl^-	I^-	Cl^-	I^-	K^+	Cl^-	I^-	Cs^+	Cl^-	Br^-
Γ_{meas}(Å)	−6.3	−4.02	−0.5	+0.55	−3.54	−3.69	−3.35	−4.95	−1.90	−4.05	−2.00	−3.10	−4.30	−1.90	−3.10	−4.25	−2.60
	±1.5	±0.15	±0.2	±0.15	±0.16	±0.13	±0.1	±0.3	±0.15	±0.15	±0.1	±0.1	±0.1	±0.1	±0.1	±0.35	±0.15
Γ_{calc}	−6.0	−3.8	−0.51	+0.56	−3.65	−3.65	−2.85	−3.83	−2.12	−3.72	−1.99	−2.85	−3.69	−2.02	−2.85	−3.22	−2.56

screening) for 1-M concentration. Furthermore, these results emphasise the fact that the surface equilibria of ions mainly result from a subtle balance between the dispersion interaction, u_i^{disp}, and the short-range potential, u_i^{extra}, while the Coulombic image charge interaction, u_i^{image}, plays a less significant role [Fig. 4(a)]. In the case of Cl^- and I^-, e.g. the dispersion forces are almost equal, and the difference in the top 5 Å is mainly due

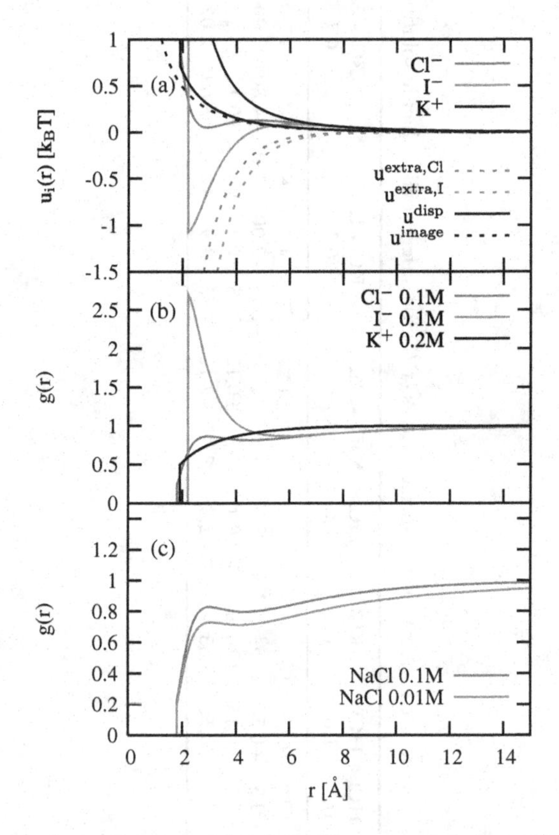

Fig. 4. Interface-ion pair potential $u_i(z)$ and concentration profiles $g(z)$ of ions obtained by fitting the effective strength of the short-range exponential potential as described in the text in order to obtain the experimental surface excesses. (a) Interface-ion potential for Cl^-, I^- and K^+ in KCl 0.1 M + KI 0.1 M mixture. The black solid line represents the dispersion interaction, u^{disp}, for Cl^-, almost equal to that for I^-. The black dashed line represents the image potential, u^{image}, which has the same strength for all ions. The red and green dashed lines represent the additional short-range exponential potential, u^{extra}, for Cl^- and I^-, respectively. (b) Concentration profiles in KCl 0.1 M + KI 0.1 M mixture. (c) Cl^- concentration profile in NaCl 0.1 M and NaCl 0.01 M.

to u_i^{extra}. The effective potential values *at contact* for Cl^- and I^- being close to each other, the potential is more attractive for the biggest I^- ion. However, the minimum in the potential is reduced by a factor of 4 due to the dispersion interaction. More generally, with the increase in ion size for the halides, we recover the Hofmeister series, $\Gamma(I^-) > \Gamma(Br^-) > \Gamma(Cl^-)$.

Knowing the effective short-range potential, the interfacial distribution of ions can be calculated. Figure 4 shows the result of such calculation for NaCl at two different concentrations and for KCl + KI mixture. As expected, the main effect of an increase in salt concentration is to reduce the repulsion of the image charge, and beyond that to wash out the profile. In the mixture, I^- is more strongly attracted to the surface than Cl^-. Interestingly, these profiles are in excellent agreement with numerical simulations for Cl^- and I^-.[16]

Grazing incidence X-ray fluorescence is able to directly determine the surface composition for concentrations ranging from 0.01 M to 1 M under ambient conditions. Coupled with competitive adsorption in mixtures of salts, this method has the unique ability to distinguish very short-range couplings at the Å level. However, it lacks depth sensitivity below about 5 nm. At solid/solution interfaces, such a sensitivity can be obtained by using the X-ray standing waves technique.

4. X-Ray Standing Waves at the Solid–Solution Interface

4.1. *Theoretical Background*

The X-ray standing waves technique (XSW, Fig. 5) uses the Bragg reflection of X-rays from a crystal or a multilayer substrate, or the interference between incident and specularly reflected plane waves at an interface in order to establish a standing waves field which can be used to probe interfaces, by recording for example the fluorescence of a given type of atom. The need for a high-flux beam with negligible divergence and a small cross section (i.e. high brightness) necessitates the use of synchrotron radiation.[30–32]

Close to a Bragg angle θ such as $\sin\theta = \lambda/2d$, with λ the wavelength, the incident and scattered travelling waves interfere to produce a standing wave field with a period equal to the spacing d of the associated scattering planes (Fig. 5). The XSW planes are parallel to the diffracting planes. The standing wave field has a penetration depth of at least microns, and extends

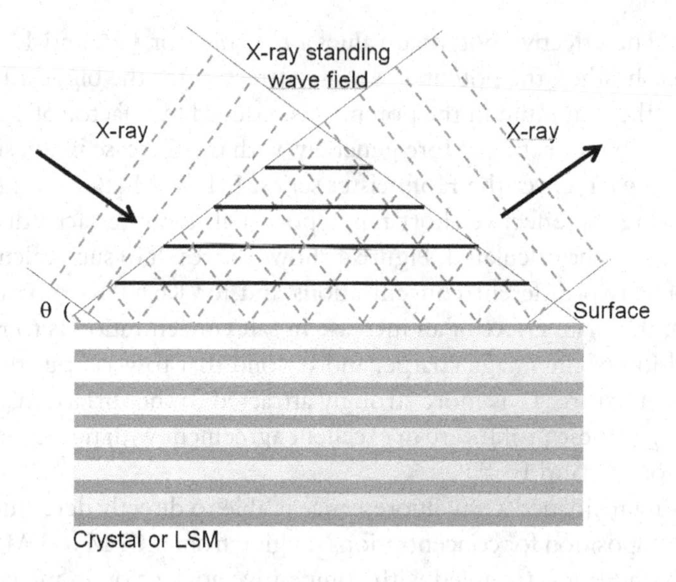

Fig. 5. Illustration of the X-ray standing wave field formed by the interference between the incident and Bragg-diffracted plane waves. Tilting angle θ through the row reflection condition causes the antinodes of the standing wave to shift in by one-half of the diffraction-plane spacing d.

well above the surface. The X-ray field experienced by the substrate or atoms close to the interface depends on their location within the standing wave field relative to the nodal planes. As the angle θ is increased from the low angle side of the Bragg peak where there is a node at the surface, the standing waves move inward until the antinode reaches the diffraction planes. As the incident angle is varied, the nodal or antinodal planes of the XSW will pass through the plane of an atom present above the surface. Since the photoelectric effect is proportional to the E-field intensity at the centre of an atom, the fluorescence emission yield from atoms above the surface will show a characteristic modulation as the substrate is rocked in angle. The phase and amplitude of the signal are respectively depending on the average position modulo the d-spacing of the layered system and the width of the distribution of the atoms. If all the atoms considered are contained in a plane parallel to the diffracting planes, the contrast of the modulation would be equal to the Bragg peak reflectivity R. If the heavy atoms are randomly distributed in a thick overlayer, the contrast of the modulation would be zero.

The fluorescence yield, normalised to the off-Bragg yield, varies as,

$$\Upsilon(\theta) = 1 + R(\theta) + 2[R(\theta)]^{1/2} f_H \cos[\phi(\theta) - 2\pi P_H], \quad (4)$$

where the reflectivity $R(\theta)$ and the XSW phase $\phi(\theta)$ are derived for the $\mathbf{H} = (hkl)$ diffraction condition from dynamical diffraction theory. The classical way of characterising the data, which is well adapted to thin adsorbate layers, is to use the coherent position, $P_H = \mathbf{H} \cdot \mathbf{r}$ for atoms located at $\mathbf{r}$, and the coherent fraction, f_H which can be obtained from a χ^2 fit of Eq. (4) to the XSW data and are model independent.

Under total external reflection condition, for $\theta = 0$, a standing-wave node lies at the mirror surface with the first antinode infinitely far above the surface. As the incident angle θ is increased through the reflection, the first antinode moves inward until it coincides with the mirror surface for $\theta = \theta_c$. Above θ_c, the first antinode remains at the mirror surface while the amplitude of the standing wave decreases very rapidly because of the drastic reduction in intensity of the specularly reflected plane wave.

An advantage of the technique is that different spectroscopies, e.g. fluorescence or photoelectrons, can be used. In case fluorescence is used, a full spectrum including fluorescence and elastic scattering is recorded for each value of the grazing angle of incidence θ as previously described for the grazing incidence X-ray fluorescence at the air–solution interface.

4.2. *Example: Surface Composition of Rb and Sr Adsorbed from Aqueous Solutions at the Rutile and Mica Water Interface*

The surface composition of Rb and Sr adsorbed from aqueous solutions at the rutile (110)-water interface has been determined using X-ray standing waves in Ref. 33. The surface used to produce the XSW was a chemo-mechanically polished synthetic single crystal of TiO_2. In order to get results representative of rutile powder, which mainly present (110) and (100) faces, a (110) surface was chosen. mM Solutions were used and a wide pH range was investigated. The sample was contained in a Kel-F cell using a 8 μm-thick Kapton film as an X-ray window, in which a thin solution layer is held against the mirror surface and a negative pressure was applied to the solution in order to maintain a film thickness of about 2 μm.[33] The XSW experiments were performed at the beamline station 12-ID-D (BESSRC-CAT) at the Advanced Photon Source at Argonne

National Laboratory. Monochromatic X-ray beams were used at two photon energies of 16.7 and 16.2 keV to measure Sr and Rb respectively, using a 20 μm × 200 μm high beam, giving a footprint of 0.17 × 0.20 mm). The SrK_α and RbK_α fluorescence emission lines at 14.2 and 13.4 keV were monitored using a Si(Li) solid state detector. The Sr^{2+} ions were observed to be at a height of 3.07(0.07) Å and that Rb^+ adsorbs at a height of 3.44 (0.03) Å.[34] Several similar experiments were further performed to investigate Rb^+, Sr^{2+}, Zn^{2+}, and Y^{3+} adsorption at rutile-water interface[34,35] or Se(IV) at the hematite(100)-water interface.[36]

Rb^+ and Sr^{2+} adsorption at the mica surface was also investigated in Ref. 37 by resonant anomalous reflectivity. It was shown that Rb^+ adsorbs in a partially hydrated state and incompletely compensates the surface charge, whereas Sr^{2+} adsorbs in both fully and partially hydrated states while achieving full charge compensation. These differences are driven by balancing the energy cost of disrupting ion and interface hydration with the electrostatic attraction between the cation and charged surface.

4.3. *Example: Determination of the Surface Composition of Electrical Double-Layer Structure at Solid–Water Interface*

We have recently investigated mixtures of 10^{-5}M solution of KCl and KI, $LaCl_3$ and KCl, and KCl and CsCl. The surface charge and the magnitude of electrostatic forces were controlled by changing the pH from 3 to 9 (via the protonation–deprotonation of silanol groups at the solid–water interface). The substrates considered were Si/Mo multilayer substrates with a 30 Å period, thus allowing one to probe the diffuse Gouy–Chapman layer. The X-ray standing waves (XSW) measurementis were performed at the BW1 beamline of the DESY synchrotron (Hamburg, Germany) with an incident beam of 8 keV. The reflectivity of the first order Bragg reflection and the fluorescence yields were simultaneously recorded. From an analysis of the fluorescence data with a homemade program similar to that described in the previous section, we were able to extract the ions fluorescence angular dependence (Fig. 6).

The fluorescence from La^{3+}, K^+ and Cl^- in a mixture of 10^{-5}M of $LaCl_3$ and KCl at pH 7 is represented in Fig. 6 together with the profiles and the Bragg peak. Obviously, there is a strong condensation of the ions at the interface, and La^{3+} is more strongly condensed than K^+. This experiment represents a first step in probing the ion distribution at

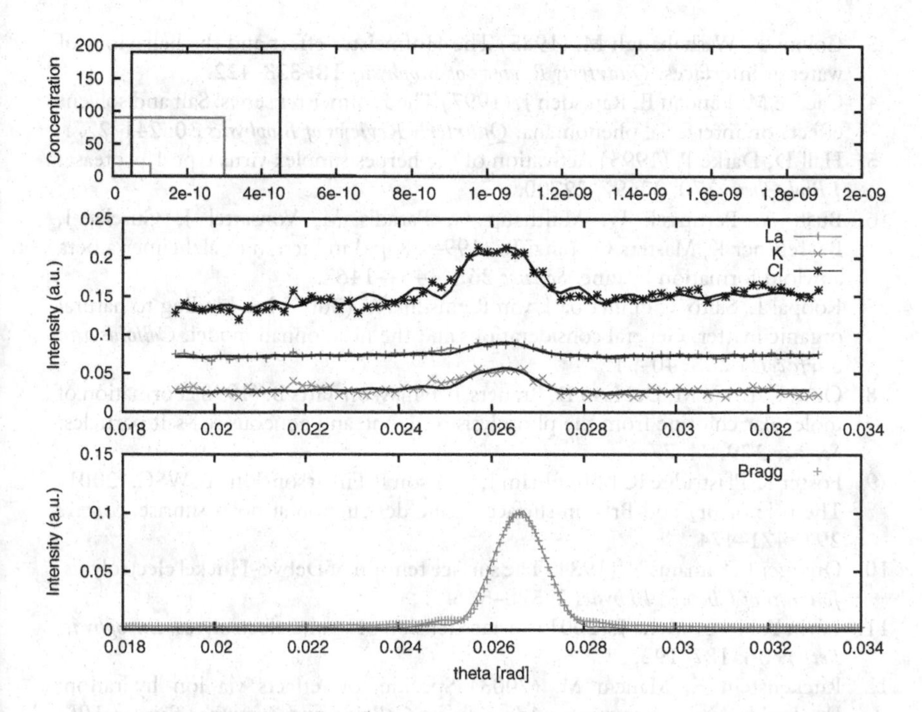

Fig. 6. Fluorescence from La^{3+}, K^{+} and Cl^{-} in a mixture of 10^{-5} M of $LaCl_3$ and KCl at pH 7 as a function of the Bragg angle. Experimental points and calculated intensity (lines). Corresponding ionic profiles (top). Experimental and calculated Bragg peak (bottom).

interfaces. More experiments and simulations are in progress in order to obtain a quantitative description of the ion distribution in various cases.

Acknowledgements

J. D. and P. V. acknowledge support of the Indo-French Centre for the promotion of advanced research and of ANR SISCI.

References

1. Craig V, Ninham B, Pashley R. (1993) Effect of electrolytes on bubble coalescence. *Nature* **364**: 317–319.
2. Weissenborn P, Pugh R. (1996) Surface tension of aqueous solutions of electrolytes: Relationship with ion hydration, oxygen solubility, and bubble coalescence. *Journal of Colloid and Interface Science* **184**: 550–563.

3. Collins K, Washabaugh M. (1985) The Hofmeister effect and the behaviour of water at interfaces. *Quarterly Reviews of Biophysics* **18**: 323–422.
4. Cacace M, Landau E, Ramsden J. (1997) The Hofmeister series: Salt and solvent effects on interfacial phenomena. *Quarterly Reviews of Biophysics* **30**: 241–277.
5. Hall D, Darke P. (1995) Activation of the herpes simplex virus type 1 protease. *J Biol Chem* **270**: 22697–22700.
6. Bush A, Pettingell W, Multhaup G, Paradis M, Vonsattel J, Gusella J, Beyreuther K, Masters C, Tanzi R. (1994) Rapid induction of alzheimer a beta amyloid formation by zinc. *Science* **265**: 1464–1467.
7. Koopal L, Saito T, Pinheiroa J, van Riemsdijk W. (2005) Ion binding to natural organic matter: General considerations and the nicadonnan model. *Colloids and Surfaces A* **265**: 40–54.
8. Oum K, Lakin M, DeHaan D, Brauers T, Finlayson-Pitts B. (1998) Formation of molecular chlorine from the photolysis of ozone and aqueous sea-salt particles. *Science* **279**: 74–77.
9. Foster K, Plastridge R, Bottenheim J, Shepson P, Finlayson-Pitts B, WSC. (2001) The role of Br_2 and BrCl in surface ozone destruction at polar sunrise. *Science* **291**: 471–474.
10. Onsager L, Samaras N. (1934) The surface tension of Debye–Huckel electrolytes. *Journal of Chemical Physics* **2**: 528–536.
11. Levin Y, Flores-Mena J. (2001) Surface tension of strong electrolytes. *Europhysics Letters* **56**: 187–192.
12. Ruckenstein E, Manciu M. (2003) Specific ion effects via ion hydration: II. double layer interaction. *Advances in Colloids and Interface Science* **105**: 177–200.
13. Edwards SA, Williams D. (2006) Surface tension of electrolyte solutions: Comparing the effects of ionic dispersion forces and solvation. *Europhysics Letters* **74**(5): 854–860.
14. Ninham B, Yaminsky V. (1997) Ion binding and ion specificity: The Hofmeister effect and onsager and lifshitz theories. *Langmuir* **13**: 2097–2108.
15. Kunz W, Belloni L, Bernard O, Ninham B. (2004) Osmotic coefficients and surface tensions of aqueous electrolyte solutions: Role of dispersion forces. *Journal of Physical Chemistry B* **108**(7): 2398–2404.
16. Jungwirth P, Tobias D. (2001) Molecular structure of salt solutions: A new view of the interface with implications for heterogeneous atmospheric chemistry. *Journal of Physical Chemistry B* **105**: 10468–10472.
17. Jungwirth P, Tobias D. (2006) Specific ion effects at the air/water interface. *Chemical Reviews* **106**: 1259–1281.
18. Liu D, Ma G, Levering L, Allen H. (2004) Vibrational spectroscopy of aqueous sodium halide solutions and air-liquid interfaces: Observation of increased interfacial depth. *Journal of Physical Chemistry B* **108**(7): 2252–2260.
19. Raymond E, Richmond G. (2004) Probing the molecular structure and bonding of the surface of aqueous salt solutions. *Journal of Physical Chemistry B* **108**(16): 5051–5059.
20. Petersen P, Johnson J, Knutsen K, Saykally R. (2004) Direct experimental validation of the Jones-ray effect. *Chemical Physics Letters* **397**: 46–50.

21. Weber R, Winter B, Schmidt P, Widdra W, Hertel I, Dittmar M, Faubel M. (2004) Photoemission from aqueous alkali-metal-iodide salt solutions using euv synchrotron radiation. *Journal of Physical Chemistry B* **108**(15): 4729–4736.
22. Ghosal S, Hemminger J, Bluhm H, Mun B, Hebenstreit E, Ketteler G, Ogletree D, Requejo F and Salmeron M. (2005) Electron spectroscopy of agueous solution interfaces reveals surface enhancement of halides: Implications for atmospheric chemistry. *Science* **307**: 563–566.
23. Daillant J, Gibaud A. (eds.) (2009) *X-Ray and Neutron Reflectivity, Principles and Applications.* Springer Verlag, Heidelberg.
24. Luo G, Malkova S, Yoon J, Schultz D, Lin B, Meron M, Benjamin I, Vanysek P, Schlossman M. (2006) Ions distributions near a liquid–liquid interface. *Science* **311**(5758): 216–218.
25. Sloutskin E, Baumert J, Ocko B, Kuzmenko I, Checco A, Tamam L, Ofer T, Gog E, Gang O, Deutsch M. (2007) The surface structure of concentrated aqueous salt solutions. *Journal of Chemical Physics* **126**: 054704.
26. van Gysel M, Lemberge P, van Espen P. (2003) Implementation of a spectrum fitting procedure using a robust peak model. *X-ray Spectrometry* **32**: 434–441.
27. van Gysel M, Lemberge P, van Espen P. (2003) Description of compton peaks in energy-dispersive X-ray fluorescence spectra. *X-ray Spectrometry* **32**: 139–147.
28. Viswanath P, Daillant J, Belloni L, Mora S, Alba M, Konovalov O. (2007) Specific ion adsorption and short-range interactions at the air aqueous solution interface. *Phys Rev Lett* **99**: 086105.
29. Mahanty J, Ninham B. (1976) *Dispersion Forces.* Academic Press, London.
30. Bedzyk M, Bommarito G, Schildkraut J. (1989) X-ray standing waves at a reflecting mirror surface. *Physical Review Letters* **62**(12): 1376–1379.
31. Bedzyk M. (1990) Measuring the diffuse-double layer at an electrochemical interface with long period X-ray standing waves. *Synchrotron Radiation News* **3**(5): 25–29.
32. Abruna H, White J, Bommarito G, Albarelli M, Acevedo D, MJB. (1989) Structural studies of electrochemical interfaces with X-rays. *Review of Scientific Instruments* **60**(7): 2529.
33. Fenter P, Cheng L, Rihs S, Machesky M, Bedzyk M, Sturchio N. (2000) Electrical double-layer structure at the rutilewater interface as observed *in situ* with small-period X-ray standing waves. *Journal of Colloid and Interface Science* **225**: 154–165.
34. Zhang Z, Fenter P, Cheng L *et al.* (2004) Ion adsorption at the rutile-water interface: Linking molecular and macroscopic properties. *Langmuir* **20**: 4954–4969.
35. Zhang Z, Fenter P, Cheng L, Sturchio N, Bedzyk M, Machesky M, Anovitz L, Wesolowski D. (2006) Zn^{2+} and Sr^{2+} asorption at the TiO_2 (110)-electrolyte interface: Influence of ionic strengh, coverage, and anions. *Journal of Colloid and Interface Science* **295**: 50–64.
36. Catalano J, Zhang Z, Fenter P, Bedzyk M. (2006) Inner-sphere adsorption geometry of se(iv) at the hematite (100) water interface. *Journal of Colloid and Interface Science* **297**: 665–671.
37. Park C, Fenter P, Nagy K, Sturchio N. (2006) Hydration and distribution of ions at the mica-water interface. *Phys Rev Lett* **97**: 016101.

Chapter 6

The Determination of Specific Ion Structure by Neutron Scattering and Computer Simulation

George W. Neilson*, Philip E. Mason† and John W. Brady†

The presence of ions in water constitutes a rich and varied environment in which many natural processes occur. This review provides results of structural studies of aqueous solutions derived from state-of-the-art neutron scattering methods. The enhanced resolution provided by methods such as Neutron Diffraction and Isotopic Substitution (NDIS) have given scientists new insights into the contrasting hydration structures of a variety of ions and small solute molecules, and crucially how these structures might affect the general properties of solutions. The discussion points out common features of ionic hydration within particular series such as the alkalis, halides and transition metals, and also indicates where significant differences in hydration structure appear. For polyatomic ions there is a clear need for additional information to interpret the neutron scattering results. Accordingly we show how computer simulation can assist in the particular case of hydration of complex ions such as guanadinium (Gdm^{+}) or thiocyanate (SCN^{-}), both of which have important consequences for protein denaturation in aqueous solution.

1. Introduction

Neutron scattering methods provide the best experimental means currently available to probe the atomic structure of aqueous solutions.[1] It can be proved that a formal mathematical (Fourier transformation) link can be formed between the neutron scattering pattern obtained experimentally and the pair radial distribution functions $g_{\alpha\beta}(r)$ of pairs of atoms α and β of the system.[2] Knowledge of these functions, either individually or as combinations $[G_{\alpha}(r)]$ specific to a particular atom (or ion), α,

*H. H. Wills Physics Laboratory, University of Bristol, Tyndall Avenue, BS8 1TL, UK.
†Department of Food Science, Cornell University, Ithaca, NY, USA.

can be used to define atomic correlation distances, coordination numbers, and the extent of local order around that atom (or ion). This information can then be used in a critical examination of results obtained from model potential calculations.[3]

Since the mid-1970s, we have been concerned with the development of a variety of X-ray and neutron scattering methods aimed at the most detailed structural description experimentally possible of interatomic structure of ions in aqueous electrolyte solution.[2] Foremost amongst these methods has been that of Neutron Diffraction and Isotopic Substitution (NDIS) — combinations of differences between the scattering patterns of isotopically labelled samples can be used to obtain information directly on all aspects of the pairwise inter-atomic structure of an aqueous electrolyte solution, including that of the water solvent itself.[4]

In this article, we present a brief outline of the methods used and introduce the structural properties which can be calculated directly from the experimental data. Several examples have been selected to demonstrate the power of these methods, with reference to studies of monatomic ions such as Ni^{2+} and Cl^- (Figs. 1–3). We also mention an extension to the NDIS method that applies to aqueous solutions of 'null' water and enables one to determine the nearest neighbour coordination of strong cations for cations which do not possess isotopes suitable for the NDIS method.[5]

In recent years and with the dramatic advances in computer power, we have developed in parallel computer simulation methods to help determine the structure around complex ions such as guanidinium (Gdm^+) or thiocyanate (SCN^-) in water[6,7] (Fig. 4). The combination of computer simulation and NDIS has enabled us to obtain reliable structural information on the structural effects of these and other ions (Fig. 5) and their potential influence in the denaturation of proteins in solution.

2. Methods

2.1. *Neutron Diffraction Based on Isotopic Substitution (NDIS)*

The main advantage of neutron methods rests on the fact that neutrons interact predominantly with atomic nuclei through the strong force and the interaction is isotropic. As a result, structural information in the form of $g_{\alpha\beta}(r)$'s is accessible directly from the experimental diffraction data.

Moreover, the strength of interaction, which is characterised by the scattering length parameter b_a, varies from isotope to isotope.[8] Therefore, two solutions with the same composition of atomic material but containing isotopically different nuclei (e.g. H for D, ^{35}Cl for ^{37}Cl, etc.) will give different neutron scattering patterns. This enables detailed information to be obtained which is specific to a particular isotopically substituted species.

If neutrons are scattered by a liquid containing N nuclear/atomic species, the intensity of the scattered neutrons, $I(q)$ is given by[9]

$$I(q) = \sum_{\alpha}\sum_{\beta} b_\alpha b_\beta \left\langle \sum_{i(\alpha)}\sum_{i(\beta)} \exp i\mathbf{q}\cdot(r_j(\beta) - \mathbf{r}_i(\alpha)) \right\rangle$$

$$= N\left[\sum_\alpha c_\alpha b_\alpha^2 + \sum_\alpha\sum_\beta c_\alpha c_\beta b_\alpha b_\beta [S_{\alpha\beta}(q) - 1]\right]$$

$$= N\left[\sum_\alpha c_\alpha b_\alpha^2 + F(q)\right], \quad (1)$$

where $c_\alpha = N_\alpha/N$. In Eq. (1), $\mathbf{r}_i(\alpha)$ denotes the position of the ith nucleus of α-type characterised by a neutron scattering length b_α, and $\mathbf{q}$ is the scattering vector whose modulus, q, for elastic scattering (i.e. $|\mathbf{k}_0| = |\mathbf{k}_1| = 2\pi/\lambda$), is given by $q = 4\pi \sin\theta/\lambda$, where θ is half the scattering angle and λ is the wavelength of the incident neutrons. The angular brackets show that an ensemble average has been taken. $F(q)$ is the (total) structure factor of the solution, and $S_{\alpha\beta}(q)$ are the *partial structure factors (p.s.fs.)*. The sum extends over all the atomic species in the liquid, and for n atomic species there will be $n(n+1)/2 S_{\alpha\beta}(q)s$, each of which can be inverted to yield $g_{\alpha\beta}(r)$ through the Fourier transformation

$$g_{\alpha\beta}(r) = 1 + \frac{V}{2\pi^2 Nr}\int dq(S_{\alpha\beta}(q) - 1)k \sin qr. \quad (2)$$

The $b_\alpha s$ are independent of q for neutrons, i.e. the scattering is isotropic. It is this property which facilitates the direct determination of solution structure by neutron diffraction. (Although the same formalism applies to the scattering of X-rays, the fact that the $b_\alpha s$ are dependent on q means that the structural information cannot be obtained directly from the scattering function.)

Thus $S_{MO}(q)$ and $S_{MH}(q)$ for an aqueous electrolyte solution, MX_n in H_2O, can in principle yield the real space distribution functions $g_{MO}(r)$ and $g_{MH}(r)$ which are necessary if a proper description of ionic hydration is to be given. Unfortunately the contribution of these partial structure factors to the scattered intensity is very small even for relatively concentrated solutions, and special procedures must be adopted for their determination.

Total neutron diffraction studies of ionic hydration in aqueous electrolytes have been carried out,[10] however due to the similarity of scattering lengths (b) of all the atomic nuclei, the structural patterns are usually less informative than those obtained from X-ray diffraction studies. Therefore, as with X-ray diffraction, models are required to assist in the interpretation of data. By contrast, the difference methods of neutron diffraction and isotopic substitution (NDIS) can be used to determine directly and without modelling individual $g_{\alpha\beta}(r)$'s or their linear combinations of the form $G_\alpha(r)$ which is specific to the substituted species α.

Specifically, the NDIS method can be illustrated by reference to an aqueous electrolyte solution of a salt (MX_n) in water (H_2O) or more preferably from a neutron scattering viewpoint heavy water (D_2O).[1,2] The first difference method applied to cations or anions by isotopic exchange M′ for M or X′ for X can be used to obtain information concerning aqua-ion structure in terms of the function $G_M(r)$ or $G_X(r)$ (Fig. 1). In mathematical terms

$$\begin{aligned} G_M(r) &= Ag_{MO}(r) + Bg_{MH}(r) + Cg_{MX}(r) + Dg_{MM}(r) + E, \\ E &= -(A + B + C + D), \end{aligned} \tag{3}$$

where $A = 2c_M c_O b_O \Delta b_M$, $B = 2c_M c_H b_H \Delta b_M$, $C = 2c_M c_X b_X \Delta b_M$, $D = c_M^2(b_M^2 - (b_{M'})^2)$, $\Delta b_M = b_M - b_{M'}$.

Furthermore, if one carries out H/D substitution of the water molecules as well as with the ions, it is possible to obtain to a good approximation the individual pair radial distribution functions (r.d.f.s) $g_{IO}(r)$ and $g_{IH}(r)$, (Fig. 2).

For example, by carrying out isotopic substitution experiments on similarly labelled salts such as $^{62}Ni^{nat}Cl_2$, $^{58}Ni^{nat}Cl_2$ and $^{nat}Ni^{37}Cl_2$, $^{58}Ni^{35}Cl_2$ in light water, it is possible essentially to separate $g_{NiO}(r)$ from $g_{NiH}(r)$ and $g_{ClO}(r)$ from $g_{ClH}(r)$, (Fig. 2).

It can also be shown that second difference experiments involving six samples with isotopically different salt isotopic exchanges of M′ for M, X′ for X, can be used to obtain the individual pair functions for the solute [$g_{MM}(r)$,$g_{MX}(r)$,$g_{XX}(r)$], (Fig. 3).

For completeness, we mention that results for three solutions with differing isotopic amounts of H_2O and D_2O will provide information on individual r.d.f.s $g_{HH}(r)$ and $g_{OH*}(r)$ of the solvent, where the asterisk

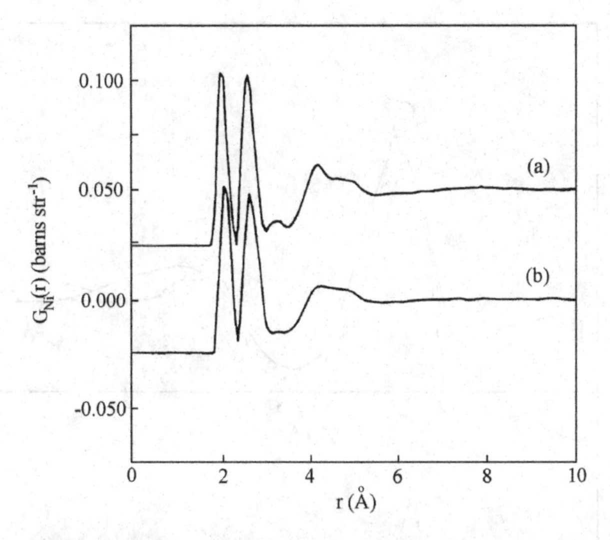

Fig. 1. The total Ni^{2+} pair distribution function, $G_{Ni}(r) = 8.86g_{NiO}(r) + 14.45g_{NiD}(r) + 1.39g_{NiN}(r) + 0.4g_{NiNi}(r) - 25.1$ mbarns str^{-1} for a 3.4 molal solution of nickel nitrate in heavy water at (a) 100 K (glass) and at (b) 270 K (liquid), (Howell, I. Ph.D. thesis, University of Bristol, 1993). The peaks at 2.08 Å and 2.56 Å represent respectively the near neighbour Ni–O and Ni–D correlations, and integration over them gives a hydration number of 6 [Eq. (4)]. Similar structures are found for ions such as Cr^{3+}, Fe^{2+} and Fe^{3+}, and many of the lanthanide ions. By contrast, Cu^{2+} shows an appreciably different first shell configuration,[45] and Zn^{2+} exhibits less well-resolved first hydration shell.[52]

on $g_{OH}(r)$ means that it is only an approximation to the true $g_{OH}(r)$; in the limit of high dilution, it will tend to the exact function as in pure water.[11] It is also worth noting that because the scattering length of the oxygen isotopes ^{16}O, ^{17}O and ^{18}O are almost the same,[8] determination of $g_{OO}(r)$ in pure water is liable to large errors, and is not currently accessible in solution.

To determine $G_M(r)$ or $G_X(r)$, it is preferable to work with solutions of heavy water (D_2O) because of the prohibitively large neutron incoherent scattering of the proton (H) spins, and the large inelastic scattering from the proton nuclei in H_2O.

The coordination number of α atoms around β is defined for the range $r_1 \leq r(\text{Å}) \leq r_2$ as

$$\bar{n}^{\alpha}_{\beta} = \pi c_{\alpha}\rho \int_{r_1}^{r_2} g_{\alpha\beta}(r)r^2\mathrm{d}r, \tag{4}$$

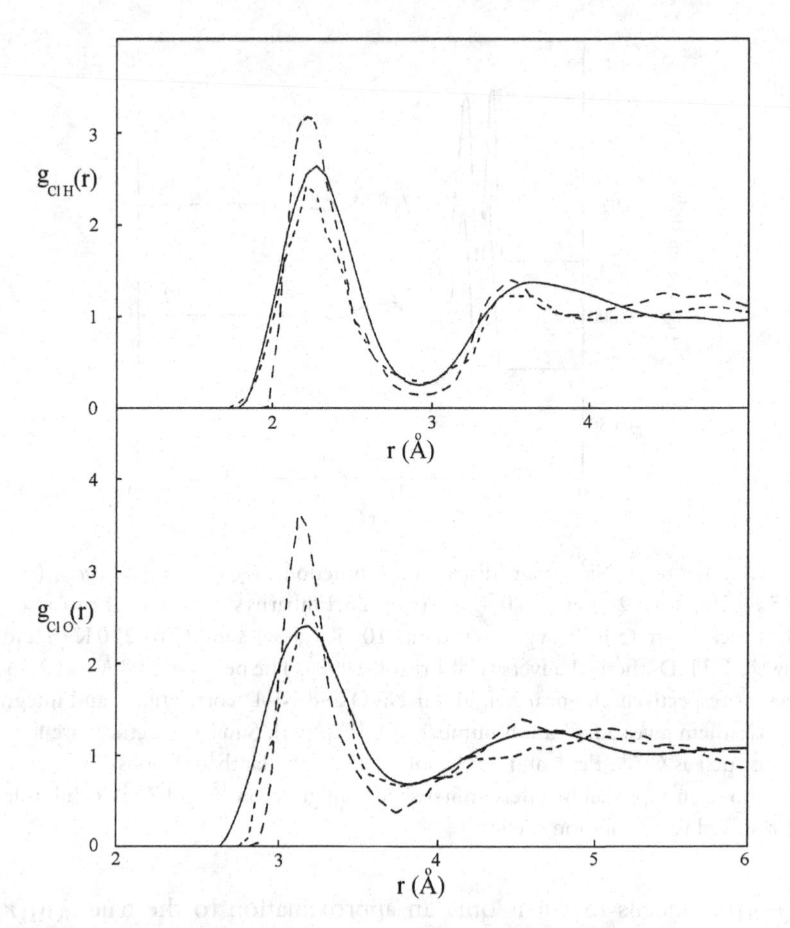

Fig. 2. The radial distribution functions $g_{\mathrm{ClH}}(r)$ and $g_{\mathrm{ClO}}(r)$ (full curve) obtained from NDIS experiments (Powell *et al.*, 1993), on an aqueous solution of two molal nickel chloride, compared with simulation results of Sprik *et al.* (1990)[12] (short dashed curves) and Dang *et al.* (1991)[13] (long dashed curves).

where ρ is the total number density in atoms per Å^{-3}. A hydration number for an ion I in solution can be defined as

$$\bar{n}_{\mathrm{I}}^{\mathrm{H_2O}} = 4\pi\rho c_{\mathrm{O}} \int_{r_1}^{r_2} g_{\mathrm{IO}}(r) r^2 \mathrm{d}r, \tag{4a}$$

or

$$\bar{n}_{\mathrm{I}}^{\mathrm{H_2O}} = 4\pi\rho c_{\mathrm{H}} \int_{r_1}^{r_2} g_{\mathrm{IH}}(r) r^2 \mathrm{d}r, \tag{4b}$$

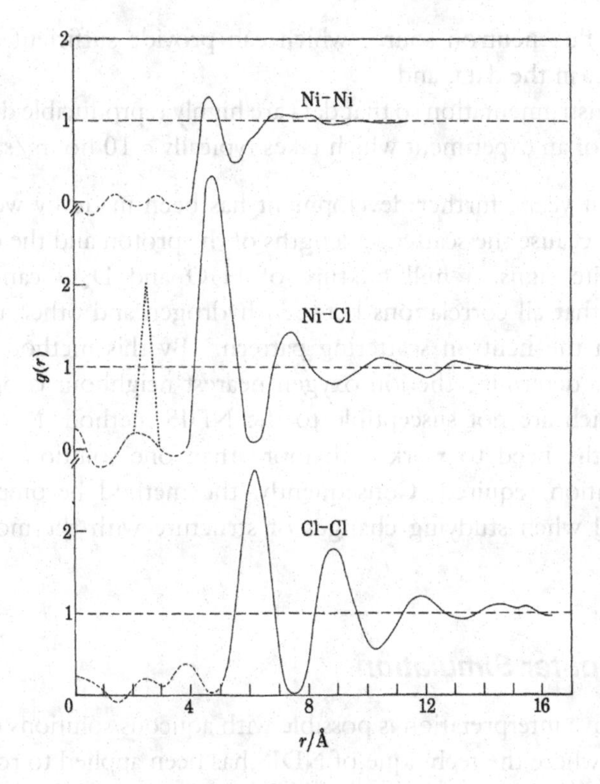

Fig. 3. The radial pair distribution functions $g_{NiNi}(r), g_{ClCl}(r), g_{NiCl}(r)$ in a 4.3 molal solution of nickel chloride in heavy water.[29] The dotted curve centred at 2.4 Å indicates the magnitude of $g_{NiCl}(r)$ expected if inner sphere complexing were to occur.

depending on which of these functions can be more readily identified. For more complex systems, NDIS methods can again be used to isolate structures around a particular species or to determine individual radial distribution functions (r.d.f.s). This procedure has been used successfully to determine the ion-specific structure in molten salts of electrolytes,[14] aqueous solutions of electrolytes,[1,2] polyelectrolytes,[15] and indeed sugar/water mixtures.[16,17]

The success of NDIS experiments depends crucially on several factors. These include

(i) high quality samples whose composition is accurately known in terms of atomic concentration and isotope content,

(ii) a high flux neutron source which can provide sufficient statistical accuracy in the data, and
(iii) stable instrumentation so that data are highly reproducible during the course of an experiment which takes typically ~10 hours/sample.

In recent years, further development has been made by working in null water: because the scattering lengths of the proton and the deuteron have opposite signs, a null mixture of H_2O and D_2O can be prepared such that all correlations between hydrogen and other atoms are absent from the neutron scattering pattern.[5] By this method we have been able to determine the ion-oxygen nearest neighbour coordination for ions which are not susceptible to the NDIS method. Moreover, it overcomes the need to work with more than one solution to extract the information required. Consequently, the method becomes particularly useful when studying changes of structure with thermodynamic state.

2.2. *Computer Simulation*

While intuitive interpretation is possible with aqueous solutions of strong electrolytes where the technique of NDIS has been applied to reduce the total number of correlations that contribute to the experimental neutron scattering, this simply is not possible with more complex systems, such as those containing polyatomic ions where many more correlations may be present. In systems containing molecular species (on which isotopic substitution has been performed), there are always strong intra-molecular correlations, even at lower concentrations.[17] These must be removed from the experimental data before anything significant can be determined about the structure of water around these solutes. Even in the simplest case, such as the highly symmetrical biologically important ion Gdm^+, the interpretation of the neutron scattering data is complicated.[18] Due to its three-fold symmetry and high solubility, it is possible to perform $^{14}N/^{15}N$ substitution experiments on solutions of this ion. By using knowledge of the covalent structure of this ion, we were able to construct a model and empirically subtract this from the experimental data, thereby enabling us to gain some insight into the hydration of this ion.[6] However, such methods are primitive compared to the sophistication of current molecular dynamics (MD) methods, which yield predictions not only for the structure of the molecules, but also of the solution itself. While such simulations provide powerful visualisation, they remain only as good as the experiments (and QM calculations) from which they are parameterised. Over the last decade, both MD and neutron scattering have matured in such a timely fashion to mutually complement each other and advance our insight into aqueous solutions. On the MD side, progress has come both from improved methodologies and significant increases in computer speed, while on the neutron scattering side this has come from a new generation of neutron scattering machines such as the D4C at ILL, typically providing several order of magnitude more counts per second and appreciably higher degrees of stability than with previous diffractometers.

While it is possible to use neutron scattering results to appraise and refine MD simulations, molecular dynamics can also be used as a heuristic test to yield the non-intuitive relationship between structures of aqueous solutions and the experimental results those structures would produce, and hence allow interpretation of the NDIS results. In this role the physical validity of the MD used to generate a three-dimensional structure is largely a secondary concern, as the significant test of this method is the comparison of the neutron scattering result to the formally correct mathematical prediction of the experimental result from the disordered structure produced by the modelling.

While it largely falls outside the scope of this review, which deals mostly with the method of NDIS applied to solutes, it should be noted that significant advances have also been made in the union of MD with NDIS results obtained by application of the method to the solvent (water).[19,20] Such NDIS results are much more complicated to interpret, and it is generally found that the direct comparison of NDIS to MD is disappointing. However, it has been found that if the comparison between simulation results and experimental neutron scattering is made on a differential measurement (e.g. two different solutions), then the agreement is excellent. In such studies it has been found that the differential measurement is significantly more complicated to interpret than a simple neutron scattering measurement or difference studies, and the use of the MD simulations for guidance is necessary to interpret the neutron scattering data. We have made extensive use of this union to highlight the aggregation/non-aggregation trends of ions in solutions such as Gdm_2SO_4/GdmSCN, Gdm_2CO_3/GdmCl, and Cs_2CO_3/$CsNO_3$.[21–23]

3. Results and Discussion

Since their implementation in the 1970s, NDIS difference methods have been used to investigate ion specific structure in aqueous electrolyte solutions for many monatomic species, and several reviews are already available in the literature.[24,25] In recent years, the NDIS methods have been applied to polyatomic ions, many of which are used in the denaturation of proteins in water. The complexity of these ions means that the neutron data are often difficult to interpret and some sort of modelling is required. Accordingly, as mentioned above we have developed computer simulation codes that provide results which can be directly compared with those obtained from NDIS. In the second part of this section, we present a few results which illustrate the power of this combined method.

3.1. *Monatomic Ions*

The NDIS difference methods were initially applied to concentrated solutions of nickel chloride in heavy water. Both Ni^{2+} and Cl^- possess isotopes of sufficient number and variation in neutron coherent scattering lengths and results were obtained for all aspects of the Ni^{2+} and Cl^- hydration as a function of concentration (Figs. 1 and 2) and with variations of temperature and pressure.[26,27] Results were also obtained to determine the extent of the preferential hydration of Ni^{2+} regarding the number of H_2O and D_2O molecules in its first hydration shell (Fig. 2); as far as it could be ascertained, at ambient conditions there is no preference for the heavier molecule over the lighter one.[28] From a complete set of six experiments on six isotopically labelled samples of a 4.3 molal solution of $NiCl_2$ in heavy water,[29] all three r.d.f.s of the solute were calculated from sets of double difference (Fig. 3).

Once the limits of the NDIS methods had been established, the structures around many other cations and anions were eventually obtained, and today there exists a substantial body of information of the structure of many cations.[2,24] Most recently this information has been augmented with the developed technique of working with salts in solutions of null water — a mixture of 63.7% H_2O and 36.3% D_2O which from a neutron scattering viewpoint contains no information on correlations between hydrogen and all other atoms in the system.[5] Experiments on such solutions[5,30] have provided information on the nearest neighbour coordination of Na^+ and Be^{2+}, and can be used to determine structure around other strong cations such as Al^{3+}, Co^{3+}, Au^{3+}, and many others.

Regarding the results themselves it is worthwhile discussing them in terms of the groups in the Periodic Table to which the ions themselves belong. For the alkali ions, results for Li^+, Na^+ and K^+ are already available in the literature.[6,31,32] Although Rb^+ and Cs^+ are not susceptible to NDIS methods, their first hydration shells have been investigated by the technique of anomalous X-ray Scattering (AXS).[33,34] As a general observation, one notes that other than for Li^+ there is no evidence for a second hydration shell around any of the other ions in this series. Also of interest is the observation that there is a clear contrast in the hydration shells around Na^+ and K^+, whereby the former appears to be much more strongly coordinated to its hydrated water molecules. The hydration numbers of these ions range from around 4 for Li^+ to around 7 for Rb^+ and Cs^+.

The hydration shells of the alkaline earth ions Be^{2+}, Mg^{2+} (isomorphic with Ni^{2+}), Ca^{2+}, and Sr^{2+} have all been studied by NDIS methods, and the results show a progressive and systematic change in the hydration structure as one might expect from arguments based on charge density.[30,35–40] The small Be^{2+} cation has a well-defined first hydration shell[33] with a hydration number of exactly 4. Results for Mg^{2+} show a well-defined first hydration shell of six water molecules[36] and further hydration shells similar to that for Ni^{2+}. The Ca^{2+} ion also has a well-defined hydration regime[37,38] with up to around nine water molecules in its first shell depending on concentration. Similarly, Sr^{2+} has a first shell of eight water molecules.[39,40]

For the transition metal ions, Cr^{3+}, Fe^{2+}, Fe^{3+}, Ni^{2+}, and Cu^{2+} have all been studied by NDIS methods, with the results attesting to the richness and uniqueness of structures these ions possess in aqueous solution.[41–48] Of particular interest for these cations are results for Cu^{2+} which are strongly dependent on counterion and concentration. The exact details of the Cu^{2+} coordination remains controversial with claims for five-fold overall coordination in copper perchlorate solutions.[46] For solutions of copper nitrate and copper perchlorate, there is no evidence for cation–anion contacts, however, in aqueous solutions of copper chloride at concentrations above three molal, there is strong evidence for copper-chloride compexation.[48]

Several lanthanide cations including Sm^{3+}, Gd^{3+} and Dy^{3+} have also been investigated by NDIS, with the results confirming their strong coordination to water and a change in first shell coordination number from around 9 to 8 in progression across the group.[49–51]

The NDIS technique has also been applied to other cations — Zn^{2+}, Hg^{2+}, Ag^{+}, and the results present a clear picture of the degree and extent of coordination around each of these ions.[52–54]

The only monatomic anion which is suitable for exploitation by NDIS is Cl^- with its two isotopes ^{35}Cl and ^{37}Cl. In aqueous solution Cl^- appears to have a structure which is remarkably independent of counterion and concentration. Specifically, results for $g_{ClO}(r)$ and $g_{ClH}(r)$ show that Cl^- has a well-defined first hydration shell and no longer-range structure (Fig. 2).[55] In the glass phase and when present in concentrated aqueous solutions containing large bio-molecules, it would appear that its nearest neighbour Cl–H correlation becomes sharper.[56,57]

At more elevated concentrations, and in the presence of Zn^{2+} or Cu^{2+} counterions, there is an appreciable reduction of coordination number to around 4. Significantly, however, the general conformation of the Cl^-–H_2O water molecules does not change with concentration.[58] The fact that such a change in coordination number does not occur in solutions of nickel chloride is probably a consequence of the effective self-screening of the relatively strong Ni^{2+} cation. The hydration of Cl^- is also sensitive to large changes in temperature and pressure.[59] Again the effect seems more pronounced for the case of relatively weaker counterions being present.

Although other ions in the halide series are not susceptible to NDIS methods, the use of AXS has provided information on the ion–oxygen coordination for Br^- and I^-.[60,32] The results show that the extent of the hydration shell is as for Cl^- limited to the first few Angstroms.

3.2. Complex Polyatomic and Hofmeister Ions

The hydration structures of several polyatomic ions such as ND_4^+, NO_3^-, NO_2^-, and ClO_4^- have already been established.[61–64] Recently NDIS methods have also been applied to more complex ions such as Gdm^+ and SCN^- (Fig. 4), which as the salt guanidinium thiocyanate is a particularly strong denaturant of proteins in solution,[65] and both ions play leading roles in the Hofmeister series.[66]

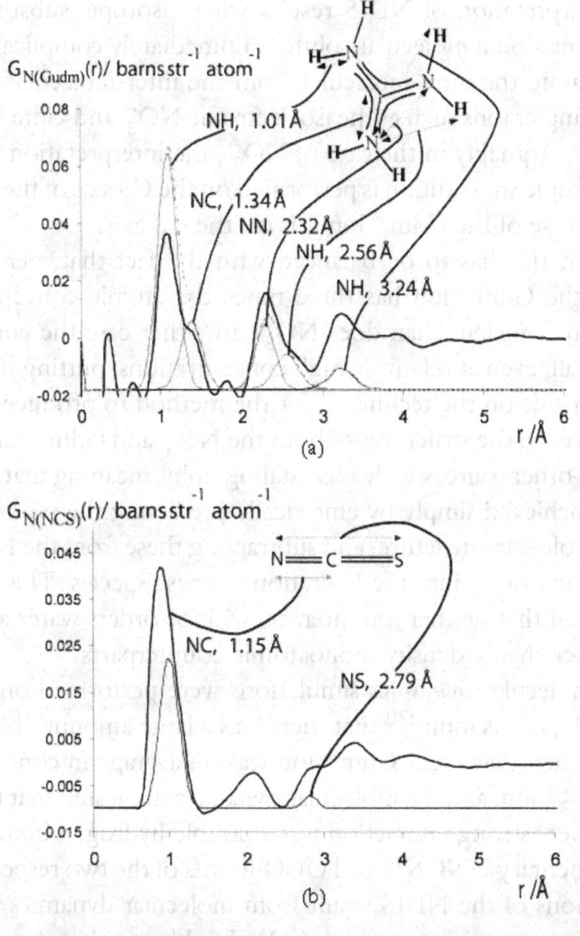

Fig. 4. (a) The thick-lined curve is the total nitrogen atom pair distribution function $G_N(\text{Gdm})(r)$ of the Gdm^+ cation in 3.33 mol of GdmCl and 55.55 mol of heavy water. $[G_N(\text{Gdm}) = 12.36g_{NH}(r) + 4.56g_{NO}(r)1.12(r) + 0.31g_{NC}(r) + 0.45g_{NCl}(r) + 18.80.]$ The dashed-line curve represents the corresponding intra-molecular part of $G_{N(\text{Gdm})}(r)$ calculated on the assumption that correlations between the nitrogen atom and other atoms in the molecule can be represented by a spread function of the form $\exp(-(r^2/w_0^2))/r$ (shown as a thin solid line, labelled with the correlation they represent) where w_0 is the full width at half height of the peak. (b) The thick-lined curve is the total nitrogen atom pair distribution function $G_{N(\text{SCN})}(r)$ of the SCN^- anion in 5.00 mol of NaSCN and 55.55 mol of heavy water. $[G_{N(\text{SCN})} = 6.21g_{NH}(r) + 2.70g_{NO}(r) + 0.33g_{NN}(r) + 0.28g_{NC}(r) + 0.15g_{NNa}(r) + 0.12g_{NS}(r) - 9.80.]$ The thin-lined curve represents the corresponding intramolecular part of $G_{N(\text{SCN})}(r)$ calculated by assuming that correlations exist between the nitrogen atom and other atoms in the molecule. See also Ref. 6.

The interpretation of NDIS results where isotopic substitution has been performed on a molecular solute is immediately complicated by the need to separate the intra-molecular from the inter-molecular structure. Even with simpler ions such as the isoelectronic NO_3^- and Gdm^+ ions, this is non-trivial. Arguably in the case of NO_3^-, the interpretation is simpler, since the isotopic substitution is performed on the C3 axis of the molecule, while in the case of the Gdm^+ ion it is off the C3 axis.

However, this has to be balanced with the fact that, per unit concentration, the Gdm^+ ion has three times the atomic concentration of the substituted nucleus than does NO_3^-. In either case the contrasts are relatively small, even at relatively high concentrations, putting high experimental demands on the technique for the method to produce insightful results. However, the structures of both the NO_3^- and Gdm^+ ions are well known from other sources such as crystallography, meaning that some success can be achieved simply by empirically predicting the peaks expected due to the molecular structure, and subtracting these from the NDIS data, allowing some insight into the hydration of these species. This has led to the conclusion that neither ion interacts with or orders water as strongly as their higher charge density monoatomic counterparts.

When molecular dynamics simulations were performed on solutions of Gdm_2CO_3, it was found[20] that there was a large amount of hetero-ion association, such that every Gdm^+ ion was on average in contact with at least one CO_3^{2-} ion, and the molecular dynamics suggested that the nature of this relatively strong interaction was a double hydrogen bond between the complementary HNCNH and OCO motifs of the two respective ions. The predictions of the NDIS result from molecular dynamics suggested that the strong ion pairing in Gdm_2CO_3 should provide a detectable signature in the experimental results. It was also predicted that the neutron result would be clearer and easier to interpret if the experiment was conducted in null water, rather than D_2O (which would be the more normal choice of solvents for such an experiment). It was found that the experimental results reproduced the most significant portion of the MD prediction that related to the ion-pairing and longer-range association of the ions in this solution (Fig. 5). Furthermore, the longer-range structure predicted by the molecular dynamics simulations led to small angle neutron scattering experiments which confirmed the simulation predictions.

These examples demonstrate the utility and power of integrating the methods of neutron scattering and molecular dynamics, in that, given

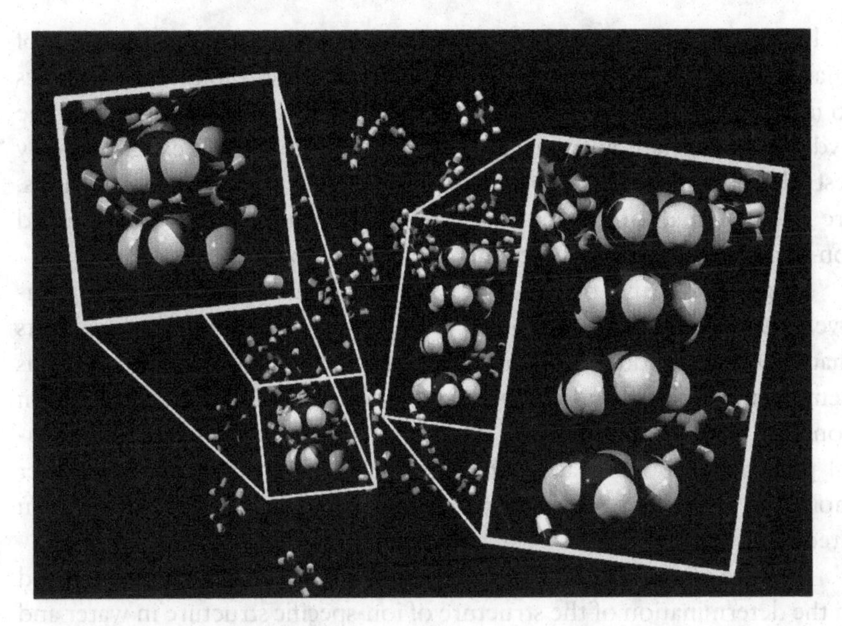

Fig. 5. A typical snapshot of the van der Waals surface representation of guanidinium ion positions taken from the MD simulation, illustrating a stacked dimer and tetramer.

an interesting suggestion of a phenomenon by molecular dynamics, the molecular dynamics can be used to suggest the NDIS experiment that would be capable of testing the veracity of the predictions.

4. Conclusions and Perspectives

Results from NDIS experiments provide information on specific ion structure at the highest level of resolution currently available. This information is not only useful as a systematic characterisation of ions within particular groups, but also can be used to assess the quality of computer simulation studies based on model potentials.[3] The information also helps in the interpretation of results obtained from spectroscopy and thermodynamic measurements.

For the case of polyatomic ions, the use of computer simulation methods has greatly helped in the analysis of data from aqueous solutions containing salts of complex ions.[15,20,21] This methodology has yielded insight into the hydration of ions that gives experimentally testable predictions into the detailed behaviour of some ions of the Hofmeister series in biological systems, such as the reversal of the denaturing effect of guanidinium by sulphate.[18] Such studies, that use molecular dynamics to interpret NDIS experimental data and thereby gain an insight into the hydration of denaturant ions, and then use this to predict experimentally testable behaviours of polypeptides and proteins in electrolyte solutions, are essential in advancing our understanding of how specific ion–ion and ion–protein interactions affect the stability of proteins.

In general terms, the flexibility of the water network (and the relatively short relaxation times of the water molecules themselves) suggests that NDIS is an ideal probe of ion–water structures. Our observations demonstrate the uniqueness of each ion in terms of its own hydration zone. By extension, our results also strongly argue against the fundamental principle of homeopathy which suggests that by some fiction water molecules are able to remember the environments in which they have been sited.

The use of neutrons will remain as the foremost experimental method in the determination of the structure of ion-specific structure in water and indeed other solutions where sufficient numbers of ions can be solvated, i.e. at ~ 0.5 molal. However, in this article we have ignored the enormous contribution that X-ray methods offer to complement the neutron results for heavier ions ($Z > 25$). Indeed as a matter of record there are various X-ray techniques such as isomorphic substitution and anomalous X-ray scattering and EXAFS spectroscopy which will continue to provide a useful and extensive probe of the structure of ionic solutions.[65] The use of EXAFS in particular will enable results to be obtained at ionic concentrations of a few millimolal.

It is clear from our observations that no one method will be sufficient to resolve structure at the required level of detail around all hydrated species. Instead one must rely on a full complement of diffraction and other techniques including computer simulation to answer the many and outstanding questions regarding the degree to which ions specifically influence the properties of aqueous electrolyte solutions.

Finally, it is clear that many challenges remain; foremost amongst these are:

(i) the determination of the three ion–ion $g_{\alpha\beta}(r)$'s in a concentrated solution of lithium chloride — the results will provide theorists with a test of the extent to which primitive models can be used to explain the behaviour of physical properties of 1-1 electrolyte in water.

(ii) a wide-ranging study of the hydration structure of Hofmeister ions such as Gdm^+, SCN^-, Cl^-, in an aqueous solution containing a well-characterised protein as it undergoes a folding/unfolding transition. This will yield insight into ion–ion and ion–protein interactions and how these affect the protein stability.

Acknowledgements

During the course of the work described, we have been fortunate in receiving funds from EPSRC to carry out investigations at sites such as ILL, Grenoble; ISIS, Chilton and the ESRF, Grenoble. We also thank our colleagues, John Enderby (Bristol University, UK) who was instrumental in the development of the NDIS methods, and Stuart Ansell (Rutherford Appleton Laboratory, UK), for the welcome advice they have provided during many discussions. The many contributions of Chris Dempsey (Biochemistry Department, Bristol University) are also gratefully acknowledged. This project was supported by grant GM63018 from the National Institutes of Health.

References

1. Neilson GW, Enderby JE. (1996) *J Phys Chem* **100**: 1317.
2. Enderby JE, Neilson GW. (1979) In: Franks F (ed.), *Water, A Comprehensive Treatise*, Vol. 6, Chap. 1, Plenum Press, New York.
3. Konesham S, Rasaiah JC, Lynden-Bell RM, Lee SH. (1998) *J Phys Chem B* **102**: 4193.
4. Botti A, Bruni F, Imberti I, Ricci MA, Soper AK. (2004) *J Chem Phys* **120**: 10154.
5. Mason PE, Ansell S, Neilson GW. (2006) *J Phys Condens Matter* **18**: 8437.
6. Mason PE, Neilson GW, Dempsey CE, Barnes AC, Cruickshank JM. (2003) *Proc Nat Acad Sci USA* **100**: 4557.
7. Mason PE, Brady JW, Neilson GW, Dempsey CE. (2007) *Biophysical Journal: Biophysical Letters*, L04–L06.
8. Sears VF. (1992) *Neutron News* **3**(3): 26.
9. Squires G. (1978) *Thermal Neutron Scattering*. Cambridge University Press.
10. Narten AH, Vaslow P, Levy HA. (1973) *J Chem Phys* **58**: 5017.
11. Tromp RH, Neilson GW, Soper AK. (1992) *J Chem Phys* **96**: 8460.
12. Sprik M, Klein ML, Watanabe K. (1990) *J Phys Chem* **94**: 6483.
13. Dang LX, Rice JE, Caldwell J, Kollman PA. (1991) *J Am Chem Soc* **113**: 2481.
14. Enderby JE, Neilson GW. (1980) *Adv Physics* **29**: 323.
15. Neilson GW, Adya AK, Ansell S. (2002) *Annu Rep Prog Chem* **98**: 1–52.

16. Liu Q, Brady JW. (1997) *J Phys Chem B* **101**: 1317.
17. Mason PE, Neilson GW, Enderby JE, Saboungi ML, Brady JW. (2005) *J Am Chem Soc* **127**: 10991–10998.
18. Mason PE, Neilson GW, Enderby JE, Saboungi ML, Dempsey CE, MacKerell AD, Brady JW. (2004) *J Am Chem Soc* **126**: 11462–11470.
19. Mason PE, Neilson GW, Barnes AC, Enderby JE, Brady JW, Saboungi ML. (2003) *J Chem Phys* **119**: 3347–3353.
20. Mason PE, Neilson GW, Enderby JE, Saboungi ML, Brady JW. (2005) *J Phys Chem B* **109**: 13104–13111.
21. Mason PE, Dempsey CE, Neilson GW, Brady JW. (2005) *J Phys Chem B* **109**: 24185.
22. Mason PE, Neilson GW, Dempsey CE, Brady JW. (2006) *J Am Chem Soc* **128**: 15136.
23. Mason PE, Neilson GW, Kline SR, Dempsey CE, Brady JW. (2006) *J Phys Chem B* **110**: 13477.
24. Neilson GW, Adya AK. (1997) *Roy Soc Chemistry Annual Repts Section C* **93**: 101.
25. Ansell S, Barnes AC, Mason PE, Neilson GW, Ramos S. (2006) *J Biophys Chem* **124**: 171.
26. Howell I, Neilson GW. (1996) *J Chem Phys* **104**: 2036.
27. de Jong PHK, Neilson GW, Bellissent-Funel M-C. (1996) *J Chem Phys* **105**: 5155.
28. Powell DH, Neilson GW, Enderby JE. (1993) *J Phys Condens Matter* **5**: 5723.
29. Neilson GW, Enderby JE. (1983) *Proc Roy Soc A* **390**: 353.
30. Mason PE, Ansell S, Neilson GW, Brady JW. (2008) *J Phys Chem B* **112**: 1935.
31. Howell I, Neilson GW. (1996) *J Phys Condens Matter* **8**: 4455.
32. Neilson GW, Skipper NT. (1985) *Chem Phys Letts* **114**: 35.
33. Ramos S, Barnes AC, Neilson GW, Capitan M. (2000) *Chemical Physics* **258**: 171.
34. Ramos S, Neilson GW, Barnes AC, Buchanan P. (2005) *J Chem Phys* **123**: 214501.
35. Marx D, Sprik M, Parinello. (1997) *Chem Phys Letts* **273**: 360.
36. Skipper NT, Neilson GW, Cummings S. (1989) *J Phys Condens Matter* **1**: 3489–3506.
37. Hewish NA, Neilson GW, Enderby JE. (1982) *Nature* **297**: 138.
38. Badyal YS, Barnes AC, Cuello GJ, Simonson JM. (2004) *J Chem Phys* **108**: 11819.
39. Neilson GW, Broadbent RD. (1990) *Chem Phys Letts* **167**: 429.
40. Ramos S, Neilson GW, Barnes AC, Capitan MJ. (2003) *J Chem Phys* **118**: 5542–5546.
41. Broadbent RD, Neilson GW, Sandstrøm M. (1992) *J Phys Condens Matter (Liquids)* **4**: 639.
42. Herdman GJ, Neilson GW. (1992) *J Phys Condens Matter (Liquids)* **4**: 649.
43. Herdman GJ, Neilson GW. (1992) *J Phys Condens Matter (Liquids)* **4**: 627.
44. Powell DH, Neilson GW. (1990) *J Phys Condens Matter* **2**: 3871.
45. Salmon PS, Neilson GW. (1989) *J Phys Condens Matter* **1**: 5291.
46. Pasquarello A, Petri I, Salmon PS, Parisel O, Car R, Toth E, Powell DH, Fischer H, Helm L, Merbach AE. (2001) *Science* **291**: 856.

47. Salmon PS, Neilson GW, Enderby JE. (1988) *J Phys C Solid State Phys* **21**: 1335.
48. Ansell S, Tromp RH, Neilson GW. (1995) *J Phys Condens Matter (Liquids)* 7: 1513.
49. Narten AH, Hahn RL. (1983) *J Phys Chem* **87**: 3193.
50. Sobolev O, Charlet L, Cuello GJ, Gehin A, Brendle J, Geoffroy N. (2008) *J Phys Condens Matter* **20**: 104207.
51. Cossy C. Barnes AC, Enderby JE, Merbach AE. (1989) *J Chem Phys* **90**: 3254.
52. Powell DH, Gullidge PMN, Bellissent-Funel MC, Neilson GW. (1990) *Mol Phys* **71**: 1107.
53. Sobolev O, Cuello CJ, Roman-Ross G, Skipper NT, Charlet LJ. (2007) *Phys Chem A* **111**(24): 5123.
54. Sandström M, Neilson GW, Johannson G, Yamaguchi T. (1985) *J Phys C Solid State Phys* **18**: L1115.
55. Cummings S, Newsome JR, Neilson GW, Enderby JE, Howe RA, Howells WS, Soper AK. (1980) *Nature* **287**: 714.
56. Ansell S, Dupuy J, Jal J-F, Neilson GW. (1997) *J Phys Condens Matter* **9**: 8835.
57. Wilson JE, Ansell S, Enderby JE, Neilson GW. (1997) *Chem Phys Letters* **278**: 21.
58. Powell DH, Barnes AC, Enderby JE, Neilson GW, Salmon PS. (1988) *Faraday Discussions* **85**: 137.
59. de Jong PHK, Neilson GW, Bellissent-Funel M-C. (1996) *J Chem Phys* **105**: 5155.
60. Ramos S, Barnes AC, Neilson GW, Lequien S, Thiaudiere D. (2000) *J Phys Condens Matter* **12**(8A): 203.
61. Hewish NA, Neilson GW. (1981) *Chem Phys Letts* **84**: 425–427.
62. Neilson GW, Enderby JE. (1982) *J Phys C Solid State Phys* **15**: 2347.
63. Kameda Y, Arakawa H, Hamgai K, Uemura O. (1992) *Bull Chem Soc Japan* **65**: 2154.
64. Neilson GW, Schiöberg D, Luck WAP. (1985) *Chem Phys Letts* **122**: 475.
65. Chomczynski P, Sacchib N. (1987) *Analytical Biochemistry* **162**: 156.
66. Baldwin RL. (11996) *Biophysical J* **71**: 2056.
67. Fulton JL, Pfund DM, Wallen SL, Newville M, Stern EA, Ma Y. (1996) *J Chem Phys* **105**: 2161.

Chapter 7

Specific Ion Effects at the Air–Water Interface: Experimental Studies

Vincent S. J. Craig* and Christine L. Henry*

Ion specificity is important in all soft matter systems at high electrolyte concentrations. Specific ion effects that influence many biological processes fall within the Hofmeister paradigm, whereby the strength of action of the anions and cations follow a well-defined order, independent of the co-ion. In contrast, the ion specificity seen when salts inhibit bubble coalescence depends on the combination of ions present. Single electrolytes at 0.1 M may inhibit bubble coalescence or have no effect, as determined by simple ion combining rules and empirical assignments (α or β) of the cation and the anion. The coalescence of mixed electrolyte systems can also be predicted by an extension of these rules. Bubble coalescence in some non-aqueous solvents can also show ion-specificity that depends upon ion combination in a way analogous to the combining rules observed in water — which suggests that this is a general property of the air–solution interface. Here, we summarise evidence that the contrasting specific ion effects observed in bubble coalescence and in other soft matter systems have the same origin: Ion specificity arises from differences in the affinity of ions for interfaces. This leads to the conclusion that combining rules seen in bubble coalescence inhibition arise from the dynamic nature of the air–solution interface.

1. Introduction

Specific-ion effects at the air–water interface are perhaps the simplest of all the systems that exhibit ion specificity in terms of number and type of chemical species involved. The change in air–water surface tension as a function of electrolyte concentration and the stability of bubbles in aqueous electrolyte solutions are both strongly dependent on the type of ions present. For example, the surface tension of water is raised upon

*Department of Applied Mathematics, Research School of Physics, Australian National University, Canberra, Australia.

the addition of NaCl but lowered upon the addition of HCl; bubble coalescence is inhibited in NaCl solutions, whereas in $NaClO_4$ solutions there is no inhibitory effect. So, as we shall see later, ion specific effects at the air–water interface are easily revealed by macroscopic measurements and give rise to a rich and complex behaviour. They can in some cases be codified, but are not understood. Our inability to explain the specific ion effects in these chemically simple systems is indicative of substantial weaknesses in the theories of colloid and interface science when describing systems with concentrated electrolytes.

Studies of the air–water interface are an experimentally appealing embarkation point for an investigation of specific ion effects. The systems are chemically simple and the influence of electrolytes is large and easily measured and gives rise to complex specific ion behaviour. So why, given the importance of the problem, has there been surprisingly little experimental work in this field? We propose two main reasons for this. First, the specific ion effects seen at the air–water interface are very different to Hofmeister effects. The influence of the ions does not follow the Hofmeister series and in general the influence of ions cannot be placed in a specific order of effectiveness. In the case of bubble coalescence, the *combination* of ions present is a more important factor. Therefore, it is not apparent how these studies connect to the much larger body of specific ion effects that fall under the Hofmeister paradigm. Secondly, there is no adequate theoretical framework for the description of specific ion effects at the air–water interface. This absence of a theoretical foundation on which ideas and new experimental investigations can be built has meant that the field has progressed sporadically and with no clear direction. This reflects the fact that the experimentalist is often unsure which direction to pursue if the effort is more ambitious than merely the collection of data. Without a theoretical framework, it has been nearly impossible to connect the fundamental properties of the ions to the available experimental parameters such as the change in gradient of the surface tension with electrolyte concentration or the inhibition of bubble coalescence. However, with recent developments in theory, simulation and highly surface sensitive spectroscopic techniques, the relationship between these macroscopic measurements and an atomic level understanding of ion specific behaviour is now on the horizon, as is the connection of this work to the wider field of Hofmeister type specific ion effects. With this in mind we now examine recent experimental work on specific ion effects on bubble coalescence and attempt to relate this to the specific ion effects revealed in surface tension measurements.

2. Bubble Coalescence Inhibition in Electrolyte Solutions

The influence of electrolytes on bubble coalescence is evident in the white-water generated when waves break on the beach. The white foam produced that lasts approximately ten seconds is largely due to the high concentration of electrolytes in the sea, see Fig. 1. The stabilising effect of electrolytes on bubbles can also be observed by comparing the results obtained upon shaking a flask containing pure water and a flask containing concentrated NaCl solution. This reveals that the influence of salt is to extend the lifetime of the bubbles at the air–water interface from about one second to about ten seconds — thus, coalescence is said to be inhibited rather than prevented by electrolytes. The same effect is manifested in three ways in a

Fig. 1. The high concentration of dissolved salts is responsible for the whitewater that is formed as waves break in seawater. The foam is short-lived, typically lasting for approximately ten seconds.

solution sparged with bubbles in a bubble column. Firstly, the inhibition of coalescence has an effect on the production of bubbles at the frit, as the bubbles are inhibited from coalescing when they snap off.[1] Secondly, the bubbles produced in the column are stabilised when the collision time is less than the coalescence time.[2] Thirdly, transient foaming is induced at the top of the liquid column.[3] Thus, the single phenomenon of bubble coalescence inhibition is manifested in the production of bubbles, the outcome of collisions between bubbles, and the lifetime of bubbles at the surface of a liquid. Therefore, bubble coalescence inhibition studies have used a variety of measures to investigate ion specificity including gas hold-up,[4] bubble production,[1] thin film drainage[5] and bubble coalescence.[2,3,6–8] Bubble coalescence has been followed by analysing the outcome of carefully controlled collisions between two bubbles[8] or by averaging the result of many bubble–bubble collisions.[2] All these approaches give results that are consistent, though the precise concentration at which the effect is manifested for a particular salt is seen to vary between investigations as it is sensitive to the size of the bubbles and the timescale of the collisions between them. In general for 1:1 electrolytes, a concentration in excess of 0.05 M is required to exhibit bubble coalescence inhibition and the full effect is manifested below 0.2 M.[2,8]

The specificity of ion effects on bubble coalescence is apparent on two levels. Most significantly some electrolytes have little or no effect on bubble coalescence whereas others are seen to significantly inhibit coalescence.[2,6,9,10] Moreover, those electrolytes that inhibit bubble coalescence do so with different levels of effectiveness.[8,9] If the conditions of the experiment are held constant, this effectiveness can be quantified by the concentration of electrolyte that gives rise to a given level of coalescence inhibition. This effectiveness is usefully characterised by the 'transition concentration', the concentration at which 50% of the full effectiveness of bubble coalescence inhibition is observed.[2]

By analysing the amount of light scattered in a gas-sparged bubble column as depicted in Fig. 2, the degree of bubble coalescence can easily be followed. In pure water the bubbles coalesce readily, grow rapidly and scatter little light. The column appears clear. In the presence of a coalescence inhibiting salt, the bubbles produced at the frit remain small, much light is scattered and the column appears white. This approach has the added advantage of being insensitive to small amounts of contaminant which are floated to the top of the column and deposited on

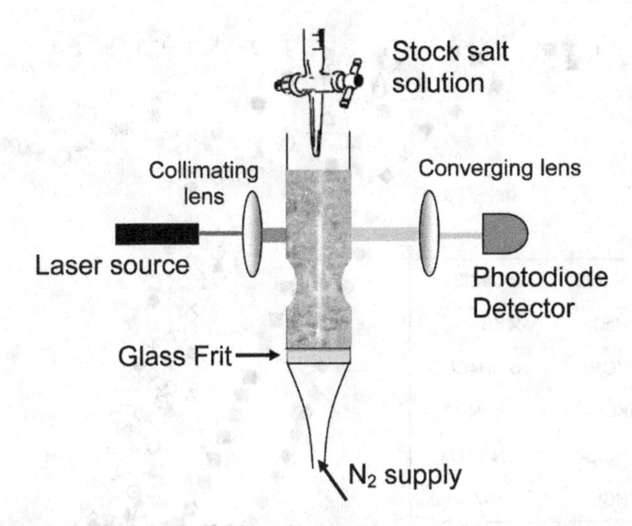

Fig. 2. Schematic of Bubble Coalescence Apparatus. Small bubbles are produced at the frit and rise in the column passing through the constriction that promotes bubble collisions. In pure water, they rapidly coalesce such that the bubbles breaking the laser beam are large and few in number. In this case, most of the light is transmitted and strikes the detector. When coalescence is inhibited, the bubbles remain small and numerous, resulting in a large amount of scattering and low light intensity at the detector.

the walls during the normal operation of the experiment. The data presented here were obtained using N_2 gas as the sparging gas. A range of other gases have been employed with similar results, though the transition concentration is altered slightly in line with the diffusivity of the gas.[2,11]

The data obtained using the sparging column, for a range of electrolytes, are shown in Fig. 3. In addition to these electrolytes that strongly inhibit bubble coalescence at moderate concentrations (≈ 0.1 M), we find that other electrolytes, previously called non-inhibiting, have little or no effect on bubble coalescence at moderate electrolyte concentrations. We shall now call these salts less inhibiting as there is some evidence that at concentrations above 1 M, they may inhibit bubble coalescence.[12] This is currently unresolved as the inhibition of coalescence at these concentrations may be influenced both by non-electrolyte and electrolyte contamination. We now turn our attention to understanding why this partitioning in the effectiveness of electrolytes exists.

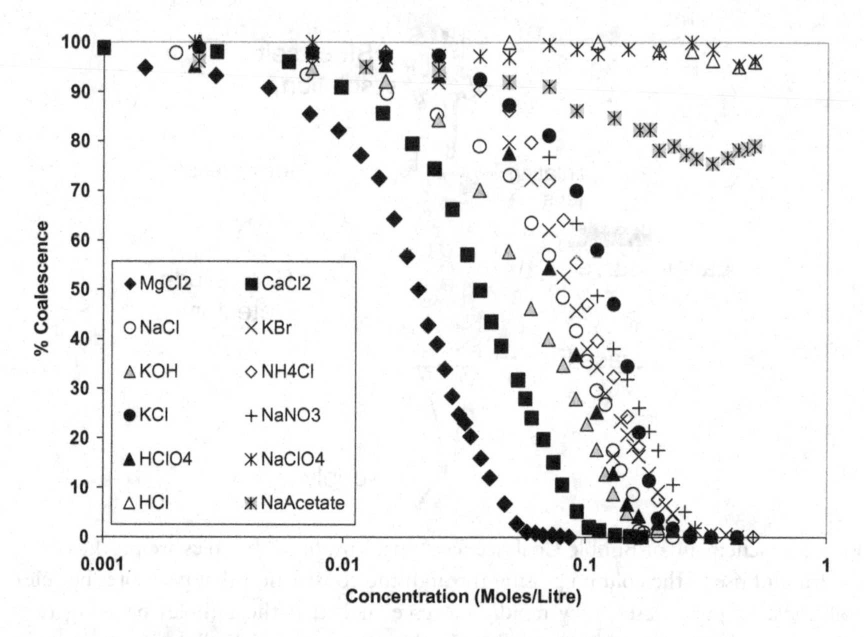

Fig. 3. Bubble coalescence data obtained using an air sparged column for a selected range of electrolytes. Three of these electrolytes, HCl, $NaClO_4$ and NaAcetate are deemed less inhibiting; the remainder are called inhibiting.

3. Predicting Bubble Coalescence Inhibition

How can we understand why some electrolytes inhibit bubble coalescence whilst others are less inhibiting, as shown in Figs. 3 and 4? In order to illustrate the ion specificity of bubble coalescence inhibition, we have selected four electrolytes, two that inhibit bubble coalescence and two that are less inhibiting. These are shown in Fig. 4. It is apparent that whether an electrolyte is inhibiting or non-inhibiting is determined by the combination of the anion and cation that make up the electrolyte. Thus a particular anion or cation cannot be said to strongly favour or disfavour bubble coalescence inhibition. The effect of different cation/anion combinations is captured in combining rules that have been developed empirically.[2,6] The development of these combining rules is illustrated in Table 1. NaCl is an electrolyte that inhibits bubble coalescence so this is indicated by a tick in the table at the intersection of the Na^+ column and the Cl^- column. Both of these ions have been arbitrarily assigned the property α. In comparison HCl has no effect on bubble coalescence inhibition, therefore a cross is

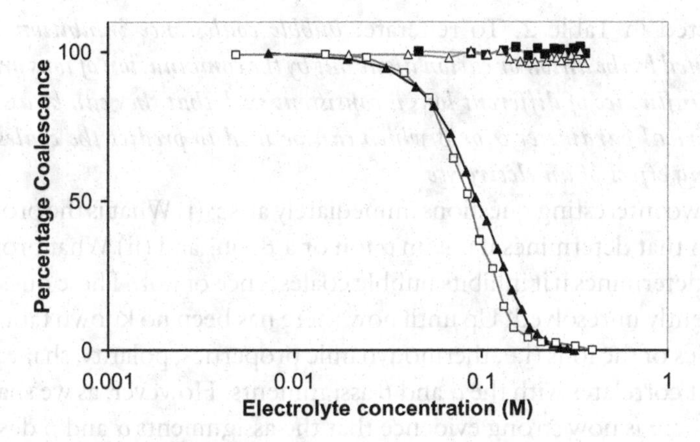

Fig. 4. Ion-specificity and combining rules in electrolyte bubble coalescence inhibition. Coalescence is plotted as a function of electrolyte concentration on a log scale. Electrolytes NaCl ($\alpha\alpha$) (▲) and $HClO_4$ ($\beta\beta$) (□) inhibit at 0.1 M, while their 'cross-products' $NaClO_4$ ($\alpha\beta$) (■) and HCl ($\beta\alpha$) (△) have no effect up to 0.5 M concentration. 100% coalescence is defined in pure water; 0% is a stable low value in inhibiting electrolytes.

Table 1. Illustration of the empirical assignment of the property α and β, which are used with the combining rules to codify bubble coalescence inhibition behaviour.

Ions	Cations	H^+	Na^+
Anions		β	α
Cl^-	α	×	✓
ClO_4^-	β	✓	×

placed in the table and the H^+ ion is designated β, to indicate that its influence is different to Na^+. Similarly, $NaClO_4$ has no effect on bubble coalescence so the ClO_4^- is also designated β to indicate that the influence of the ClO_4^- ion is different to Cl^-. However, we find that $HClO_4$ inhibits bubble coalescence and therefore a tick is placed in the table. Thus, $\alpha\alpha$ or $\beta\beta$ combinations of ions correspond to ticks (bubble coalescence inhibition) and $\alpha\beta$ or $\beta\alpha$ combinations are less inhibiting.

These empirically determined combining rules can be applied to a wide range of electrolytes and they are found to work without exception, as depicted in Table 2. To reiterate, *bubble coalescence inhibition is not determined by the anion or cation alone but by the combination of ions present, and the influence of different ions is consistent such that they can be assigned an empirical parameter α or β which can be used to predict the coalescence inhibiting effect of an electrolyte.*

So two interesting questions immediately arise: (i) What is the property of an ion that determines if it is an α ion or a β ion? and (ii) What property of a salt determines if it inhibits bubble coalescence or not? These questions are currently unresolved. Up until now there has been no known tabulated properties of the ions (i.e. thermodynamic properties, polarity, shape, size, etc.) that correlates with the α and β assignments. However, as we shall see below, there is now strong evidence that the assignments α and β describe how the ions arrange themselves within the air–water interface and it is this arrangement of ions within the interface that controls bubble coalescence inhibition.

4. Bubble Coalescence in Mixed Electrolytes

Whilst a deep understanding of bubble coalescence in single electrolyte systems is lacking, it is clear that the combining rules can be used to predict whether a particular electrolyte will inhibit bubble coalescence or not. Many practical electrolyte systems contain more than one type of anion or cation and therefore the question arises as to whether the influence of mixed electrolytes on bubble coalescence can also be predicted. Consider a solution containing both HCl and $NaClO_4$, both of which individually have no effect on bubble coalescence. What influence on bubble coalescence should we expect? One might predict that as these electrolytes individually have no effect, the net effect might also be minor. However, the same solution can be obtained by using $HClO_4$ and NaCl, both of which do inhibit bubble coalescence. Thus, coalescence inhibition cannot be predicted based on a simple extrapolation from the behaviour of the component electrolytes. The influence of a range of equimolar mixtures of electrolytes on bubble coalescence is shown in Fig. 5. The electrolyte mixtures behave similarly to single electrolytes in that some inhibit bubble coalescence and others are less inhibiting.[13]

Let us consider the ion assignments made for single electrolytes in an effort to describe the behaviour of mixed electrolyte systems. Pairs of electrolytes can be categorised based on these assignments and they give rise

Table 2. Codification of Bubble Coalescence in electrolyte solutions by the empirical assignment of the property α or β and the use of combining rules. (After Craig *et al.*,[2] with additional data.)

Ions		**Li^+**	**Na^+**	**K^+**	**Cs^+**	**Mg^{2+}**	**Ca^{2+}**	**NH_4^+**	**H^+**	**$(CH_3)NH_3^+$**	**$(CH_3)_2NH_2^+$**	**$(CH_3)_3NH^+$**	**$(CH_3)_4N^+$**
Assignment		α	α	α	α	α	α	α	β	β	β	β	β
OH^-	α		✓	✓					×				
Cl^-	α	✓	✓	✓		✓	✓		×	×	×	×	×
Br^-	α		✓	✓	✓				×				×
NO_3^-	α	✓	✓	✓			✓		×				
SO_4^{2-}	α	✓	✓	✓		✓			×				
$(COO_2)^{2-}$	α			✓					×				
IO_3^-	α		✓										
ClO_3^-	β		×										
ClO_4^-	β		×			×		×	✓				
CH_3COO^-	β		×	×	×	×		×	✓				✓
SCN^-	β		×										

✓ = inhibit coalescence; × = no inhibition; $\alpha\alpha, \beta\beta$ = ✓; $\alpha\beta, \beta\alpha$ = ×.

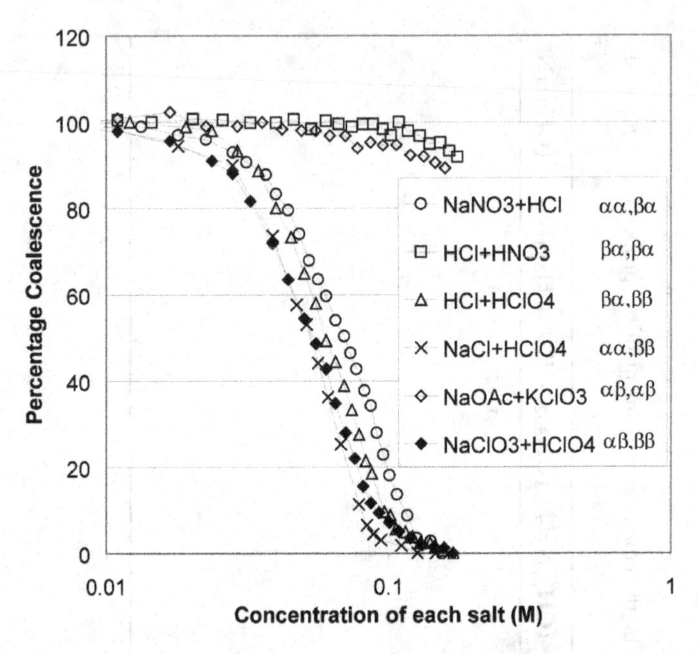

Fig. 5. Bubble coalescence behaviour of a selection of mixed electrolytes.[13] The empirical assignments used in the combining rules for single electrolytes are shown for each electrolyte combination. As with single electrolyte systems, some combinations exhibit a sharp transition to bubble coalescence inhibition and others do not.

to nine different classes of mixed electrolytes (i.e. $\alpha\alpha + \beta\beta, \alpha\alpha + \alpha\beta$, etc.). We adopt the convention of putting the cation assignments first. It is found that in terms of their influence on bubble coalescence and their influence on surface tension, electrolyte mixtures that are in the same class behave similarly.[13] This indicates that the mixed electrolytes are appropriately categorised based on the combination of α and β ions. This is shown in Table 3 where the categories are listed and their influence on bubble coalescence is shown.

How can the behaviour captured in Table 3 be encoded? Inspection reveals that the only categories that have no influence on bubble coalescence are those where all the anions are of one type and all the cations are of the other type. That is the $\alpha\beta$, $\alpha\beta$ and $\beta\alpha$, $\beta\alpha$ categories. Thus we can use the table to make the prediction that any system that has ions that can make an $\alpha\alpha$ combination or a $\beta\beta$ combination will inhibit bubble coalescence.

Table 3. Summary of bubble coalescence inhibition by mixtures of electrolytes categorised using the empirically assigned properties α and β. The cation is listed first.

Categorisation based on empirical assignments	Coalescence inhibition observed
$\alpha\alpha, \alpha\alpha$	✓
$\alpha\alpha, \alpha\beta$	✓
$\alpha\alpha, \beta\alpha$	✓
$\alpha\alpha, \beta\beta$ or $\alpha\beta, \beta\alpha$	✓
$\alpha\beta, \alpha\beta$	×
$\alpha\beta, \beta\beta$	✓
$\beta\alpha, \beta\alpha$	×
$\beta\alpha, \beta\beta$	✓
$\beta\beta, \beta\beta$	Not done*

*No suitable combinations are available.

An attempt to rationalise the behaviour of both single electrolytes and mixed electrolyte systems has been made based on a proposal of Marcelja[14] regarding the locations of individual ions within the interfacial region. Recent work has shown that the concentration profile of ions within the interfacial region is not monotonic but is considerably more complicated and that some ions are actually concentrated at the surface whilst others are depleted at the surface.[15–18] Marcelja suggested that for bubble coalescence inhibition to occur, an ion that was at the surface needed to be paired with an ion that was withdrawn from the surface.[14] The suggestion is that if α cations are withdrawn from the surface, β cations are at the surface and conversely α anions are at the surface and β anions are withdrawn from the surface. Thus in the Marcelja paradigm, an $\alpha\alpha$ or $\beta\beta$ combination describes a pair of ions that are separated in the surface region, whereas in an $\alpha\beta$ combination both ions are withdrawn from the interface and in a $\beta\alpha$ combination both ions preferentially inhabit the interface. The appeal of this explanation is that it encompasses the predictive combining rules for the single electrolytes. Moreover, it can be extended to mixtures of electrolytes by making the assumption that as long as ion separation at the interface exists, other ions have only a minor contribution.[13] That is if the mixture contains a sufficient concentration of ions that can form either an $\alpha\alpha$ or $\beta\beta$ pair, then bubble coalescence inhibition will be observed.

However, the correlation between the assignments of α and β and the expected interfacial location of the ions predicted by simulation and spectroscopic studies is not good. For example, the simulation studies predict large differences in the surface affinity of the highly polarisable Br^- ion and the less polarisable Cl^- ion, whereas they have a similar influence on bubble coalescence inhibition and are both designated α anions. Whilst some of the more complex ions are difficult to simulate, it appears unlikely that this disagreement will be resolved merely by improvements in simulations. This strongly suggests that the Marcelja model which fits with the combining rules needs to be modified in order to accurately describe the relationship between the ion assignments (α and β) and bubble coalescence inhibition.

5. Bubble Coalescence in Non-Aqueous Solvents

A great deal of literature is related to specific ion effects in aqueous systems, yet despite this, the influence of the solvent is often ignored or treated as secondary to the roles of the ions. Alternatively, many see the role of water to be special or even unique, given the small size of the molecule, its ability to form a hydrogen-bond network and the strength of water as a solvent. This is particularly the case in biological systems where water is seen as having a central role in all the interactions. Therefore, it is of some interest to examine specific ion effects in non-aqueous solvents. In regard to bubble coalescence inhibition of electrolytes, investigating other solvents may provide some insight into the mechanism of coalescence as different solvents have different physical properties, such as surface tension, viscosity and density, and the interaction forces across a thin film should differ with the dielectric, optical and structural properties of the solvent. Additionally, different solvents will have different chemical interactions with the ions and in particular different interfacial affinities. Given the multitude of possible means by which a solvent could influence bubble coalescence inhibition by electrolytes, it is reasonable to expect that the solvent will play a significant role.

Specific ion effects on bubble coalescence in non-aqueous solvents have only recently been investigated.[19] Four solvents were employed, dimethylsulfoxide (DMSO), formamide, methanol, and propylene carbonate. Note that only high dielectric solvents are suitable for these investigations as they are able to solvate a satisfactory range of salts sufficiently. Bubble coalescence inhibition due to added electrolyte was exhibited in all four solvents, showing that the phenomenon is not unique to water, but rather may prove to be universal. In both methanol and DMSO, all the electrolytes added were found to inhibit bubble coalescence; this included electrolytes that have no effect on bubble coalescence in aqueous systems. Therefore we can state that at least some ions behave differently in these systems and given that all salts were found to be effective, the combining rules are not applicable in these solvents. In formamide, the influence of electrolytes on bubble coalescence was remarkably similar to that found in water, as shown in Fig. 6. Some salts showed strong inhibition of bubble coalescence and others showed no effect or an effect that was only elicited at a much higher concentration. Thus for formamide, a table similar to Table 2 above for water can be constructed and the ion assignments are found to be the same as those in water, see Table 4.

The influence of electrolytes on bubble coalescence in propylene carbonate (PC) revealed electrolytes with both inhibiting and non-inhibiting or weakly inhibiting character, however in this case the influence of electrolytes differed from those found in water. A table could also be constructed for PC whereby the coalescence behaviour was described by

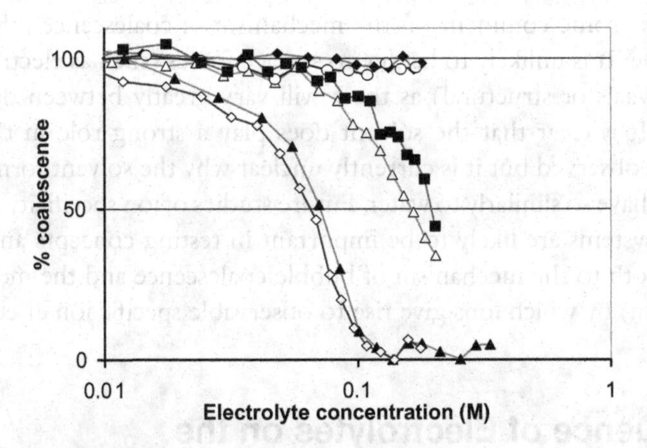

Fig. 6. Inhibition of bubble coalescence by selected electrolytes in formamide solutions as a function of concentration. NaI (◇), LiCl (▲), $LiClO_4$ (△), $(CH_3)_4NBr$ (■), CH_3COONa (♦), CH_3COONH_4 (○). 100% coalescence is defined in pure formamide, 0% coalescence is a stable, low-voltage signal in inhibiting electrolytes.

Table 4. Codification of bubble coalescence inhibition by electrolytes in formamide.

Ions		Li^+	Na^+	NH_4^+	$(CH_3)_4N^+$
Assignment		α	α	α	β
Cl^-	α	✓			
Br^-	α				×
I^-	α	✓	✓		
ClO_4^-	β	×			
CH_3COO^-	β		××	××	×[a]

✓ indicates bubble coalescence inhibition. $\alpha\alpha$ and $\beta\beta$ salts = ✓. × and ×× indicate partial inhibition and no inhibition, respectively, relative to pure formamide. $\alpha\beta$ and $\beta\alpha$ salts = × or ××.
[a]Tetramethylammonium acetate is believed to be affected by contamination.

empirical assignment of the terms α and β to different ions and the use of the combining rules, but in this case some of the ions such as ClO_4^- and SCN^- required different assignments from those in the aqueous system.

For those electrolytes that readily inhibited bubble coalescence, a lower concentration was required in the non-aqueous solvents, see Figs. 3 and 6. Also, the transition concentration did not follow the same trend in the different solvents. Thus the four solvents studied, whilst all exhibiting bubble coalescence inhibition, showed a diverse range of behaviour. This allows some comment on the mechanism of coalescence inhibition to be made. It is unlikely to be due to surface forces (such as electrostatic, van der Waals or structural) as these will vary greatly between different solvents. It is clear that the solvent does play a strong role in the ion-specificity observed but it is currently unclear why the solvent formamide should behave so similarly to water. Future studies of ion specificity in non-aqueous systems are likely to be important in testing concepts and ideas relating both to the mechanism of bubble coalescence and the molecular mechanisms by which ions give rise to observable specific ion effects.

6. Influence of Electrolytes on the Hydrodynamic Boundary Condition

The mobility of the solvent–vapour interface may play a significant role in the bubble coalescence process.[20] A mobile interface is expected for a pure liquid in contact with a gas and this will lead to lower hydrodynamic drag and faster drainage. The adsorption of a small amount of surfactant is known to alter the boundary such that the no-slip boundary condition is appropriate.[21] This raises the possibility that electrolytes may also alter the boundary condition from slip to no-slip. The effect this would have on bubble coalescence may be manifested at two stages. Firstly, immobile interfaces reduce film drainage velocity during the collision of two bubbles; and secondly, once a film has formed then a no-slip boundary condition will inhibit the growth of capillary waves that lead to rupture of the film.

A convenient means to study the boundary condition at the solution–gas interface is determination of the terminal rise velocity of a small spherical bubble. A slip boundary condition results in a 50% greater rise velocity, for a given bubble size in a given solvent.[22] The main challenge in these studies is ensuring that the system is free of even tiny amounts of surface active impurities. The influence of a range of inhibiting and non-inhibiting electrolytes on bubble terminal rise velocity has been investigated and no influence was found on the hydrodynamic boundary condition.[23] It is therefore tempting to discard any possible influence of the boundary condition, however bubble rise studies probe an isolated interface and the possibility remains that for two interfaces interacting hydrodynamically, the effect may be different.

7. Bubble Coalescence and Surface Tension of Electrolyte Solutions

There are numerous ways in which surface tension can be measured. The technique that is perhaps the most suitable for electrolyte solutions is the Maximum Bubble Pressure technique,[24] as the continuous generation of fresh interface assists in acquiring data and is not influenced by low levels of contamination. Typically a bubble is produced from a fine capillary every five to ten seconds and as it grows the pressure is monitored. When the pressure reaches a maximum, the radius of the bubble is equivalent to the inside radius of the hydrophilic capillary tip. At this time the radius of the bubble and the internal pressure are known, therefore the surface tension can easily be calculated using the Laplace equation. This bubble then detaches and a new bubble is created, allowing for easy repeat measurements of surface tension at a freshly generated interface.

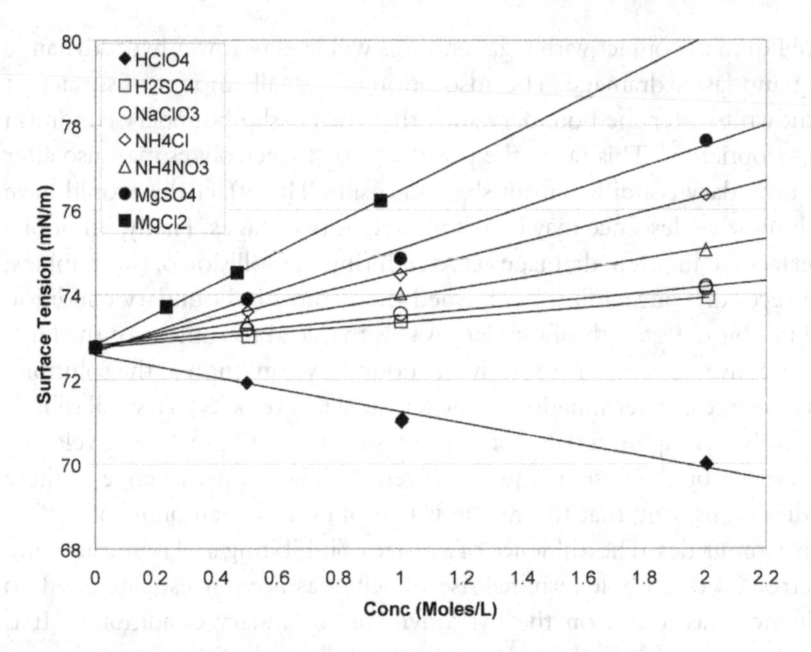

Fig. 7. Surface tension versus concentration for a selection of electrolytes. Most of the simple electrolytes exhibit a linear relationship. When measurements at a range of concentrations are made, this can be used to assess the accuracy of the individual surface tension measurements and improve the evaluation of the surface tension gradient. For single electrolytes, those with a small gradient in surface tension with concentration are found to have no effect on bubble coalescence ($NaClO_3$ and H_2SO_4 are the two examples of non-inhibiting electrolytes shown in the plot).

For most electrolytes, the surface tension changes linearly over a wide concentration range and unambiguous ion specific effects are seen as shown in Fig. 7. These effects are yet to be adequately described theoretically (note for some electrolytes at millimolar concentrations, there is a significant deviation known as the Jones–Ray effect, but this effect is small and only evident at low concentrations). This allows the gradient of the surface tension with bulk concentration to be easily determined. It has long been known that electrolytes that have a higher surface tension gradient inhibit bubble coalescence at lower concentrations. More recently it has been shown that there is a correlation between the transition concentration and the inverse of $(\frac{d\gamma}{dc})^2$ and that the value of $(\frac{d\gamma}{dc})^2$ is low for the non-inhibiting electrolytes.[25,26] The inverse of $(\frac{d\gamma}{dc})^2$ is plotted against the transition concentration of a range of electrolytes in Fig. 8.

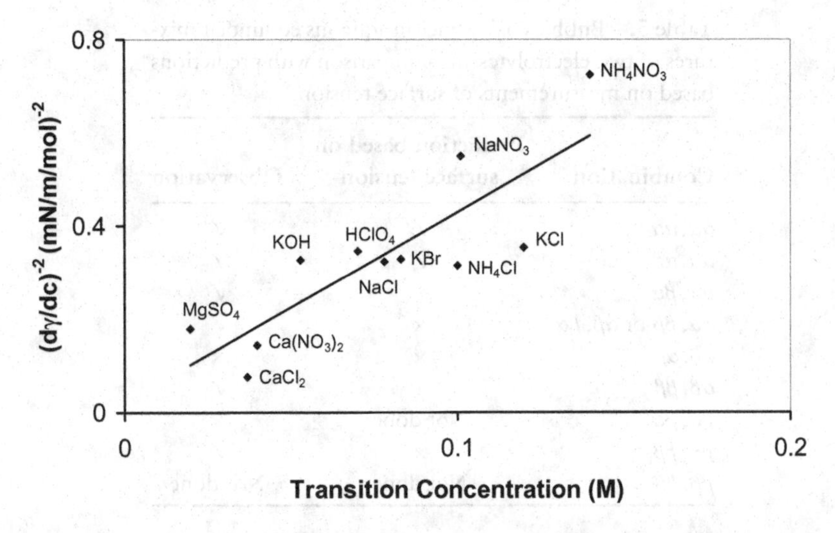

Fig. 8. The transition concentration of a range of electrolytes is compared to $\frac{1}{(\frac{d\gamma}{dc})^2}$. Note that sodium acetate has an ordinate of ~0.50, but is non-inhibiting and is therefore not shown.

Why should the surface tension be related to bubble coalescence inhibition? A possible mechanism is the influence of the Gibbs elasticity[27] on the stability of the liquid film separating two bubbles. The stability of foam films formed in the presence of surfactant is aided by the Gibbs elasticity that arises due to the difference in surfactant concentration between the surface and the bulk. If the film is disturbed, the generation of non-equilibrated interface results in an interfacial elasticity that opposes the disturbance. The magnitude of the effect is proportional to the square of the interfacial surface tension gradient with regard to concentration.[28] Gibbs elasticity also applies to species that are depleted from the interface such as electrolytes. However, for electrolytes the surface tension gradients are much smaller and consequently the Gibbs elasticity is expected to be very much smaller. Furthermore, as the effect is dynamic, rapidly diffusing species such as small ions can limit the effectiveness of the Gibbs elasticity. The Gibbs elasticity of electrolyte interfaces has not been measured directly. Nonetheless the correlation with $(\frac{d\gamma}{dc})^2$ suggests that the Gibbs elasticity could be controlling bubble coalescence. However, an extensive study using equimolar mixtures of pairs of electrolytes (see Fig. 5) found no correlation at all between the magnitude of the surface

Table 5. Bubble coalescence in aqueous equimolar mixtures of two electrolytes and comparison with predictions based on measurements of surface tension.

Combination	Prediction based on surface tension	Observation
$\alpha\alpha, \alpha\alpha$	✓	✓
$\alpha\alpha, \alpha\beta$	✓	✓
$\alpha\alpha, \beta\alpha$	×	✓
$\alpha\alpha, \beta\beta$ or $\alpha\beta, \beta\alpha$	×	✓
$\alpha\beta, \alpha\beta$	×	×
$\alpha\beta, \beta\beta$	×	✓
$\beta\alpha, \beta\alpha$	Not done	×
$\beta\alpha, \beta\beta$	×	✓
$\beta\beta, \beta\beta$	Not done	Not done

tension gradient and bubble coalescence inhibition.[13] Based on observations in single electrolytes, it was predicted that electrolyte mixtures with a $(\frac{d\gamma}{dc})^2 < 1$ (mN/m/M)^2 would not inhibit bubble coalescence. This prediction was compared to coalescence inhibition measurements, as summarised in Table 5.

So whilst the propensity for an electrolyte to inhibit bubble coalescence correlates well with the inverse of $(\frac{d\gamma}{dc})^2$ for solutions of single electrolyte, this correlation is not observed for solutions with two electrolytes.[13] This could indicate that the correlation between surface tension gradient and the inhibiting effect of single electrolytes is a secondary effect, in that the surface tension gradient itself is not important but it may be correlated with an unidentified feature that truly controls bubble coalescence.

The effect of an inhibiting electrolyte on bubble coalescence is to extend the lifetime of the thin aqueous film separating two bubbles. Even for a strongly inhibiting electrolyte, the thin film separating two bubbles will rupture in less than ten seconds and in the absence of electrolyte, this rupture typically occurs in less than one second.[29] It should therefore be remembered that coalescence is a dynamic process and hence equilibrium values of the surface tension may not be appropriate. Currently, there is no dynamic surface tension data available for electrolyte solutions. The acquisition of such data would be challenging as the influence of electrolytes on surface tension is small (and the dynamic influence will be even smaller) and the diffusion of electrolyte is rapid, thus highly accurate measurements at short time scales are required. However, one can surmise that as the time scale is reduced, the surface tension value will approach that of pure water. It is possible that for single electrolytes the dynamic surface tension is reflected in the gradient of the equilibrium surface tension with concentration, whereas for the mixed electrolytes this correlation may be destroyed by the presence of multiple ion types, each of which will have different diffusion rates. Alternatively, the static surface tension gradient could be correlated with another, important dynamic property of the interface. It has been shown that in surfactant-stabilised thin films, the film stability can be related to the dilational surface modulus ε, a complex viscoelastic quantity that reflects the surface tension response to a transient change in surface area during film deformation[30,31] This dynamic surface rheology may also be important in electrolyte stabilisation of thin films. As the dilational surface modulus is related to the change in surface tension in response to a change in surface area, it is possible that for simple electrolyte solutions, it is correlated with the equilibrium surface tension gradient.

If we accept that a dynamic process controlling bubble coalescence is reflected in measurements of the equilibrium surface tension only when a single solute is present, then we can make a connection between specific ion effects in bubble coalescence and the Hofmeister series by using the Solute Partitioning Model of Pegram and Record.[32] By applying a thermodynamic analysis of surface tension data, they have determined single ion coefficients that describe the relative concentration of the ion at the interface compared to bulk. An ion partition coefficient of less than one indicates that the ion concentration at the interface is less than in bulk and a value greater than one corresponds to an accumulation at the interface relative to the bulk. The single-ion partition coefficients are found to be independent of bulk concentration and additive for the different ions.

If the ions are ordered from the highest to lowest values of the single-ion partition coefficients a series can be derived for both the cation and anions describing their affinity for the interface. For the anions the series is $ClO_4^- \approx SCN^- > ClO_3^- >$ Acetate$^- > I^- > NO_3^- > Br^- > Cl^- > F^- > CO_3^{2-} > SO_4^{2-}$.[32–34] With the exception of acetate, this series agrees with the Hofmeister series. The ions that are more strongly excluded from the interface, as measured by the single-ion partition coefficients, are those that are the most effective Hofmeister ions (i.e. strongly salt-out proteins, etc.).

The surface tension gradient is found by Pegram and Record to be a function of the individual ion partition coefficients. Significantly, the magnitudes of the ion partition coefficients for the air–water interface correspond to the empirically assigned ion parameters α and β as shown in Table 6 for the anions and Table 7 for the cations. The values of these coefficients suggest that an α cation or α anion is withdrawn from the interface and a β cation or β anion is adsorbed to the interface. Note that for the anions, this is the reverse of the original suggestion of Marcelja.[14] Thus it suggests that it is not the separation of ions within the interfacial region as originally proposed by Marcelja that is important for

Table 6. Anion partition coefficients at the air–water interface.

Ion	$K_{p,i} \pm$ s.d.[a]	Ion assignment
SO_4^{2-}	0.00 (defined)	α
F^-	0.53 ± 0.02	α
OH^-	0.58 ± 0.04	α
Cl^-	0.69 ± 0.04	α
Br^-	0.86 ± 0.08	α
NO_3^-	0.98 ± 0.09	α
I^-	1.18 ± 0.12	α
CH_3COO^-	1.30 ± 0.05	β
ClO_3^-	1.44 ± 0.03	β
SCN^-	1.64	β
ClO_4^-	1.77 ± 0.04	β

[a]Data taken from Pegram and Record (2007).[32]

Table 7. Cation partition coefficients at the air–water interface.

Ion	$K_{p,i} \pm \sigma$[a]	Ion assignment
Na^+	0.00 (defined)	α
Cs^+	0.01 ± 0.04	α
Li^+	0.08 ± 0.21	α
K^+	0.12 ± 0.08	α
NH_4^+	0.25 ± 0.07	α
H^+	1.50 ± 0.04	β

[a]Data taken from Pegram and Record (2007).[32]

bubble coalescence inhibition, but the accumulation of both ions at the interface or the removal of both ions from the interface that results in bubble coalescence inhibition. This model also accommodates the recent simulation and experimental data on the partitioning of ions within the interface.[18,35,36]

We have tested the Pegram and Record ion partition coefficients as predictors of surface tension gradients for mixed electrolytes, and found that the values of surface tension gradient predicted by the ion partition coefficients are in quite good agreement with the measured surface tensions (see Fig. 9). This is significant as the surface tension gradients of the mixed electrolyte systems *do not* correlate with their bubble coalescence inhibiting properties.

Up until now the parameters of α and β have been assigned empirically using the combining rules. Once an ion had been assigned a property, the combining rules could be used to propagate predictions for bubble coalescence behaviour for other ion pairings. Now the assignment of the parameters α or β can be made to an ion based on the ion partition coefficients, without recourse to an empirical assignment. This shows that ion partition coefficients when used to assign the parameters α and β can be married with the combining rules to effectively predict bubble coalescence inhibition.

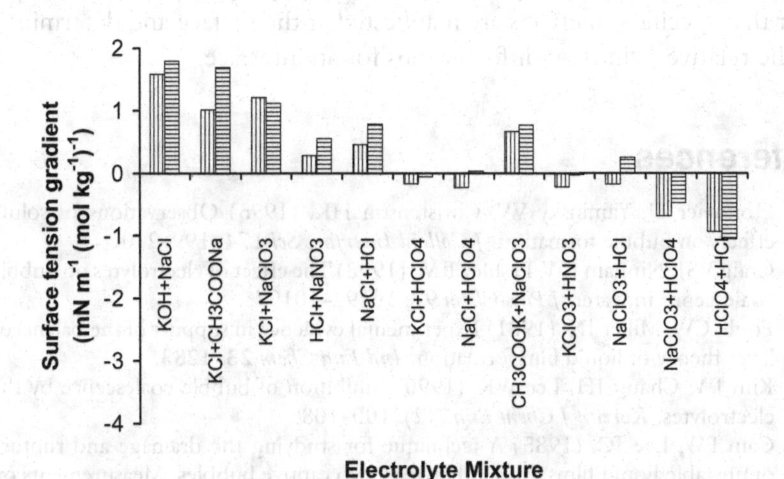

Fig. 9. Comparison of measured surface tension gradients[13] for equimolar mixtures of pairs of electrolytes with calculated values of the surface tension gradient using the ion partition coefficients of Pegram and Record.[32]

Also, as the ion partition coefficients apply to mixed electrolyte systems and the behaviour of mixed electrolyte systems is not predicted by measurements of the surface tension, we can state that the partitioning of the ions at the interface controls bubble coalescence through a mechanism other than the surface tension.

8. Conclusion

The influence of electrolytes on bubble coalescence in both aqueous and non-aqueous systems is ion specific. The available evidence suggests that ion specific effects at the air–water interface are strongly correlated with the partitioning of ions in the interfacial region. Ion partition coefficients determined from surface tension data correlate well with the empirically assigned α and β parameters used to predict bubble coalescence inhibition. Bubble coalescence is inhibited when there is an accumulation of both ions at the interface or a removal of both ions from the interface, though the precise mechanism of inhibition remains elusive.

The ion partition coefficients are also correlated with the Hofmeister series (with some exceptions such as acetate). Therefore, there is now a substantiated link between the ion specific effects observed at the air–water interface and the more ubiquitous Hofmeister-type specific ion effects. It is clear that specific ion effects are manifested at the surface and determined by the relative affinity of different ions for an interface.

References

1. Hofmeier U, Yaminsky VV, Christenson HK. (1995) Observations of solute effects on bubble formation. *J Colloid Interface Sci* **174**: 199–210.
2. Craig VSJ, Ninham BW, Pashley RM. (1993) The effect of electrolytes on bubble coalescence in water. *J Phys Chem* **97**: 10192–10197.
3. Foulk CW, Miller JN. (1931) Experimental evidence in support of the balanced-layer theory of liquid film formation. *Ind Eng Chem* **23**: 1283.
4. Kim JW, Chang JH, Lee WK. (1990) Inhibition of bubble coalescence by the electrolytes. *Korean J Chem Eng* 7(2): 100–108.
5. Cain FW, Lee JC. (1985) A technique for studying the drainage and rupture of unstable liquid films formed between two captive bubbles: Measurements on KCl solutions. *J Coll Interface Sci* **106**(1): 70–85.
6. Craig VSJ, Ninham BW, Pashley RM. (1993) Effect of electrolytes on bubble Coalescence. *Nature* **364**: 317–319.

7. Deschenes LA, Barrett J, Muller LJ, Fourkas JT, Mohanty U. (1998) Inhibition of bubble coalescence in aqueous solutions. I. Electrolytes. *J Phys Chem B* **102**: 5115–5119.
8. Lessard RR, Zieminski SA. (1971) Bubble coalescence and gas transfer in aqueous electrolytic solutions. *Ind Eng Chem Fundam* **10**(2): 260–269.
9. Foulk CW. (1924) Foaming of boiler water. *Ind Eng Chem* **16**(11): 1121–1126.
10. Craig VSJ. (2004) Bubble coalescence and specific-ion effects. *Curr Opin Colloid Interface Sci* **9**: 178–184.
11. Pashley RM, Craig VSJ. (1997) Effects of electrolytes on bubble coalescence. *Langmuir* **13**: 4772–4774.
12. Christenson HK, Bowen RE, Carlton JA, Denne JRM, Lu Y. (2008) Electrolytes that show a transition to bubble coalescence inhibition at high concentrations. *J Phys Chem C* **112**: 794–796.
13. Henry CL, Dalton CN, Scruton L, Craig VSJ. (2007) Ion-specific coalescence of bubbles in mixed electrolyte solutions. *J Phys Chem C* **111**: 1015–1023.
14. Marcelja S. (2006) Selective coalescence of bubbles in simple electrolytes. *J Phys Chem B* **110**: 13062–13067.
15. Jungwirth P, Tobias DJ. (2001) Molecular structure of salt solutions: A new view of the interface with implications for heterogeneous atmospheric chemistry. *J Phys Chem B* **105**: 10468–10472.
16. Jungwirth P, Tobias DJ. (2002) Ions at the air–water interface. *J Phys Chem B* **106**(25): 6361–6373.
17. Jungwirth P, Tobias DJ. (2006) Specific ion effects at the air–water interface. *Chem Rev* **106**: 1259–1281.
18. Jungwirth P, Winter B. (2008) Ions at aqueous interfaces: From water surface to hydrated proteins. *Annu Rev Phys Chem* **59**: 343–366.
19. Henry CL, Craig VSJ. (2008) Ion-specific influence of electrolytes on bubble coalescence in non-aqueous solvents. *Langmuir* **24**: 7979–7985.
20. Coons JE, Halley PJ, McGlashan SA, Tran-Cong T. (2005) Bounding film drainage in common thin films. *Colloids and Surface A: Physicochem Eng Aspects* **263**: 197–204.
21. Danov KD, Valkovska DS, Ivanov IB. (1999) Effect of surfactants on the film drainage. *J Colloid Interface Sci* **211**: 291–303.
22. Parkinson L, Sedev R, Fornasiero D, Ralston J. (2008) The terminal rise velocity of 10–100 μm diameter bubbles in water. *J Coll Interface Sci* **322**: 168–172.
23. Henry CL, Parkinson L, Ralston JR, Craig VSJ. (2008) A mobile gas–water interface in electrolyte solutions. *J Phys Chem C* **112**: 15094–15097.
24. Fainerman VB, Miller R. (2004) Maximum bubble pressure tensiometry — an analysis of experimental constraints. *Adv Colloid Interface Sci* **108–109**: 287–301.
25. Weissenborn PK, Pugh RJ. (1995) Surface tension and bubble coalescence phenomena of aqueous solutions of electrolytes. *Langmuir* **11**: 1422–1426.
26. Weissenborn PK, Pugh RJ. (1996) Surface tension of aqueous solutions of electrolytes: Relationship with ion hydration, oxygen solubility and bubble coalescence. *J Colloid Interface Sci* **184**: 550–563.
27. Gibbs JW. (1928) *The Collected Works*, pp. 300–315. Longmans, Green and Co., New York.

28. Christenson, HK, Yaminsky, VV. (1995) Solute effects on bubble coalescence. *J Phys Chem* **99**: 10420.
29. Li D, Liu S. (1996) Coalescence between small bubbles or drops in pure liquid. *Langmuir* **12**: 5216–5220.
30. Andersen A, Oertegren J, Koelsch P, Wantke D, Motschmann H. (2006) Oscillating bubble SHG on surface elastic and surface viscoelastic systems: New insights in the dynamics of adsorption layers. *J Phys Chem B* **110**: 18466–18472.
31. Koelsch P, Motschmann H. (2005) Relating foam lamella stability and surface dilational rheology. *Langmuir* **21**: 6265–6269.
32. Pegram LM, Record MT Jr. (2007) Hofmeister salt effects on surface tension arise from partitioning of anions and cations between bulk water and the air–water interface. *J Phys Chem B* **111**: 5411–5417.
33. Pegram LM, Record MT Jr. (2008) Thermodynamic origin of Hofmeister ion effects. *J Phys Chem B* **112**: 9428–9436.
34. Pegram LM, Record MT Jr. (2008) Quantifying accumulation or exclusion of H^+, OH^-, and Hofmeister salt ions near interfaces. *Chem Phys Letts* **467**: 1–8.
35. Garrett BC. (2004) Ions at the air/water interface. *Science* **303**: 1146–1147.
36. Petersen PB, Saykally RJ, Mucha M, Jungwirth P. (2005) Enhanced concentration of polarizable anions at the liquid water surface: SHG spectroscopy and MD simulations of sodium thiocyanide. *J Phys Chem B* **109**: 10915–10921.

Part C

NEWEST RESULTS FROM THEORY AND SIMULATION

Chapter 8

Ion Binding to Biomolecules

Mikael Lund*, Jan Heyda† and Pavel Jungwirth†

We investigated specific anion binding to basic amino acid residues as well as to a range of patchy protein models. This microscopic information was subsequently used to probe protein–protein interactions for aqueous lysozyme solutions. Using computer simulations to study both atomistic and coarse grained protein molecules, it is shown that the ion–protein interaction mechanism as well as magnitude is largely controlled by the nature of the interfacial amino acid residues. Small anions interact with charged side-chains via ion-pairing while larger, poorly hydrated anions are attracted to nonpolar residues due to a number of solvent-assisted mechanisms. Taking into account ion and surface specificity in a mesoscopic model for protein–protein interactions, we investigated the association of the protein lysozyme in aqueous solutions of sodium iodide and sodium chloride. As observed experimentally, it is found that 'salting out' of lysozyme follows the reverse Hofmeister series for pH below the iso-electric point and the direct series for pH above.

1. Introduction

The stability of protein solutions is governed not only by the macromolecular net charge, salt concentration and valency, but also by the chemical nature of the dissolved ions.[1] Traditionally, the latter falls under the category of Hofmeister or ion-specific effects which in recent years has seen an appreciable renaissance — both from experimental and theoretical perspectives. In Hofmeister's original studies,[2] ions were arranged according to their ability to precipitate or 'salt-out' egg white proteins. For the anions, which we focus on here, the following order was found: $F^- > CH_3COO^- > Cl^- > NO_3^- > Br^- > I^- > SCN^-$. The effects

*Department of Theoretical Chemistry, University of Lund, P.O.B. 124, SE-22100 Lund, Sweden.

†Institute of Organic Chemistry and Biochemistry v.v.i., Czech Academy of Sciences, Flemingovo nam. 2, 16610 Prague, Czech Republic.

of cations are generally less pronounced. The situation is, however, more complicated in that the Hofmeister ordering is in fact dependent on the solution pH and the protein iso-electric point, pI. For example, second virial coefficient measurements of lysozyme, which has pI in the range of 10 to 11, showed that this protein follows the reverse Hofmeister series at low pH and that anions such as iodide and thiocyanate very effectively induce protein association.[3] In a systematic study using small-angle X-ray scattering, it was demonstrated[4] that a Hofmeister reversal for pH $<$ pI is observed not only for lysozyme but for a range of small proteins including α-crystallins, γ-crystallins, ATCase and Brome Mosaic Virus.

Recent studies indicate that the ion specificity of protein association to a large extent is governed by interactions between salt ions and the macromolecular surface.[5] Then for pH $<$ pI, any absorption of anions will effectively reduce the repulsion between the cationic biomolecules and thus assist complexation. The reverse is true for negatively charged proteins (pH $>$ pI) where binding of anions will increase the repulsion and thus stabilise the solution. From the above mentioned experiments,[3,4] we thus conclude that anion binding to proteins follows the reverse Hofmeister series — i.e. thiocyanate binds more strongly than fluoride irrespective of the net charge.

It has been shown[6] that combining Poisson–Boltzmann continuum electrostatics with ion-specific dispersion forces between salt ions and spherical macro-ions, the Hofmeister reversal can be obtained. In the dispersion framework the larger ions such as iodide and thiocyanate interact stronger with the (averaged) macromolecular surface than does, for example, chloride. While dispersion is indeed ion specific (see Chap. 11 by Boström *et al.*), other important mechanisms are also at play. Protein surfaces are far from uniform and consist rather of an intricate network of polar and nonpolar groups to which salt ions have widely different affinities. For example, large anions are attracted to hydrophobic interfaces via surface-modified solvation and polarisation.[7,8] Direct ion-pairing[9,10] between salt particles and charged surface groups also gives rise to ion specific phenomena.[11] In this chapter we will focus mainly on the contributions from these mechanisms using a combination of atomistic and mesoscopic simulation techniques.

2. Ion-Pairing

2.1. *Ion-Pairing and Simple Electrolytes*

Shortly after the introduction of the Debye–Hückel (DH) theory, it was recognised that bulk properties of strongly associating ions are difficult to reproduce. This, in 1926, led Bjerrum to introduce the concept of 'ion-pairing' where the association of two ions is described by an equilibrium constant.[9,12] While this partitioning of bound and unbound states is always subject to an arbitrary definition — and invented to mend an incomplete microscopic description — the concept of ion-pairing remains useful for understanding ion-specific effects. In the following, we outline the basic concepts for bulk electrolyte solutions and, later, show that these apply to proteins as well.

One way of probing the interactions between solvated ions is to study aqueous bulk electrolyte activity coefficients, γ. For the most common choice of reference state, the mean activity coefficient is a measure of the excess free energy, $\mu = k_B T \ln \gamma$, of transferring a solvated salt pair from an infinite dilution to a solution with a finite salt concentration. In the present context, we are interested in the ion *specificity* and therefore — for a fixed salt concentration — define a free energy of exchanging one counter ion with another. For example,

$$\Delta\mu_{\mathrm{KCl}\rightarrow\mathrm{KI}} = \mu_{\mathrm{KI}} - \mu_{\mathrm{KCl}} = k_B T \ln(\gamma_{\mathrm{KI}}/\gamma_{\mathrm{KCl}}).$$

The range of activity coefficient measurements is vast and we can thus use the change in excess chemical potential as a direct measure of relative ion specificity for a large number of species. In Fig. 1 we show the chemical potential change for exchanging iodide with fluoride [Fig. 1(a)] and potassium with sodium [Fig. 1(b)]. From these experimentally obtained data, it is clear that the larger alkali cations like Cs^+ and Rb^+ prefer iodide over fluoride while the opposite is true for the smaller sodium ion. Potassium shows an equal preference for both halides. Likewise, we also observe that acetate — a useful proxy for acidic amino acid side chains — prefers sodium over potassium. Interestingly, the free energy difference varies almost perfectly linearly with the solute concentration, i.e. $\partial\Delta\mu/\partial c = \mathit{constant}$. By plotting the excess chemical potential difference, one can argue that we have subtracted the generic DH-type, salt screened contribution and are left with the specific part only.

Let us note that the DH theory and the underlying primitive model of electrolytes are appropriate for the long-ranged nature of the electrostatic

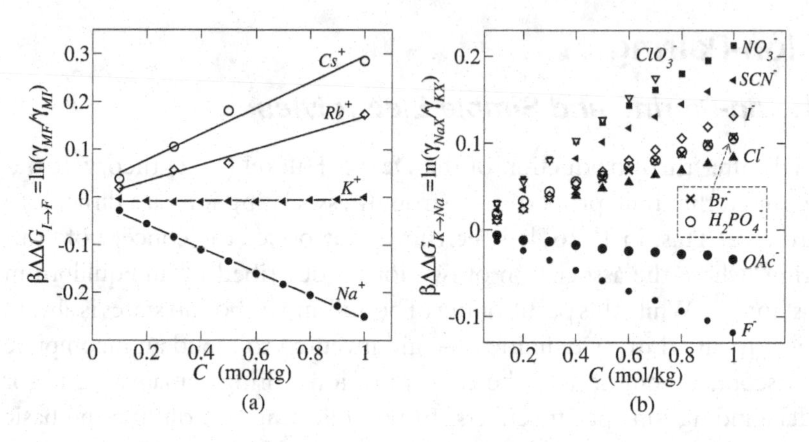

Fig. 1. (a) Measured change in excess chemical potential of exchanging iodide with fluoride in a range of alkali metal solutions.[20] (b) As (a), but for the exchange of potassium with sodium for a number of common monovalent anions.

interactions — both between ions and between ions and water molecules. It fails, however, to describe the *short-range* oscillations connected with ion-specific hydration and, to capture ion-specific effects, the original model needs extensions.

The activity coefficient is of course a macroscopic average of the excess interactions in the solution and as such does not provide a direct picture of the molecular level mechanisms. On empirical grounds it has been suggested[13] that ions with matching water affinities — which according to the Born solvation energy translate to 'matching sizes' — tend to prefer each other. Generally, this notion is in good agreement with the data shown in Fig. 1 where small–small or large–large pairs are preferred over large–small combinations.

To unravel the molecular level mechanisms, calculations have provided important insight.[10] Classical Molecular Dynamics (MD) simulations of pairs of ions in a molecular solvent as well as *ab initio* calculations in a continuum solvent show that the principle of matching sizes indeed has a molecular explanation. Examining the radial distribution functions (see Fig. 2) between sodium and fluoride — two rather small ions — we note that the large peak corresponds to an attractive configuration of a contact ion pair. At larger separations there is a second peak, corresponding to a solvent-separated pair. The attraction between the larger caesium ion and fluoride is much less pronounced since forming a contact ion pair involves a shared solvation shell and — due to the size differences between Cs^+

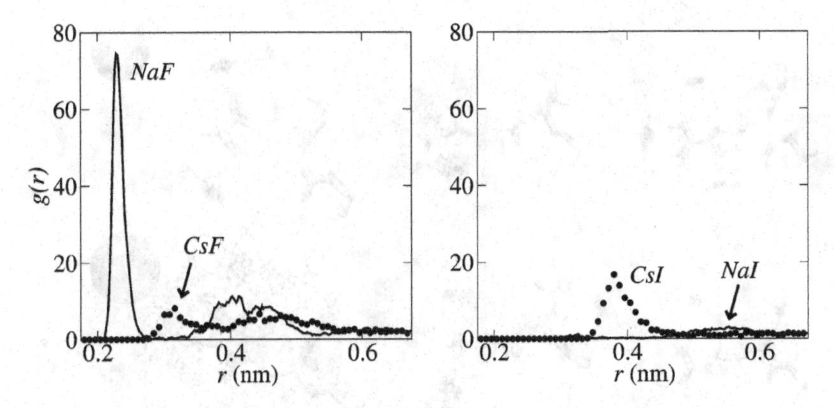

Fig. 2. Pair-wise radial distribution functions obtained from MD simulations (SPC/E water, 298 K, 1 atm) for halides of sodium and caesium.

and F^- — this is an unfavourable configuration. As expected, the situation is reversed when fluoride is exchanged with the larger iodide anion. These observations are in qualitative agreement with the monovalent activity coefficient data shown in Fig. 1, and provide a molecular interpretation of the empirical idea of matching water affinities.

2.2. *Ion Binding to Amino Acid Residues*

We now concentrate on ion-pairing in more biologically relevant systems. Simulation work on proteins, oligopeptides, and amino acid residues has shown that sodium and potassium bind differently to biomolecules.[11,14] In general, the affinity of sodium to protein surfaces is more than twice that of potassium. The cationic attraction stems mainly from pairing with acidic carboxyl groups on glutamate and aspartate, and with backbone carbonyl groups. In the following we will focus on complementary *anion* binding which generally exhibits stronger ion specific behaviour than the cations.

Let us first ask ourselves how halides bind to isolated positively charged amino acid residues such as arginine, lysine, and protonated histidine. To obtain a molecular-level insight of the ion-specific interactions, we investigate the spatial distributions of halides around the positively charged amino acid regions (see Fig. 3). This type of analysis provides information about both the strength of interaction as well as the spatial configuration. On the one hand, fluoride exhibits a strong affinity for positively charged groups,

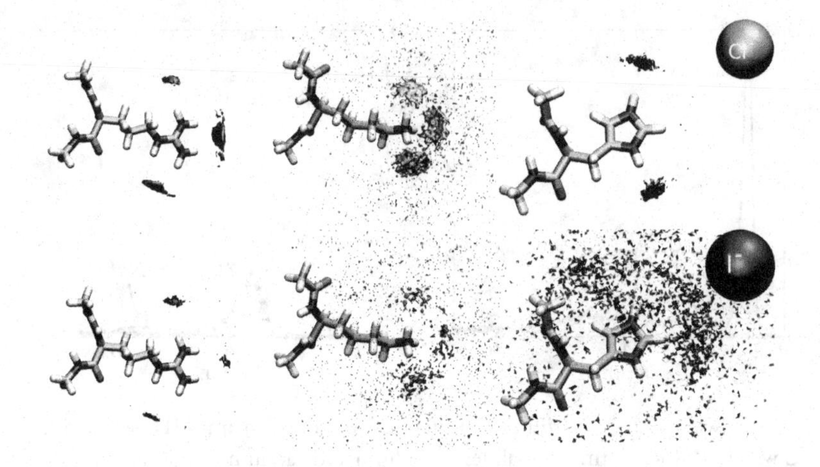

Fig. 3. Density plots showing the distributions of chloride and iodide ions around aqueous methyl-terminated basic amino acids. From left to right: arginine, lysine, and histidine.

while that of the larger halides is much weaker. On the other hand, iodide, and to a lesser extent also bromide and chloride, are weakly attracted to nonpolar regions as well as to amide hydrogens on the backbone.

Quantitatively, fluoride interacts most strongly with the guanidinium group of Arg, followed by the imidazolium group of (protonated) His, and the ammonium group of Lys. This affinity ordering is observed when using both non-polarisable and polarisable force fields, being weaker — by about a factor of two — in the latter case. For non-polarisable force fields, fluoride is the clear winner among the halides for all positively charged amino acid residues, with the dominant part of the halide–amino acid affinity being due to the charged side chain groups. Upon including the polarisation term, the interaction of F^- with the amino acids is *reduced*. This is opposite to the larger halides where polarisation effects tend to enhance the affinity for nonpolar regions. This leads to a situation where the overall affinity for the amino acid surface can be higher for iodide than for fluoride. This is caused by a sizable propensity of the former ion for the interface between water and nonpolar groups when using a polarisable force field.

The computational results show that the halide interactions with positively charged amino acid residues are *local* and generally not overwhelmingly strong. This is well documented by the fact that even fluoride anions, which exhibit the strongest interaction, frequently exchange positions in the vicinity of the amino acid and in the bulk. Therefore, additivity can be invoked and analogous ion-specific behaviour of halides can be expected at surfaces of aqueous proteins. The overall ion specific effect will then be a net result of nonpolar attractions and direct ion-pairing with positively charged side chains. In the former case, large, soft anions such as iodide win, while for the latter case, small anions like fluoride dominate. In addition, interactions with the backbone amide hydrogens should be considered as these have a considerable preference for larger anions.

3. Nonpolar Attraction

In the previous sections, we discussed how ion pairing influences bulk electrolyte properties as well as controls where and how strongly ions bind to complex biomolecules. We also mentioned that large, 'soft' anions can bind to aliphatic regions and we now focus on this mechanism that — perhaps counter intuitively — causes certain ions to be attracted to nonpolar or hydrophobic molecular regions.

Experimental as well as theoretical studies of anions close to the vapour–water and molecular interfaces reveal an appreciable ion specific segregation. In particular large, 'soft' and poorly hydrated ions such as iodide and thiocyanate are attracted to the interface[7] while small, 'hard' and well-solvated species are repelled from it. The former observation contradicts the traditional dielectric continuum picture within which a generic, solvated ion close to a low dielectric interface will experience a repulsive force due to partial dehydration.[15] Let us revise the mechanisms with which an ion may interact with a nonpolar interface:

(i) Desolvation or loss of ion–dipole energy as described in classical electrostatics by a reaction field.
(ii) Association of poorly solvated species is induced due to a reduction of the ordered water network surrounding these.[16]
(iii) Aligned water molecules near the nonpolar interface set up an electric field that leads to induced dipole interactions with polarisable ions.[17]
(iv) Solvent–solute, solute–solute and solvent–solvent dispersion interactions can lead to both an attraction and also a repulsion of ions (see Chap. 11 by Boström *et al.*)

How much each of these interactions (i)–(iv) contributes to the potential of mean force depends to a large extent on the ion type. Let us first consider a small ion, characterised by a high surface charge density and a small polarisability (both static and dynamic). In other words, the ion is strongly hydrated and (i) will be the dominant interaction type upon moving the ion towards a nonpolar surface. A large ion has opposite characteristics and, consequently, interactions (ii)–(iv) become increasingly important.

3.1. *An Idealised Bio-Colloid*

While the attraction of large anions to water–vapour interfaces is now well established, the situation for complex molecular surfaces is less scrutinised. Solvated, globular proteins are mainly hydrophilic in nature but an appreciable number of nonpolar residues can be present even at the solvent-exposed molecular surface (see Fig. 4). With the situation at the water–vapour interface in mind indeed it is imaginable that large anions exhibit a similar attraction to such nonpolar surface patches.

To decipher the interaction mechanisms of small and large anions with biomolecular interfaces, it is at least initially advantageous to look at a simplified system that captures the essential physics. With the risk of dismaying readers with a biochemical background, we will now construct an artificial 'bio-colloid' — a simple nano-sphere with distributed cationic surface charges (see Fig. 4). While we do not claim that this toy-model[8] is an adequate proxy for a real protein, it has some appealing advantages in that

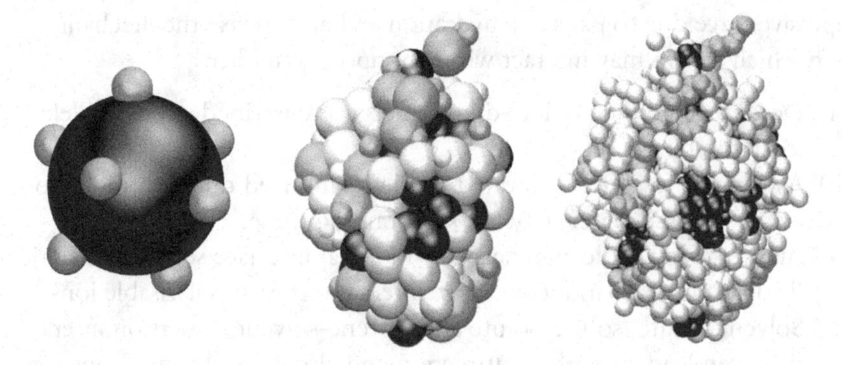

Fig. 4. Three macromolecular models with hydrophobic (black) and cationic surface groups (grey). Left: A nonpolar sphere with positively charged patches.[8] Middle: Lysozyme coarse grained to the amino acid level.[18] Right: Lysozyme in atomistic detail (hydrogens not shown).[19]

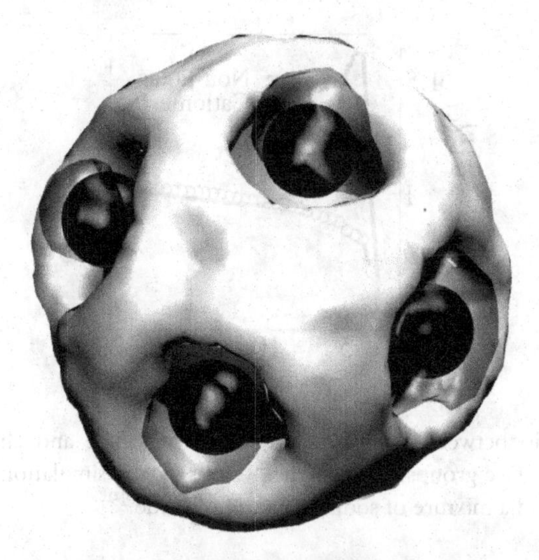

Fig. 5. Spatial iso-density plots of iodide (white) and fluoride (grey) around a spherical bio-colloid with cationic surface patches (black spheres). Results from MD simulations in explicit solvent (SPC/E) and a mixture of sodium iodide and fluoride.[8]

one can control the hydrophobicity/hydrophilicity and, due to the simple surface topology, it can be readily scrutinised. Figure 5, which shows where fluoride and iodide bind to the surface of such a bio-colloid, reveals some interesting features:

- Fluoride (small and well hydrated) binds exclusively to the discrete surface charges.
- Iodide (large and poorly hydrated) is found on the remaining nonpolar patches.

Hence, from this simplified model one can extract that specific ion binding is governed by a delicate balance between interactions with charged and nonpolar surface patches on the macromolecule. The former groups attract ions via direct pairing while the latter do so via a range of solvent-mediated interaction mechanisms (discussed above).

3.2. *Binding to a 'Real' Protein*

To confirm that different ions also segregate on complex molecular surfaces according to the distribution of charged and nonpolar patches, we

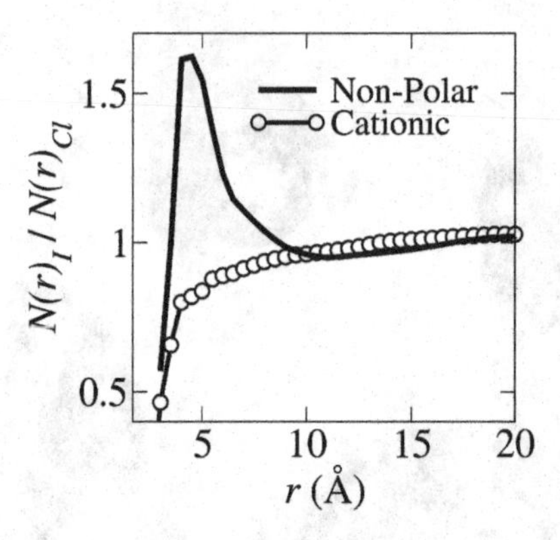

Fig. 6. The ratio between the cumulative sums of iodide and chloride around polar/nonpolar surface groups on lysozyme. Results of MD simulations in an explicit solvent (POL3) and a mixture of sodium chloride/iodide.[19]

now substitute the above 'bio-colloid' with a real protein in full atomistic detail (see Fig. 4, right) and redo the analysis. Since the protein surface is immensely more complex (as well as fluctuating), the binding cannot as easily be represented using spatial density plots as in Fig. 5. Instead, one can average the number of ions encountered in non-spherical shells around polar and nonpolar patches. The result is shown in Fig. 6. As in the simplified model, iodide (large and poorly hydrated) indeed prefers nonpolar patches while chloride (small and well hydrated) prefers cationic groups.

Coming full circle and relating to experiments, the affinity for large ions to nonpolar molecular regions is also manifested in bulk electrolyte solutions. Figure 7 shows the growing preference for iodide over chloride

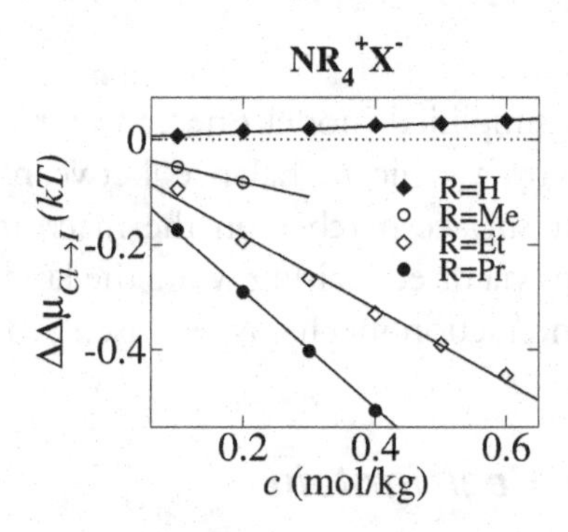

Fig. 7. Excess chemical potential difference for exchanging chloride with iodide in solutions of tetra-alkylammonium halides of increasing chain lengths. Results from experimentally determined activity coefficients.[20,21]

when increasing the chain length of the tetra-alkylammonium cations. In contrast, for the bare ammonium ion, there is a slight preference for the smaller chloride anion.

4. Protein Aggregation

Having dissected the ways in which ions bind to biomolecular surfaces, we attempt for predict ion specific phenomena in protein solutions. In Hofmeister's original work,[2] salts were ranked according to their ability to precipitate egg white proteins. Since then, many more experiments have been conducted with a rather general conclusion for protein–protein association[4]:

- Cationic proteins (pH < pI) usually follow the *reverse* Hofmeister series.
- Anionic proteins (pH > pI) usually follow the *direct* Hofmeister series.

Hence at low pH, lysozyme with pI = 10 to 11 will associate more in the presence of iodide than of chloride. This is manifested in the second virial coefficient, B_2, that can be measured using scattering techniques.[3] The thermodynamic virial coefficient is defined as an integral over the angularly averaged potential of mean force between two protein molecules, $w(r)$:

$$B_2 = -2\pi \int_0^\infty (e^{-w(r)/kT} - 1)r^2 \mathrm{d}r.$$

We now perform Monte Carlo simulations of *two* lysozyme molecules, coarse-grained to the amino acid level (see Fig. 4, middle). Due to the nano-scale length scales involved, we invoke the dielectric continuum model for water and thereby average out all structural features of the solvent. It is, however, precisely water structuring that gives rise to the ion specificity mentioned in Sec. 3 and we therefore need to include this implicitly.[17,18] Applying effective potentials between hydrophobic surfaces and ions, obtained from explicit solvent simulations, it is possible to account for the solvent-assisted interaction between ions and nonpolar amino acid residues (i.e. ALA, LEU, VAL, ILE, PRO, PHE, MET, TRP). In this framework, iodide is attracted to these groups while sodium and chloride are repelled. The specificity of anion binding to the *charged* residues is included through the distance of closest approach in the water-screened Coulomb potential. A straightforward improvement would be

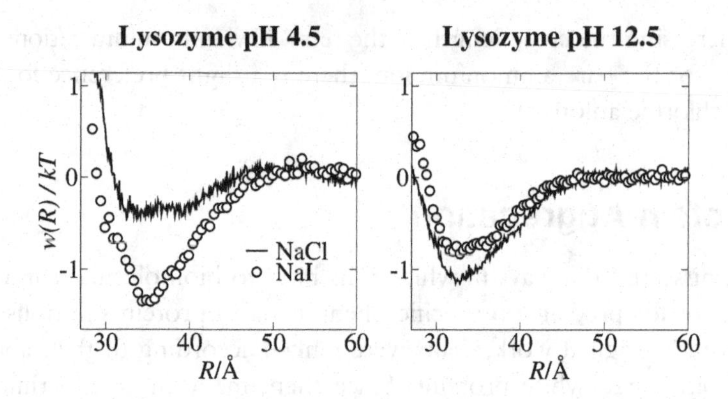

Fig. 8. Simulated potential of mean force between two mesoscopic lysozyme molecules in aqueous solutions of NaCl and NaI at low and high pH. Simulated in a continuum solvent with explicit salt- and counter-ions.[18]

to invoke a more realistic pair-correlation function, obtained from explicit solvent simulations.

As shown in Fig. 8, at low pH the inter-protein potential of mean force, $w(r)$, is more attractive in solutions of sodium iodide than in sodium chloride. This can be attributed to the fact that iodide absorbs to the nonpolar patches, thus decreasing the effective charge on the cationic proteins (pH < pI). Changing the solution pH from 4.5 to '12.5', the iso-electric point of lysozyme is exceeded and the protein net-charge goes from $+9$ to -5.5. The quotation marks are used to signal that strong alkaline conditions are hardly practical for protein solutions. Our simulations predict — in accordance with measurements[4] — that salting-out of anionic proteins indeed follows the direct Hofmeister series while the reverse is true for cationic ones.

Note that the Hofmeister reversal has also been observed in calculations of spherical-macroions via inclusion of specific dispersion interactions between ions and the macromolecules.[6] In that and other studies,[22,23] the macromolecular surface is represented by a uniform, spherical object akin to the renowned DLVO theory.[24,25] This implies that co-solutes experience the macromolecule as a spherical average of the actual, complex topology as found in real proteins. Depending on which properties one is after, this high degree of coarse graining may provide sufficient insight, although it seems less applicable for complex biomolecules where ion specificity is augmented by a subtle balance between ion-pairing and nonpolar interactions.

5. Summary

We have seen how ion specific binding to colloids, amino acid residues as well as complex proteins is governed by two main mechanisms that target distinct surface groups on the macromolecule:

- Ion-pairing, and
- Nonpolar, solvent mediated interactions.

Interestingly, both mechanisms are manifested in bulk solution properties such as the mean activity coefficient. This hints — and is indeed supported by molecular simulations — that ion specificity is controlled by a subtle combination of generic intermolecular interactions, effective for a large range of molecular dimensions.

Protein–protein association is governed by salt type and concentration as well as the solution pH. Taking into account the patchiness of protein surfaces, the experimental observations can be reproduced using coarse grained molecular simulations. In particular, ion binding to nonpolar patches leads to the observed pH-dependent reversal of the Hofmeister series.

References

1. Baldwin RL. (1996) How Hofmeister ion interactions affect protein stability. *Biophys J* **71**: 2056–2063.
2. Hofmeister F. (1888) Untitled. *Arch Exp Pathol Pharmakol (Leipzig)* **24**: 247–260.
3. Piazza R, Pierno M. (2000) Protein interactions near crystallisation: A microscopic approach to the Hofmeister series. *J Phys Condens Matter* **12**: A443–A449.
4. Finet S, Skouri-Panet F, Casselyn M, Bonnet F, Tardieu A. (2004) The Hofmeister effect as seen by SAXS in protein solutions. *Current Opinion in Colloid & Interface Science* **9**: 112–116.
5. Zhang Y, Cremer P. (2006) Interactions between macromolecules and ions: The Hofmeister series. *Model systems/Biopolymers* **10**: 658–663.
6. Boström M, Tavares FW, Finet S, Skouri-Panet F, Tardieu A, Ninham BW. (2005) Why forces between proteins follow different Hofmeister series for pH above and below pI. *Biophys Chem* **117**: 217–224.
7. Jungwirth P, Tobias DJ. (2006) Specific ion effects at the air/water interface. *Chem Rev* **106**: 1259–1281.
8. Lund M, Vácha R, Jungwirth P. (2008) Specific ion binding to macromolecules: Effects of hydrophobicity and ion pairing. *Langmuir* **24**: 3387–3391.
9. Bjerrum N. (1926) Untersuchungen Uber Ionenassoziation. I. *Kgl. Danske Vid. Selsk. Mat.-fys. Medd.* **7**: 1–48.

10. Jagoda-Cwiklik B, Vácha R, Lund M, Srebro M, Jungwirth P. (2007) Ion pairing as a possible clue for discriminating between sodium and potassium in biological and other complex environments. *J Phys Chem B* **111**: 14077–14079.
11. Vrbka L, Vondrasek J, Jagoda-Cwiklik B, Vácha R, Jungwirth P. (2006) Quantification and rationalization of the higher affinity of sodium over potassium to protein surfaces. *Proc Natl Acad Sci USA* **103**: 15440–15444.
12. Marcus Y, Hefter G. (2006) Ion Pairing. *Chem Rev* **106**: 4585–4621.
13. Collins KD. (2004) Ions from the Hofmeister series and osmolytes: Effects on proteins in solution and in the crystallization process. *Methods* **34**: 300–311.
14. Dzubiella J. (2008) Salt-specific stability and denaturation of a short salt-bridge-forming alpha-helix. *J Am Chem Soc* **130**: 14000–14007.
15. Böttcher C. (1973) *Theory of Electric Polarization*. Elsevier.
16. Chandler D. (2005) Interfaces and the driving force of hydrophobic assembly. *Nature* **437**: 640–647.
17. Horinek D, Netz R. (2007) Specific ion adsorption at hydrophobic solid surfaces. *Phys Rev Lett* **99**: 226104.
18. Lund M, Jungwirth P, Woodward CE. (2008) Ion specific protein association and hydrophobic surface forces. *Phys Rev Lett* **100**: 258105.
19. Lund M, Vrbka L, Jungwirth P. (2008) Specific ion binding to nonpolar surface patches of proteins. *J Am Chem Soc* **130**: 11582–11583.
20. Robinson RA, Stokes RH. (1959) *Electrolyte Solutions*. Butterworths Scientific Publications, London.
21. Lindenbaum S, Boyd GE. (1964) Osmotic and activity coefficients for the symmetrical tetraalkyl ammonium halides in aqueous solution at 25°C. *J Phys Chem* **68**: 911–917.
22. Tavares FW, Bratko D, Blanch HW, Prausnitz JM. Ion-specific effects in the colloid–colloid or protein–protein potential of mean force: Role of salt-macroion van der Waals interactions. *J Phys Chem B* **108**: 9228–9235.
23. Boström M, Tavares FW, Bratko D, Ninham BW. (2005) Specific ion effects in solutions of globular proteins: Comparison between analytical models and simulation. *J Phys Chem B* **109**: 24489–24494.
24. Derjaguin BV, Landau L. (1941) Theory of the stability of strongly charged lyophobic sols and of the adhesion of strongly charged particles in solutions of electrolytes. *Acta Phys Chim URSS* **14**: 633–662.
25. Verwey EJW, Th. (1948) *Theory of the Stability of Lyophobic Colloids*. Elsevier Publishing Company Inc., Amsterdam.

Chapter 9

Ion-Specificity: From Solvation Thermodynamics to Molecular Simulations and Back

Joachim Dzubiella*, Maria Fyta*, Dominik Horinek*, Immanuel Kalcher*, Roland R. Netz* and Nadine Schwierz*

Molecular dynamics simulations based on classical force fields have become the standard tool for the modelling of ion specificity in bulk and at interfaces, but the choice of the non-bonding force parameters, the so-called force fields, is a subtle issue. We discuss how thermodynamic solvation properties in the infinite-dilute limit can be used to construct optimised non-polarisable force-fields for alkali metal cations and halide anions. Next, ion specificity in bulk and at interfaces is studied. In bulk, we obtain osmotic coefficients directly from molecular simulations and discuss the relation between ion–ion correlations and the resultant osmotic thermodynamics. At the air–water interface we determine the single-ion potential-of-mean-force in the infinite dilution limit. In agreement with previous polarisable force-field simulations, the larger halide ions are less repelled from the interface and iodide is even enhanced. We conclude that the surface activity of iodide is due to the ion hydrophobicity. Combining the simulation results with a generalised Poisson–Boltzmann approach, the interfacial tension increment at one molar salt concentration is predicted in quantitative agreement with experiments. The general trend, in agreement with the Hofmeister series, is: the smaller the ion, the larger the interfacial tension increment, indicative of stronger ion repulsion from the interface. On solid hydrophobic surfaces, the behaviour is similar to the air–water interface. On solid hydrophilic surfaces, on the other hand, the Hofmeister series is reversed and small ions show enhanced binding to polar surface groups.

*Physics Department T37, Technical University Munich, 85748 Garching, Germany.

1. Introduction

Computer simulation techniques have come a long way and are now at a level where the folding dynamics of small proteins can be accurately simulated, including small librational degrees of freedom, water solvation effects as well as large-scale refolding events.[1] Yet, the at first glance simple problems of ion specificity, i.e. the systematic differences between the effects that different 'simple' ions have on physicochemical properties of complex as well as simple aqueous solutions, still pose considerable problems to the simulation community.[2,3] This is definitely true for, e.g. the ion-specific action of complex biological constructs such as ion-channels.[4] But even the apparently simple problems of predicting the surface tension of water in the presence of different salts[5–9] or the ion-specific osmotic and activity coefficients of salt solutions[7,10,11] are far from being completely understood in terms of atomistic simulations. The reasons are basically four-fold:

(i) Simulations of aqueous ionic systems reach the limits of present computer power because they require large system sizes in order to minimise finite-size-correction effects in connection with long-ranged Coulombic interactions. Such simulations are slow because time scales of ion-pair formation are often in the nanosecond range.[12]

(ii) It is often non-trivial to extract thermodynamic properties from simulation trajectories. While it is fairly straightforward to obtain a rough idea of the adsorption characteristics of different ions at interfaces from simulation snapshots or ionic density profiles, to really nail down the interfacial tension one has to integrate the ionic surface excess over the salt bulk concentration according to the Gibbs adsorption isotherm, or one has to accurately determine the anisotropy of the pressure tensor.[6,13] Along the same lines, while it is straightforward to read off the trends of ion-pair formation from the interionic radial distribution function obtained from a simulation of a moderately concentrated salt solution, the calculation of the osmotic coefficient exactly from the distribution functions requires a lot more, namely an integral of the compressibility over the salt concentration.[11] Why should a simulator care about thermodynamic properties such as heat of solution, heat of hydration, interfacial tension, osmotic and activity coefficients at all? This is because consistent and reproducible data exist for almost a century for thermodynamic properties,[14,15] while experimental techniques for extracting ionic distribution functions in bulk[16] or at interfaces[17–20] are still at the forefront of experimental development and still give rise to debate and interpretations.[21,22]

(iii) It is often said that numerical simulations correspond to computer experiments. In fact, it transpires that for molecular dynamics simulations, where Newton's equations of motion for all constituent atoms are solved based on an energy functional that is typically constructed by pairwise Lennard–Jones potentials, the force-fields employed, i.e. the energy and distance parameters entering the individual Lennard–Jones terms, play the role of the experimental parameters. Having said that, it is often unclear what a meaningful comparison between a simulation result and an experimental result is and what one really learns from, e.g. a perfect agreement between simulation and experiment: does perfect agreement merely mean that the MD force fields have been adjusted correctly, or does it mean that the simulation model reflects some underlying truth beyond what has been put into the list of force-field parameters?

(iv) Finally, many ionic properties that exhibit ion-specific behaviour result from the near-cancellation of large numbers. To give an explicit example, which we will later discuss in more detail, the heat of solution of a salt, H_{sol}, is the enthalpy difference between the aqueous solution at infinite dilution and the solid crystal and is a measure of the heat adsorbed from the surrounding during dissolution.[14,15] The enthalpies of a dilute salt solution and of a salt crystal are both typically of the order of 1000 kJ/mol (measured with respect to an ionic gas). The difference between the heats of solution, H_{sol}, is merely of the order of a few tens of kJ/mol and can be positive or negative. Positive values of H_{sol} mean that the dissolution process is endothermic; negative values mean the process is exothermic. The same cancellation of large numbers occurs for the maximal solubility of salts in water, which depends on the difference of the Gibbs free energy in the crystal and in the solvated state. Cancellation is also at work for ion-pair binding in water, where a fraction of the strongly bound water molecules in the solvation shells of a cation and an anion are released and a contact pair is formed. Back to simulations, this means

> that in order to get ionic properties right that depend on the difference between two states (which is the usual situation), one has to have a very good understanding of both ionic states, or one has to rely on cancellation of errors, which often works but as often fails.

In the past, many heuristic rules that classify ionic specificity have emerged.[23,24] As discussed in more length in other chapters of this book, seemingly unrelated properties such as surface tension, solution viscosity, heat of solution, threshold concentration to precipitate certain proteins, depend on the ion type in a universal ordering, the so-called Hofmeister series.[27] This also means that when different ionic properties are set in relation to each other, universal master curves are obtained. A particularly useful abstraction is the law of matching water affinities,[23,28] which in a slight variant is known as the rule 'like seeks like'.[24] In simple words, ions that have similar size and opposite charge will attract each other and thus have low solubility and high heat (i.e. endothermic) of solution. Examples are LiF with a maximal solvated mole ratio of $\theta_{max} = 0.0019$ at room temperature (meaning 0.0019 mol of salt per mol of water in a saturated solution) and a heat of solution of $H_{sol} = 4.7\,kJ/mol$, or CsI characterised by $\theta_{max} = 0.053$ and $H_{sol} = 33.3\,kJ/mol$. On the other end of the spectrum are asymmetric salts such as LiI with $\theta_{max} = 0.22$ and $H_{sol} = -63.3\,kJ/mol$ or CsF with $\theta_{max} = 0.43$ and $H_{sol} = -36.9\,kJ/mol$.[14,15] The wisdom is that highly asymmetric salts consist of one strongly and one weakly hydrated ion, so that the resulting attraction in solution is weak since the strongly hydrated ions prefer to keep their hydration shells rather than directly engaging with the partner ion. If, on the other hand, both ions are small and strongly hydrated, the ion-pair interaction will be strong enough to deprive both ions of their hydration shells, and if both ions are large and weakly hydrated in the bound ion-pair, a direct ion–ion contact is also easily formed. Later in this chapter, we will take the abstraction a level further and argue that the rule 'like seeks like' is expected to hold whenever an ionic property depends on the difference between a single ion property (e.g. hydration enthalpy of single ions) and an ion-pair property (e.g. lattice enthalpy of sublimation) as for example is the case for the heat of solution H_{sol}.

The first grand success of simulations in the field of ion specificity was the work by Jungwirth and Tobias,[29,30] who showed that large anions (such as iodide) show a tendency to adsorb to the air–water interface, in striking contradiction to the classical picture according to which ions are unequivocally repelled from low-dielectric surfaces due to image-charge repulsion. In the initial report, the adsorption of the larger halides was seen with polarisable force fields, not with standard non-polarisable force fields. This led to the notion that it is the polarisability of the large ions that is responsible for their surface affinity. Such a notion is impossible to prove or falsify, since there is no way to disentangle the effects of ion size, ion polarisability, ion dispersion interaction and so on. We will show in a later section that a properly optimised non-polarisable ionic force field yields a surface affinity of iodide in very good agreement with experimental surface tension results. This result does not prove the irrelevance of polarisability effects in simulations or in the experimental reality, but it demonstrates how difficult it is to draw conclusions about the underlying mechanisms of ion specificity from simulations.

2. 'Like Seeks Like': Single-Ion Versus Ion-Pair Interactions

Many ionic properties show extremal behaviour for salts that consist of ions of roughly equal sizes. Let us consider the heat of solution H_{sol}, which is particularly easy to discuss since it corresponds to the enthalpy difference between a salt crystal and the infinitely dilute salt solution. For a given cation, let us say sodium, the heat of solution shows a maximum in between sodium chloride and sodium fluoride. For the slightly larger potassium ion, the maximum has moved to potassium iodide, while for rubidium the maximum is not observed for any halide ion, as shown in Fig. 1. Since the heat of solution H_{sol} is related to the lattice enthalpy H_{latt} and the ionic hydration enthalpy H_{hyd} via

$$H_{sol} = H_{hyd} - H_{latt}, \tag{1}$$

this directly means that at the maximum in H_{sol}, the lattice energy is least favourable compared to the hydrated state. Neglecting the entropic contribution to the Gibbs free energy (which is a fairly good approximation), this would also suggest that the maximum in the heat of solution corresponds to a situation of minimal water solubility, as roughly borne out by experiments.[14,15] Many more analogies and similar extremal properties could be quoted.[23,28] The question naturally arises what the cause of that

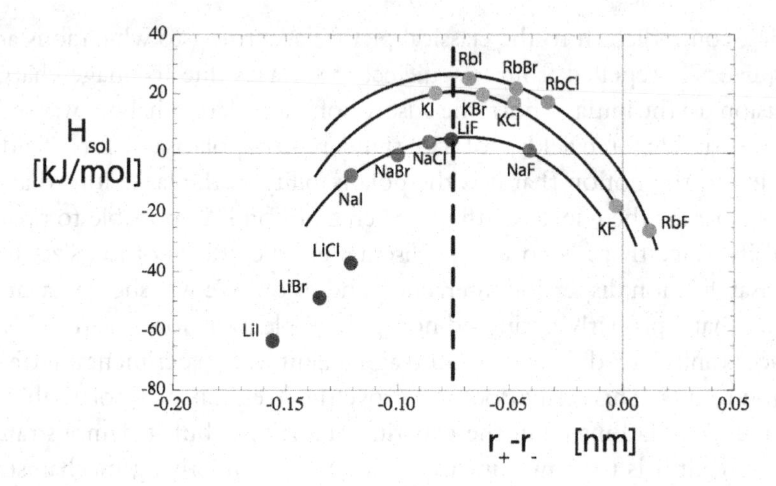

Fig. 1. Heat of solution of different alkali-halide salts, plotted as a function of the difference between the Pauling radii given in Table 1. A maximum in the heat of solution is observed for a given cation when the anion radius is larger by about $r_0 =$ 0.08 nm. Data are from Ref. 33, and the lines are parabolic fits.

behaviour is and what the simplest rational explanation could be. To that end we consider the simplest model for the heat of solution, based on pure electrostatics. The philosophy behind our model is not to describe experiments quantitatively, but rather to elucidate the mathematical features of the paradigm phrased as 'like seeks like'. The potential energy of a salt crystal is dominated by the Coulomb interactions, which can be written for a monovalent salt as[31]

$$U_{\text{latt}} = -[138.8\,\text{kJ/mol}]\,\alpha/(r_+ + r_-), \tag{2}$$

where α is the Madelung constant which for an arrangement on a simple cubic lattice takes the value $\alpha = 1.748$ and the unit of length is nm. The interionic distance is denoted by $(r_+ + r_-)$ and arbitrarily split up into the contributions of the individual ionic radii. The hydration energy is according to the Born dielectric self energy given by[14,15]

$$U_{\text{hyd}} = -[138.8\,\text{kJ/mol}](1/r_- + 1/(r_+ + r_0))/2, \tag{3}$$

and reflects the difference in dielectric self energy between the aqueous solution and the vacuum. For the cationic self energy, a length scale of the order of the OH bond length in water, r_0, has been added, since cations interact mostly with the water-oxygen which has a non-vanishing radius.

For the anion no such radius is added, since anions mostly interact with the water-hydrogens which have negligible radii. Neglecting the vibrational potential energy in the lattice and the pressure contribution to the enthalpy, which are of the order of $RT = 2.5\,\text{kJ/mol}$ and thus negligible, furthermore neglecting entropic contributions (which can be shown to be small), we approximate $H_{\text{hyd}} = U_{\text{hyd}}$ and $H_{\text{latt}} = U_{\text{latt}}$ and using Eq. (1) we arrive at

$$H_{\text{sol}} = 138.8 \frac{\text{kJ}}{\text{mol}} \left[\frac{\alpha}{r_+ + r_-} - \frac{1}{2(r_+ + r_0)} - \frac{1}{2r_-} \right]. \tag{4}$$

Introducing the short-hand notation $R = r_+ + r_-$ for the sum of the ionic radii and $\delta = r_+ - r_-$ for the difference between the radii, the expression (4) for the heat of solution can be rewritten as

$$H_{\text{sol}} = 138.8 \frac{\text{kJ}}{\text{mol}} \left[\frac{\alpha}{R} - \frac{2(R + r_0)}{(R + r_0)^2 - (\delta + r_0)^2} \right] \tag{5}$$

and thus clearly exhibits for fixed radii sum R a maximum in the heat of solution at a difference in radii corresponding to $\delta = r_+ - r_- = -r_0$. The numbers that come out from this simplistic theory for the lattice energy and also for the energy of hydration are not particularly good. This is not surprising, the literature is full of correction terms to our Ansatz that are not necessarily small.[32] In fact, our argument goes beyond the mere description of experimental data but rather aims at elucidating the very mathematical basis of the notion of 'like seeks like': it transpires that whatever terms we add in Eq. (4), they will correspond to sums of single-ion terms that depend on the ionic radii r_+ and r_- separately, and to interaction or ion-pair terms that depend on the sum $R = r_+ + r_-$. One can show that for a large class of functional forms of the single-ion and ion-pair energies, the difference between them will exhibit an extremum at a radii difference corresponding roughly to $\delta = r_+ - r_- = -r_0$. For the maximal solubility or for the activity coefficients, the actual position of the extremum, i.e. the matching ion pairs, will shift, but the general trend remains the same. The extremal behaviour of H_{sol} is exemplified in Fig. 1 for alkali-halide salts using the Pauling ion radii, c.f. Table 1, and $r_0 = 0.08\,\text{nm}$.

Table 1. Collection of Pauling radii, r, used for the heat-of-solution data analysis in Fig. 1 and for the surface charge distributions in Fig. 2, taken from Ref. 34. Also shown is σ, the surface charge density derived from the ion radii.

Ion	r (Å)	σ (e/Å^2)	Ion	r (Å)	σ (e/Å^2)	Ion	r (Å)	σ (e/Å^2)
Li^+	0.60	0.041	Mg^{2+}	0.65	0.076	F^-	1.24	−0.052
Na^+	0.95	0.025	Ca^{2+}	0.99	0.050	Cl^-	1.80	−0.025
K^+	1.33	0.018	Sr^{2+}	1.33	0.035	Br^-	1.98	−0.020
Rb^+	1.48	0.015	Ba^{2+}	1.35	0.034	I^-	2.25	−0.016
Cs^+	1.69	0.013						

3. Matching Surface Densities for Complex Ions

In the preceding section, we discussed the heat of solution of alkali-halide salts and found a maximum in H_{sol} for matching radii (modulo a constant offset r_0 that takes care of the increased solvation radii of the alkali ions). Our simple electrostatic model shows that such an extremal behaviour is expected based on single-ion and ion-pair electrostatic contributions to the energy difference between a crystal and the infinitely dilute solution. The straightforward generalisation to the situation of ions of different valencies consists not in comparing the radii but the surface charge densities. In Ref. 28 it has been argued that the water affinity, i.e. the hydration enthalpy of single ions, is an even more appropriate scale on which to compare different ions with each other. For the present purpose we stick to the surface charge density as an appropriate scale of comparing ion-pairing affinities, which is in line with the water-affinity argument since the surface charge density and the heat of hydration of single ions are monotonically related (at least if one excludes too exotic ions).

The generalisation to complex, non-spherically symmetric ions, however, poses a severe problem: a simple ion such as formiate has a neutral part and a negatively charged part, and simply dividing the net charge of the ion by the total surface area is most likely not a very accurate representation of the actual surface-charge density distribution. To clarify the situation, we calculate surface charge density distributions of various ions using the quantum chemistry program Turbomole 5.7.[35] The ionic geometry is optimised employing a density-functional-theory (DFT) calculation, and solvation effects are incorporated using the conductor-like screening model (COSMO).[36] In this model, a solvent accessible surface (SAS) is defined and the outer dielectric medium is approximated as a perfect conductor. The polarisation charge distribution on the SAS is a measure of the inverse surface charge distribution of the molecule under study. We thoroughly tested the influence of different DFT methods, basis sets and SAS discretisation effects.[37] Our results reported are in the limit where they are essentially converged and independent of the methods used. The obtained charge density distributions for all studied ions are shown in Fig. 2.

Note that the charge distributions are normalised such that when one integrates over the SAS the actual ion charge is obtained. Also shown are the surface charge densities of a few spherical ions. Note that for the cations the Pauling radii from Table 1 have been increased by a constant offset $r_0 = 0.08\,\text{nm}$ in accordance with the discussion of the results displayed in Fig. 1. For the convenient discussion of ion pairing, we plot anion and cation surface charge densities with opposite signs, i.e. we invert the charge densities of all anions. First of all, it is seen that for the most important biological anions, i.e. CH_3–COO^- and $H_2PO_4^-$, the charge distribution is broad, ranging from neutral to a maximal density corresponding to a bromide ion. The matching of the $H_2PO_4^-$ ion charge density seems to be equally good with the Na^+-ion and with the K^+-ion, which is in line with the not very different water-solubility of K_2HPO_4 and Na_2HPO_4.[28] For the carboxylate ions the situation is less simple, since we find an essentially bimodal surface charge distribution with pronounced contributions from neutral surface patches and charged surface patches. It has been argued that the carboxyl ion is a hard ion, based on the reversal of the activity coefficients comparing Br^- and CH_3–COO^- ions with the set of alkali ions.[25] The experimental evidence is clear; we simply show here that the hard character of the carboxyl group does not seem to follow naively from its surface charge distribution — most likely a distinct mechanism is at work.

4. Single Ion Thermodynamics from Atomistic Simulations

Our discussion up to this point has been overly simplistic and neglected water structural effects, entropy and dispersion interactions, polarisability effects, etc. Classical atomistic simulations with explicit solvent are a

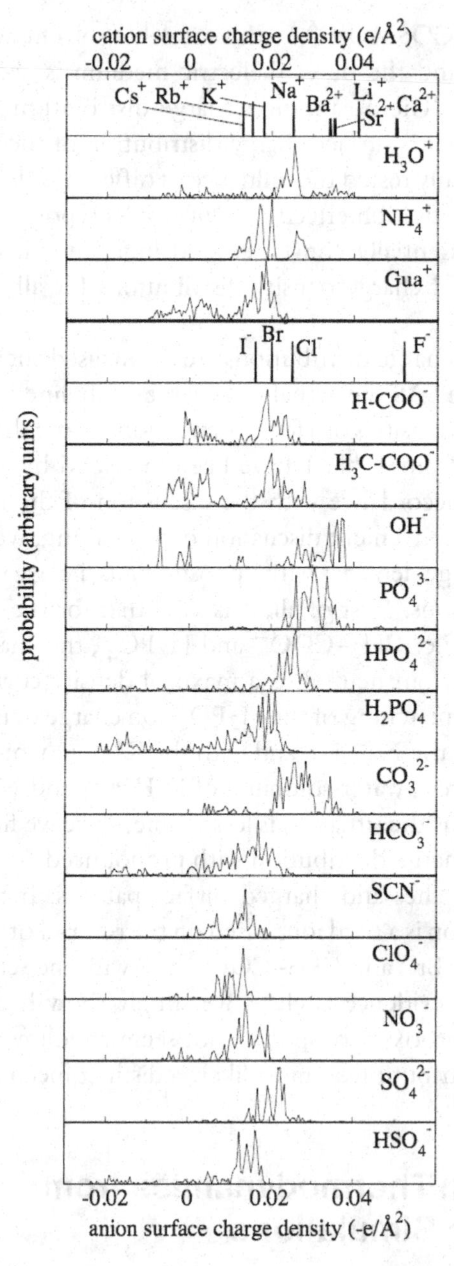

Fig. 2. Surface charge density distributions of various ions. The charge densities of cations and anions are shown with different signs, so that matching surface charge densities are located vertically underneath each other. Note that for the cations the radius offset of $r_0 = 0.08$ nm has been added to the Pauling radii tabulated in Table 1.

powerful approach for the modelling of ion solvation, because no assumptions on the type and strength of solvent-mediated forces have to be made. However, explicit-solvent simulations require the definition of all forces acting between the individual atoms, and reliable results can only be obtained if the underlying force field parameters are realistic. In MD simulations, the pair potentials between two particles i and j, V_{ij}, are modelled as a sum of the Coulomb interaction, which is free of adjustable parameters for mononuclear ions, and a Lennard–Jones (LJ) term, which accounts for van der Waals forces,

$$V_{ij}(r_{ij}) = \frac{1}{4\pi\varepsilon_0}\frac{q_i q_j}{r_{ij}} + 4\varepsilon_{ij}\left[\left(\frac{\sigma_{ij}}{r_{ij}}\right)^{12} - \left(\frac{\sigma_{ij}}{r_{ij}}\right)^{6}\right], \tag{6}$$

where q_i is the charge of atom i, and r_{ij} is the distance between the atoms. The parameter ε_{ij}, the Lennard–Jones interaction strength, and σ_{ij}, the corresponding Lennard–Jones diameter between atoms i and j, are free to be optimised to reproduce certain physical properties in combination with a selected water model and simulation setup. Out of the multitude of physical properties of solvated ions, the free energy is of key importance since it determines density distributions in equilibrium. Many parameter sets for the simulation of solvated ions are available in the literature; Figure 3 shows the ion–water (IW) parameters of a few widely used force fields.[38–41] We also show a new ion parameter set[42] which will be briefly explained further below and which is listed in Table 2. From Fig. 3 it can be seen that parameters for the same ion type differ significantly between different parameterisations. Unless one assumes that a large parameter degeneracy exists, i.e. many different parameter sets describe ionic properties equally well, this variation of parameters for the description of the same ion casts doubt on their performance in ionic simulations. In fact, in Fig. 4 we show results for the solvation free energy and entropy of alkali and halide ions, compared to experimental results.

The systematic differences between theoretical and experimental solvation data in Fig. 4 can be partly traced back to the experimental problem of splitting the free energy and entropy of neutral ion pairs into the single ion values. Single ion solvation free energies rely on an additional extrathermodynamic assumption, which is usually the solvation free energy of the proton, $\Delta G_{\text{solv}}(H^+)$. Unfortunately, there is no single, generally accepted value for $\Delta G_{\text{solv}}(H^+)$ and current estimates vary by more than 50 kJ/mol. In Fig. 4 the dashed lines show baselines that are shifted by

Table 2. Optimized Lennard–Jones parameters σ_{IW} and ε_{IW} for anions and cations obtained from the optimisation of the sums of the solvation energy and entropy of cations and the chloride reference ion, $\Sigma_{\Delta G}$ and $\Sigma_{\Delta S}$, and of the difference between solvation energy and entropy of anions and the chloride reference ion, $\Delta_{\Delta G}$ and $\Delta_{\Delta S}$. The reference parameters for Cl^- are $\sigma_{IW} = 0.378$ nm and $\varepsilon_{IW} = 0.52$ kJ/mol, for which we obtain $\Delta G_{solv} = -306$ kJ/mol and $\Delta S_{solv} = -197$ J/molK. The columns labelled [1] show the simulation results, the columns labelled [2] show the reference solvation free energies and entropies, which are taken from the compilation of Marcus.[34]

Ion	σ_{Iw} (Å)	ε_{Iw} (kJ/mol)	$\Delta_{\Delta G}$ (kJ/mol) [1]	$\Delta_{\Delta G}$ (kJ/mol) [2]	$\Delta_{\Delta S}$ (J/mol K) [1]	$\Delta_{\Delta S}$ (J/mol K) [2]
F^-	3.30	0.55	−124	−125	−86	−62
Cl^-	3.78	0.52	0	0	0	0
Br^-	4.00	0.37	27	26	17	16
I^-	4.25	0.32	65	64	40	39
Ion	**σ_{Iw} (Å)**	**ε_{Iw} (kJ/mol)**	**$\Sigma_{\Delta G}$ (kJ/mol) [1]**	**$\Sigma_{\Delta G}$ (kJ/mol) [2]**	**$\Sigma_{\Delta S}$ (J/mol K) [1]**	**$\Sigma_{\Delta S}$ (J/mol K) [2]**
Li^+	3.02	0.02	−827	−828	−259	−217
Na^+	3.49	0.02	−721	−722	−218	−186
K^+	3.85	0.02	−651	−651	−190	−149
Cs^+	4.17	0.02	−605	−605	−172	−134
Li^+	2.27	1.00	−827	−828	−254	−217
Na^+	2.65	1.00	−721	−722	−219	−186
K^+	2.97	1.00	−651	−651	−209	−149
Cs^+	3.25	1.00	−606	−605	−191	−134
Li^+	2.32	0.6	−829	−828	−268	−217
Na^+	2.70	0.65	−722	−722	−217	−186
K^+	3.03	0.65	−650	−651	−182	−149
Cs^+	3.30	0.65	−605	−605	−160	−134

the values of a reference ion, for which we choose chloride as parameterised by Dang.[44–46] Note that the shift is done with an opposite sign for cations and anions, so that the solvation energies and entropies of ion pairs are not affected. This procedure is equivalent to looking at the sum of the solvation energies/entropies of neutral salt pairs (close to what is done experimentally). Still, in these graphs it is seen that for some of the commonly used ionic force fields, the solvation free energy exhibits

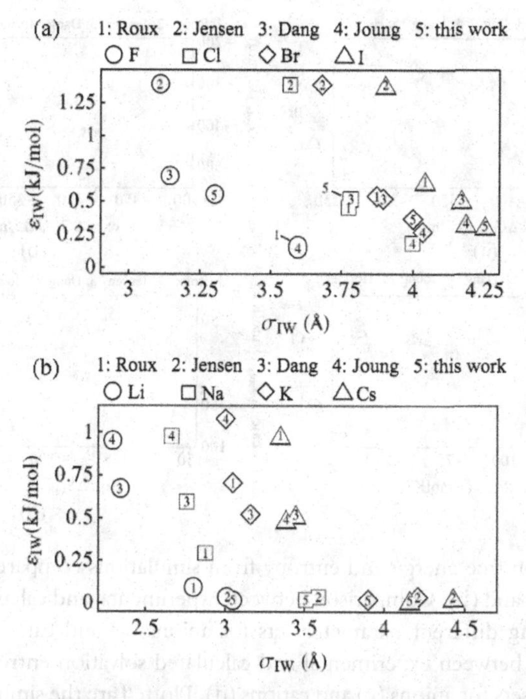

Fig. 3. Ion parameters from literature. (a) anions, (b) cations. Shown are the parameter sets of Roux,[38] Jensen,[39] Dang[44–46] (summarised by Rajamani *et al.*[40]) and Joung.[41] Different symbols represent different ions. The numbers inside every symbol indicate the parameter set to which the parameters belong. Taken from previous work.[42]

unacceptably high deviations from experiment. These deviations are especially dangerous in situations where an ion is equilibrated between solution and a surface-adsorbed state or the interior of a protein in simulations, as is often encountered in biological situations.

In order to unambiguously fix two force field parameters, in principle there is the need to decide on two experimental observables that the simulation should reproduce. In fact, we base our parameter optimisation on the free energy and entropy of hydration, for which reliable experimental data exist.[34] This choice is partly based on the diffuse notion that the solvation free energy and entropy are important for the ion-specific effects encountered in aqueous solution. We also calculate the solvated ionic radius as a consistency check (and find good results for the effective size in the anionic case even without optimising for it). Here we

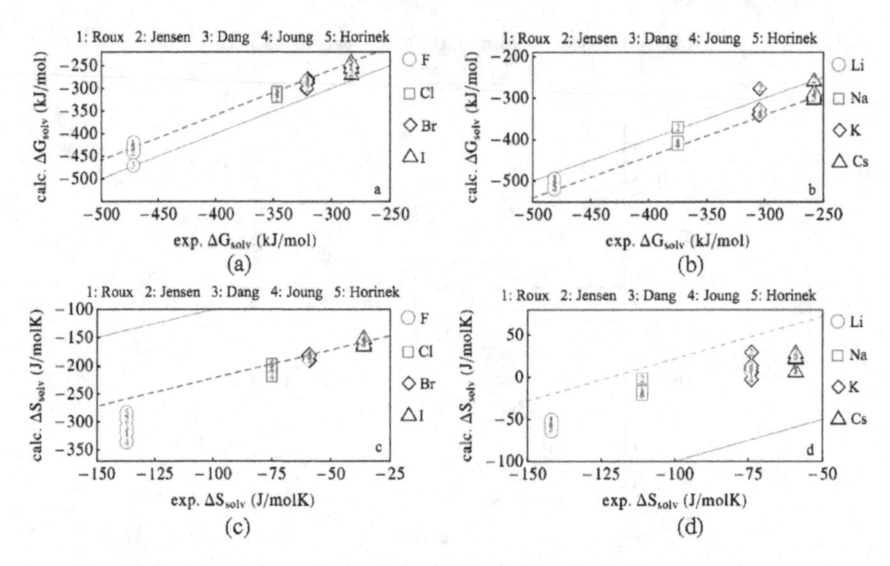

Fig. 4. Solvation free energy and entropy from simulations compared with experimental values. (a) and (b): Comparison between experimental and calculated solvation free energies using different parameter sets for anions (a) and cations (b). (c) and (d): Comparison between experimental and calculated solvation entropies using different parameter sets for anions (c) and cations (d). Plotted are the simulated solvation free energies/entropies versus the experimental solvation free energies/entropies. The experimental solvation free energies and entropies are taken from Marcus,[34] leading to the solid lines. The dashed lines show the experimental single-ion properties shifted in such a way that (i) the shifts of an anion and a cation cancel each other, and (ii) the Dang chloride ion[45] exactly reproduces the experimental data.

describe how to develop force field parameters for halide anions and alkali cations based on SPC/E water.[42] The free energy of solvation is determined by thermodynamic integration of a single ion immersed in a box of water molecules. The entropy of solvation is determined by calculating the internal energy of solvation and subtracting it from the free energy of solvation.[42] Figure 5 shows the free energy of solvation of cations and anions as a function of the LJ radius σ_{IW} for various fixed values of the LJ interaction strength ε_{IW}. It is seen that the free energy is generally negative (favourable) but increases with increasing radius, which means that the electrostatic Born self energy favours a small ion radius. The dependence on the LJ interaction strength is less pronounced, but suggests that the LJ interaction mainly comes in with its repulsive part, so that the solvation free energy goes up as ε increases.

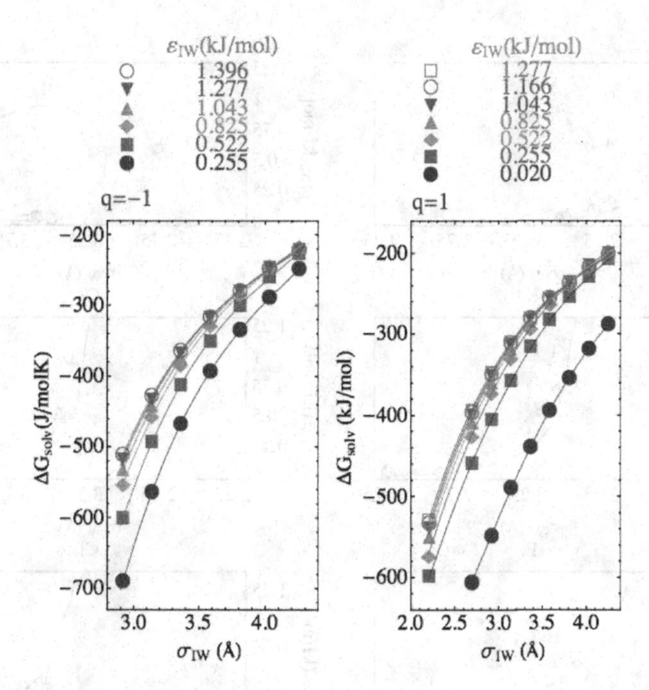

Fig. 5. Solvation free energy for anions (left) and cations (right) as a function of the ion–water Lennard–Jones radius σ_{IW} for various values of the LJ interaction strength ε_{IW}. Taken from previous work.[42]

From many such simulations, we construct the surfaces of solvation free energy, ΔG_{solv}, solvation entropy, ΔS_{solv}, and of the position of the first maximum in the ion–oxygen radial distribution function, R_1, as a function of the Lennard–Jones parameters σ_{IW} and ε_{IW}. From those surfaces, we determine contour lines of constant ΔG_{solv}, ΔS_{solv}, and R_1. In Fig. 6 we show those contour lines for the experimental values of all alkali and halide ions. For the halide ions except fluoride, we find that all three lines more or less cross in one region, which allows to determine an optimal set of force-field parameters for those ions. This thermodynamically optimised parameter set is denoted as '5'. All other force-field sets are those displayed in Figs. 3 and 4. For fluoride, no crossing between the solvation free energy and entropy is observed. Here we choose a point on the line of the experimental solvation free energy that is closest to the line of experimental solvation entropy. For the cations, the situation is dramatically different: no crossing for any ion is observed. We therefore

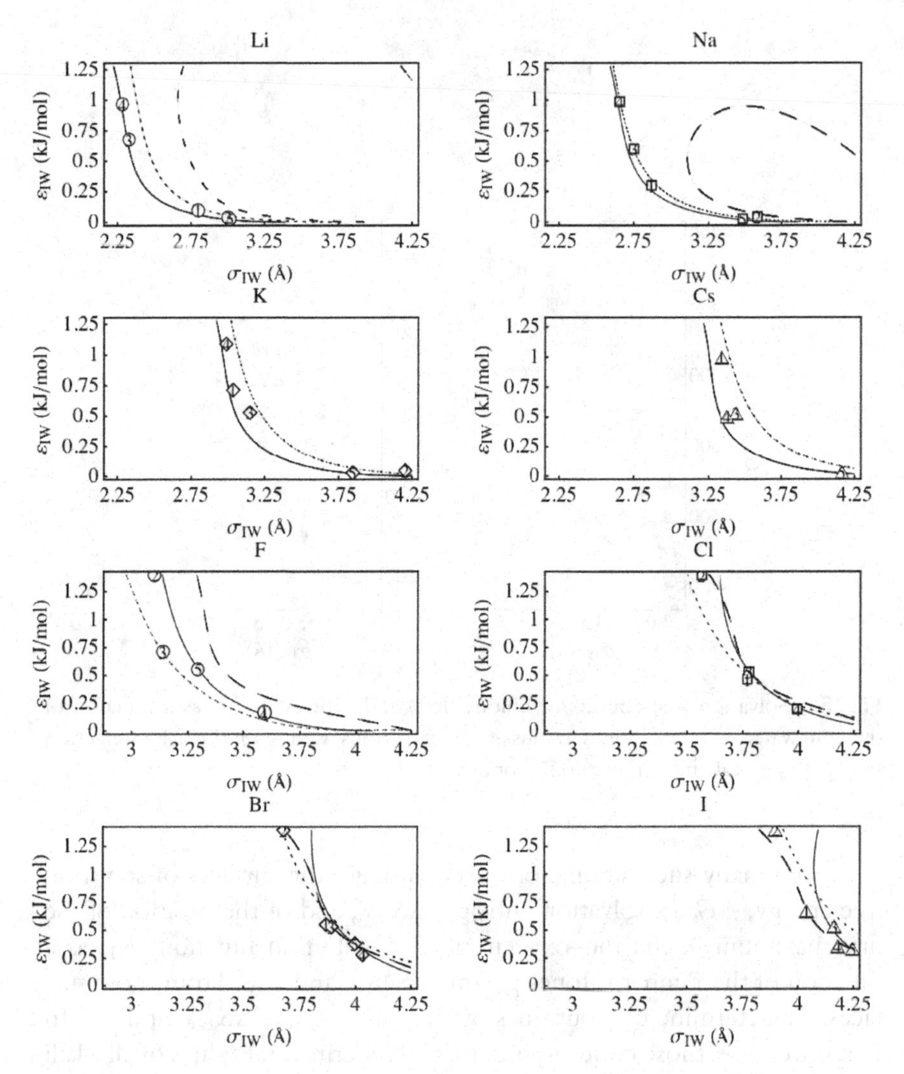

Fig. 6. Contour lines of the experimental solvation free energies (solid lines), solvation entropies (broken lines), and the ionic solvation radius (dotted lines) for the halide and alkali ions. The various force fields are denoted by symbols.

choose points on the solvation free energy line at constant values of the LJ parameter ε_{IW}. We list three of those sets in Table 2. Based on the single-ion properties studied in this section, there is no way of telling which set is superior to the others. As we will discuss in the following section, the ion-pair distribution function, however, discriminates very well between the three cationic parameter sets.

5. Ion-Pair Thermodynamics from Atomistic Simulations

In this section, we discuss the correlations and effective interactions between ions in aqueous solution. For this purpose we typically look at simulation boxes with about 2000 water molecules and add between 1 and 200 salt pairs in the solution. The main output from the simulation is the radial distribution function $g_{ij}(r)$ between the ions. In order to obtain good statistics for further analysis, long simulation runs of about 200 ns are needed. The potential of mean force (PMF) follows by Boltzmann inversion from the radial distribution function according to

$$w_{ij}(r) = -k_B T \ln[g_{ij}(r)], \tag{7}$$

and contains multi-body effects due to water and ionic correlations. For large separations and small salt concentrations ρ, one expects $w_{ij}(r)$ to be accurately represented by the Debye–Hückel limiting law

$$w_{ij}^{\mathrm{DH}}(r) = e^2 z_i z_j \exp[-\kappa(\rho)r]/(4\pi\varepsilon(\rho)r), \tag{8}$$

where $\kappa(\rho)$ is the ion-density dependent screening length and $\varepsilon(\rho)$ is the ion-density-dependent dielectric constant. We suggest the decomposition

$$w_{ij}(r) = w_{ij}^{\mathrm{DH}}(r) + V^{\mathrm{sr}}(r) \tag{9}$$

in order to extract the short-ranged, ion-specific potential $V^{\mathrm{sr}}(r)$. In Fig. 7 we plot the rescaled PMF $\log[rw_{ij}(r)/k_B T]$ which according to Eq. (8) should decay as $\log[rw_{ij}(r)] = -\kappa(\rho)r$ plus a constant for large separations. The slope determines the screening constant $\kappa(\rho)$ and the intercept with the vertical axis allows to extract the ion-density dependent dielectric constant $\varepsilon(\rho)$. We thus can extract the short-ranged potential $V^{\mathrm{sr}}(r)$ from the full PMF $w_{ij}(r)$ via the definition in Eq. (9). The inset in Fig. 7 demonstrates that $V^{\mathrm{sr}}(r)$ is largely concentration independent and thus corresponds to the true interionic short-ranged part of the pair potential, and can be estimated even from simulations at rather elevated concentrations up to $\rho = 0.5$ mol/l, in which case statistics are much improved.

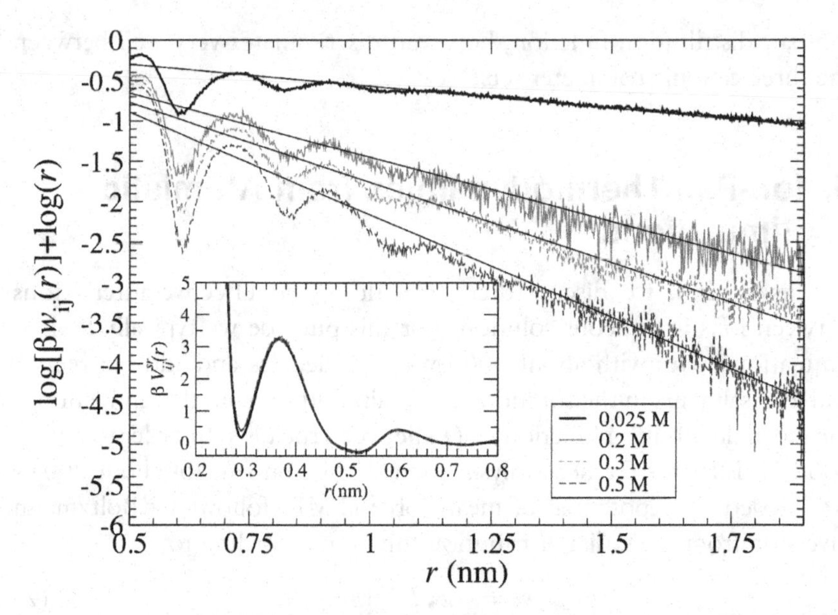

Fig. 7. Rescaled effective potential $w_{ij}(r)$ between Na^+ and Cl^- for different salt concentrations ρ (see legend). From the fit to a linear function (solid lines), the inverse screening length $\kappa(\rho)$ and concentration dependent dielectric constant $\varepsilon(\rho)$ is extracted. In the inset we show the short-ranged potential $V_{ij}^{sr}(r)$ as defined via Eq. (9) which is independent of density within the considered density range. Taken from previous work.[11]

Figure 8 shows anionic–cationic short-ranged potentials for a few different salts. To repeat: those short-ranged potentials include all interactions except the Debye–Hückel interaction which has been separated out. As such, the potentials mainly reflect the distance-dependent mutual perturbation of the ionic hydration shells. All potentials show a steep increase at contact, while the contact distance increases with increasing bare ion size, i.e. for the chloride salts the contact distance increases in the order LiCl < NaCl < KCl < CsCl as expected. All potentials show two local minima, one minimum at contact and a second minimum corresponding to a solvent-separated ion pair. For some potentials even a third local minimum is discernable, which corresponds to two ions, each surrounded by a complete hydration shell. In line with the law of matching surface charge densities mentioned in the previous section, the size-asymmetric ion pairs LiCl and NaCl show the most favourable interaction at the second, solvent-separated minimum since the cations resist stripping off their hydration

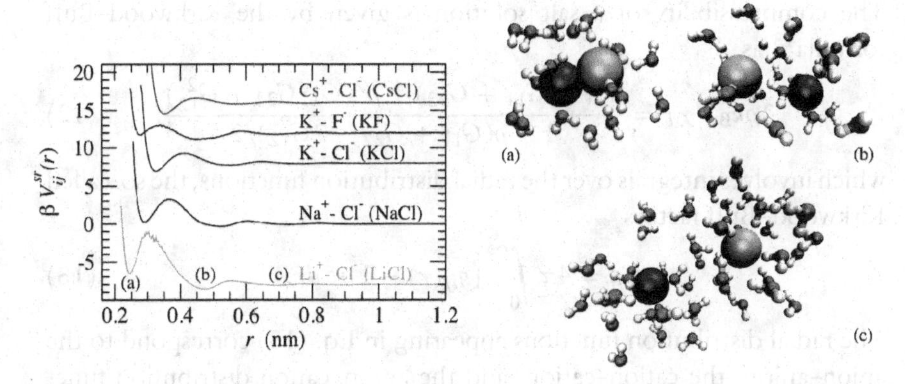

Fig. 8. Left: Short-ranged pair potentials $V_{ij}^{\mathrm{sr}}(r)$ in units of thermal energy $k_{\mathrm{B}}T = 1/\beta$ for a few different ion pairs. The curves are shifted vertically for better visibility. Right: Simulation snapshots for hydrated Li^{+} (blue sphere) and Cl^{-} (green sphere) pairs corresponding to the (a) first, (b) second, and (c) third minimum in the short-ranged part of the PMF.

shells. For the slightly larger potassium ion, the situation is reversed and the primary minimum corresponding to an ion pair in direct contact is preferred (as for the caesium ion).

This picture suggests that the more size-symmetric ion pairs such as KCl or NaCl should exhibit stronger attraction in solution than the size-asymmetric salt LiCl. This local argument should also have a bearing on integral thermodynamic properties such as osmotic coefficients, activity coefficients, maximal solubilities or heats of solution (compare Fig. 1). Most of these properties are difficult to obtain from simulations. The osmotic coefficient ϕ is comparably straightforward to calculate. It is related to the osmotic pressure Π via

$$\phi(\rho) = \beta\Pi/(2\rho), \tag{10}$$

and measures the deviation from the osmotic pressure of an ideal solute. Values of ϕ larger than unity reflect effective repulsion between ions; values smaller than unity signal attraction between ions. The osmotic coefficient ϕ is related to the isothermal compressibility χ_T via Eq. (10) and

$$\chi_T^{-1} = \rho\frac{\partial\Pi}{\partial\rho}. \tag{11}$$

The compressibility of a salt solution is given by the Kirkwood–Buff formula[43] as

$$2\rho k_B T \chi_T = \frac{1 + \rho(G_{11} + G_{22}) + \rho^2(G_{11}G_{22} + G_{12}^2)}{1 + \rho(G_{11} + G_{22} - 2G_{12})/2}, \qquad (12)$$

which involves integrals over the radial distribution functions, the so-called Kirkwood–Buff factors

$$G_{ij} = 4\pi \int_0^\infty [g_{ij}(r) - 1] r^2 \, dr. \qquad (13)$$

The radial distribution functions appearing in Eq. (13) correspond to the anion–anion, the cation–cation, and the anion–cation distribution functions which are implicitly dependent on the salt density. In practice, the compressibility is obtained for a few concentrations, augmented by the exactly known low density behaviour, and then numerically integrated according to Eq. (11) to obtain the osmotic coefficient.

Figure 9 shows the osmotic coefficients obtained from simulations (data points) for the same salt types for which the short-ranged potentials $V^{sr}(r)$ are displayed in Fig. 8. The trend observed in $V^{sr}(r)$ is fully reflected by the osmotic coefficients: the strongly hydrated Lithium ion, which only shows weak contact-pair binding but pronounced solvent-mediated ion binding, has the highest osmotic coefficient of all ions considered. The larger the cation gets, the smaller the osmotic coefficient becomes, indicating less osmotic pressure and thus more attraction in solution. The roles of potassium and caesium are interchanged if one compares our simulations with the experiments, otherwise the agreement between experiments and simulations is very good for the chloride salts. All ion parameters used correspond to the Dang parameters[44–46] (as summarised by Rajamani *et al.*[40]) which are reproduced in Table 3 and are quite similar to the optimised force fields discussed in the previous section.

For the KF salt, marked deviations between experimental and simulated osmotic coefficients are visible. We noted in the previous section that for the cations some flexibility in the force field parameters exists, as the solvation free energy of a single cation only fixes one parameter. In fact, one could in principle choose any force-field set on the line of constant solvation free energy in Fig. 6. So the question is whether one can fix the deficiency of the osmotic coefficient for KF by adjusting the force field while keeping the solvation free energy of K constant. In Fig. 10 we show a few radial distribution functions for fixed fluoride force field and four

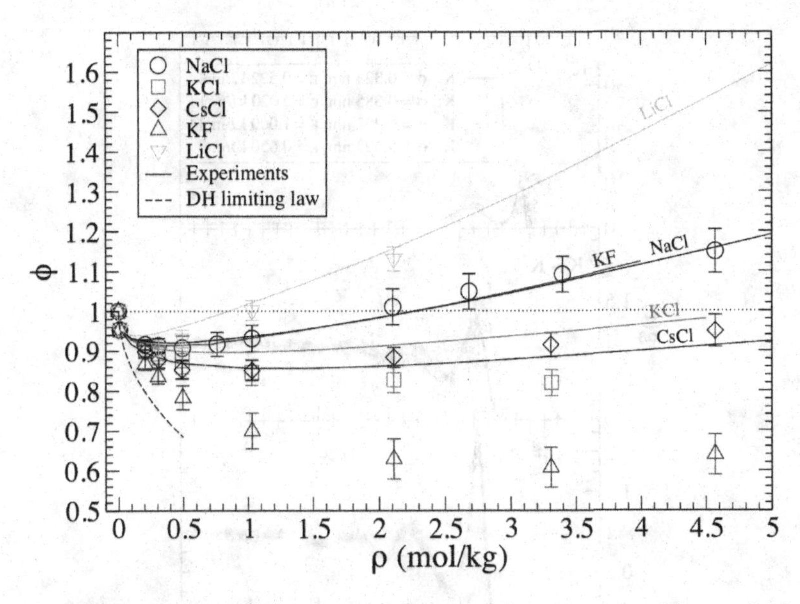

Fig. 9. The osmotic coefficient ϕ as a function of the salt concentration ρ obtained via a density integral over the compressibility according to Eqs. (10–13) for a few different salts compared to experimental values. The DH limiting law is also shown as a black dashed line. Taken from previous work.[11]

Table 3. Dang parameters[40,44–46] σ_{IW} and $\varepsilon_{\mathrm{IW}}$ used for the ions in the osmotic coefficient calculations.

Ion	σ_{IW} (nm)	$\varepsilon_{\mathrm{IW}}$ (kJ/mol)	Charge q/e	Reference
Li^+	0.2337	0.6700	+1	(44)
Na^+	0.2876	0.5216	+1	(45)
K^+	0.3250	0.5216	+1	(45)
Cs^+	0.3526	0.5216	+1	(45)
Cl^-	0.3785	0.5216	−1	(45)
F^-	0.3143	0.6999	−1	(46)
SPC/E				
O	0.3166	0.6500	−0.8476	
H	—	—	+0.4238	

different potassium force fields. Three of those K force fields are the optimised force fields shown in Table 2, while the remaining force field is from Dang[40,44–46] as displayed in Table 3. The main message of Fig. 10 is that indeed the potassium-fluoride distribution function depends sensitively on

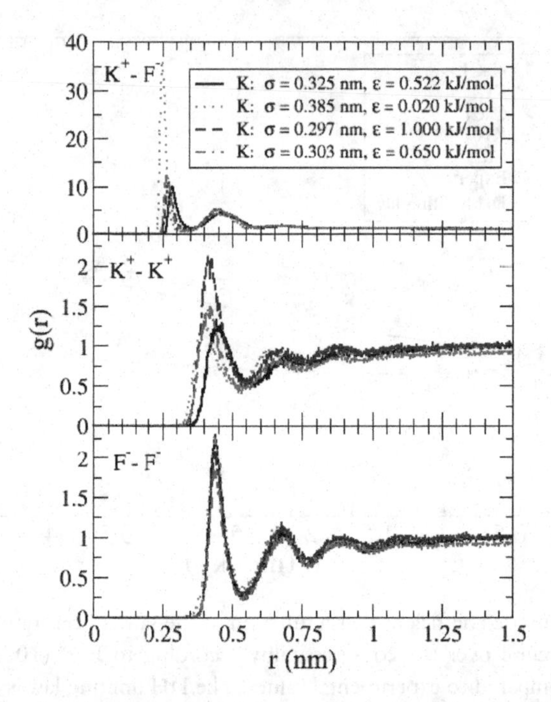

Fig. 10. Radial distribution functions for a KF solution of concentration 0.3 mol/l for fixed fluoride force field as given in Table 2 and for 4 different K force fields, namely the three optimised force fields as given in Table 2 and the Dang force field[40,44–46] given in Table 3 and used for the osmotic coefficient results shown in Fig. 9. Pronounced deviations in the K-F distribution function are observed.

the force field variation, while single-ion properties [such as solvation free energy but also the potassium-water distribution function (data not shown here)] are invariant for all force fields considered in Fig. 10. This means, maybe not surprisingly, that the single-ion solvation properties in a simulation do not fully specify the ion-pair properties. Turning this around, it means that good ionic force field should be obtainable by optimising both single-ion *and* ion-pairing properties.[11]

6. Interface Ion Thermodynamics from Simulations

So far we have discussed ion-specificity in the bulk. Hofmeister effects are most often concerned with interfacial behaviour, to which we now turn.

The classical Hofmeister effect describes the interaction between two proteins as a function of salt type and salt concentration.[26,27] Let us for the sake of simplicity assume that the two interacting proteins are identical so that we are dealing essentially with the interaction between two similar peptidic surfaces. To first approximation the interaction can be separated into contributions arising from electrostatic and depletion/adsorption effects. To understand this statement, let us confine ourselves to neutral surfaces. If the surface repels cations and anions with comparable strength, the interfacial tension will go up and consequently a rather short-ranged attraction will be felt as the surfaces get in close contact. This is commonly denoted as depletion-induced attraction.[47] Conversely, if the surface attracts cations and anions with similar strength, the interfacial tension goes down and a short-ranged repulsion between the surfaces will result; this corresponds to adsorption-induced repulsion. If, on the other hand, a surface interacts differently with anions and cations, in addition to the depletion/adsorption interaction, an effective surface charge builds up, which will lead to quite long-ranged electrostatic repulsion between similar surfaces. All these effects can be quite conveniently classified by looking at the interfacial tension and the surface potential of a single surface, which will be the subject of this section (interactions between surfaces are discussed also in the contributions by Boström[48] and Jungwirth[30]).

Let us consider the simplest surface that shows ion-specific adsorption, namely the water–air interface. In a by now classical series of papers, Jungwirth and co-workers have shown that iodide ions do adsorb at the air–water interface, in strong contrast with the traditional view.[3,29] Those simulations were performed with polarisable force fields, while the non-polarisable force fields employed at that time did not show adsorption of iodide. It was concluded that the polarisability plays a dominant role in the adsorption mechanism. Let us reconsider that problem using our novel thermodynamically optimised force fields discussed in the earlier section. We show results for the potential of mean force of a single ion at an air–water interface, calculated using umbrella sampling[49] and the WHAM method.[50]

In Fig. 11 we first show the PMFs for akali ions for the three different force-fields listed in Table 2. Although there are distinct differences between the three different figures, the trends are the same: lithium is the least strongly repelled ion; the other ions show roughly the same repulsion. In Fig. 12 we show the PMFs for the halide ions. There the

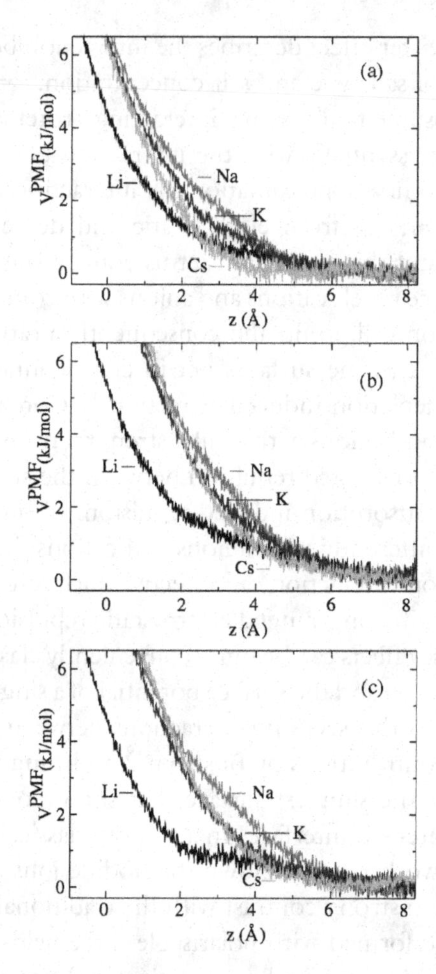

Fig. 11. Potentials of mean force for alkali ions close to the air–water interface. The simulations are performed for the three optimised ion parameter sets shown in Table 2 for (a) $\varepsilon_{IW} = 0.02$ kJ/mol, (b) $\varepsilon_{IW} = 1.00$ kJ/mol, and (c) $\varepsilon_{IW} = 0.65$ kJ/mol. All cation potentials of mean force are repulsive. For all three parameter sets, the repulsion of Li^+, the smallest cation, is significantly weaker than that of the other cations. For sodium, the second smallest cation, the strongest repulsion from the air–water interface is observed.

result is very different: the largest ion iodide is attracted to the interface, while the smaller ions are repelled. The force fields used are the non-polarisable force-fields shown in Table 2. The trend is very similar to the PMFs obtained by Dang using polarisable force fields[2] and to the

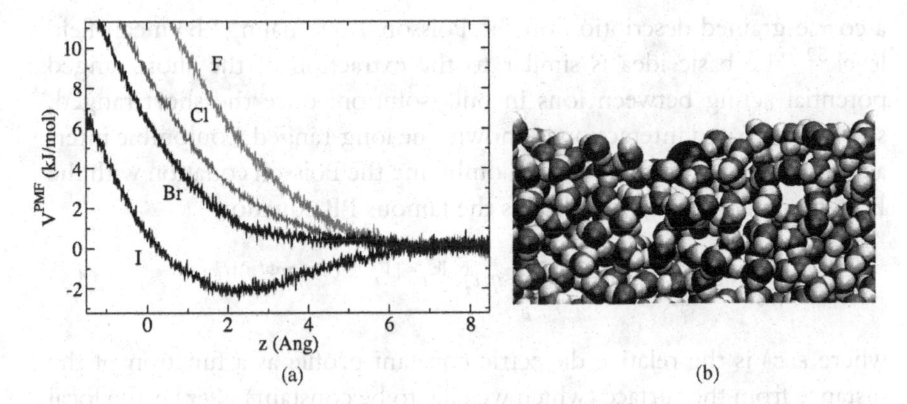

Fig. 12. (a) Potentials of mean force for halide ions at the air–water interface using the non-polarisable optimised ion parameter set shown in Table 2. There is a clear ordering of the potentials of mean force: the largest anion, iodide, is attracted to the interface; the smaller anions are increasingly repelled from the interface with decreasing ion size. (b) Snapshot of an iodide ion at the air–water interface.

density distributions obtained by Jungwirth and co-workers,[3] though in the present case the polarisability is not included but the remaining two parameters in the force field are systematically optimised based on experimental thermodynamic single-ion data. Though it is difficult to draw definite conclusions from the results shown here, we can safely state that the non-polarisable force fields for iodide used in past simulations studies that did not show interface affinity simply had too small a radius. The larger the ion, the stronger the hydrophobic effect which pushes the ion to the interface.[51] One can produce an independent argument in favour of the proposition that polarisability is not the main factor for driving ions to the interface: as has been argued a while ago, the excess static polarisability of iodide is negative, meaning that a chunk of water of the same volume as an iodide ion has a higher polarisability than iodide.[52] So it seems unlikely that it would be the polarisability of iodide that drives it into a high-electric field region like the air–water interface.

How can the putative surface affinity of certain ions be tested experimentally? As mentioned in the Introduction, direct measurements of surface ion concentrations are difficult to perform. Integrated knowledge is available from surface tension data, for which, on the other hand, simulations are cumbersome. We employ here a multi-scale modelling approach, where the PMFs obtained from MD simulations are fed into a coarse-grained description on the Poisson–Boltzmann (PB) mean-field level.[52] The basic idea is similar to the extraction of the short-ranged potential acting between ions in bulk solution: once the short-ranged, solvent-mediated interaction is known, the long-ranged Coulombic interactions can be separately added. Combining the Poisson equation with the Boltzmann equation, one obtains the famous PB equation

$$\varepsilon_0 \frac{d}{dz}\varepsilon(z)\frac{d}{dz}\Phi(z) = -\sum_i q_i c_i^{\mathrm{bulk}} e^{-[V_i^{\mathrm{PMF}}(z)+q_i\Phi(z)]/k_B T}, \tag{14}$$

where $\varepsilon(z)$ is the relative dielectric constant profile as a function of the distance from the surface (which we take to be constant), $\Phi(z)$ is the local electrostatic potential profile, q_i is the charge of the ith ion, c_i^{bulk} is its bulk concentration, and V_i^{PMF} is the PMF of the ith ion. The concentration profile of the ith ion follows as

$$c_i(z) = c_i^{\mathrm{bulk}} e^{-[V_i^{\mathrm{PMF}}(z)+q_i\Phi(z)]/k_B T}. \tag{15}$$

The surface excess of ionic species i is defined as

$$\Gamma_i = \int_{-\infty}^{z_{\mathrm{GDS}}} c_i(z)dz + \int_{z_{\mathrm{GDS}}}^{\infty} [c_i(z) - c_i^{\mathrm{bulk}}]dz, \tag{16}$$

where the Gibbs-dividing-surface position z_{GDS} is defined by the requirement that the surface excess of water itself vanishes. Finally, the Gibbs equation relates the surface tension change $\Delta\gamma$ due to added solute to the concentration integral over the surface excess,

$$\Delta\gamma = -k_B T \int_0^{c^{\mathrm{bulk}}} \sum_i \Gamma_i(c)/c\, dc. \tag{17}$$

In Fig. 13 we show the surface tension increment of a one molar solution of sodium halide, based on two different force fields for the sodium ion (open symbols) and compared to the experimental data (filled circles). The experimental trend is nicely reproduced: the larger the anion, the smaller the surface tension increment (except for the smallest ions, fluoride, for which the experimental surface tension is smaller than for chloride). The difference between the two sodium force fields is small, meaning that the force fields given in Table 2 are quite robust. Interestingly, the adsorption strength of iodide seems to be overestimated by the simulations, meaning that the surface affinity of iodide is quite a small effect.

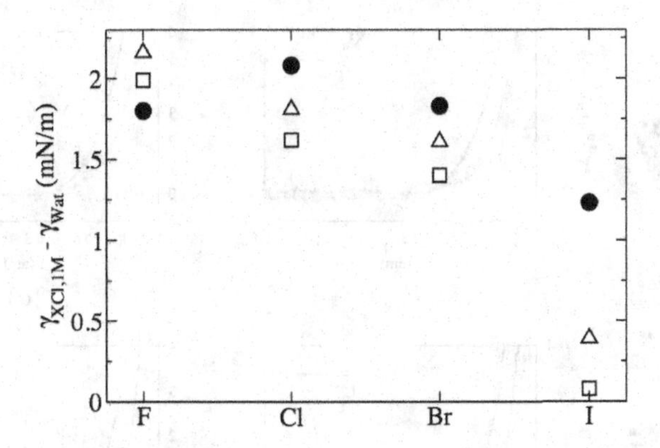

Fig. 13. The difference in surface tension $\Delta\gamma$ of a 1-M sodium halide solution and pure water. The filled circles show experimental data from Ref. 54, the open symbols show our predictions obtained with the potentials of mean force of sodium and the halides. The sodium ion is described by the $\varepsilon_{IW} = 0.02\,\text{kJ/mol}$ parameters (squares) or the $\varepsilon_{IW} = 0.65\,\text{kJ/mol}$ parameters (triangles). The qualitative trend in the experimental data is reproduced; the surface tension increase becomes smaller for the series Cl–Br–I.

We now turn to more biologically relevant surfaces, namely soft surfaces made of grafted and tightly packed polymer chains, so-called self-assembled monolayers (SAMs).[55–57] In Figs. 14(b) to 14(c) we show ionic PMFs on a hydrophobic SAM, where the end group is a methyl group. The obtained PMF structure is similar to the air–water interface, i.e. the large iodide ion is attracted to the surface, while the smaller halide ions and the cation (in this case Na^+) are repelled from the surface. Figure 14(c) shows the fitted PMFs that eventually are employed in the PB calculations. Experimentally, it is known that Hofmeister effects on hydrophilic surfaces are reversed with respect to hydrophobic surfaces.[58] Indeed, the results in Figs. 14(d) to 14(f) show that the affinity of the halide ions on a hydrophilic surface is opposite to the hydrophobic surface: the small fluoride ion is least repelled from the surface, while the larger iodide is more repelled. This salient behaviour can be best understood by considering the perturbation of the hydrogen-bonding network as ions go to the different surfaces.

Figure 15 shows snapshots of the water hydration layer around the different ions considered in Fig. 14 at a distance $z = 0.575\,\text{nm}$ from the hydrophobic surface [which corresponds to the distance of strongest attraction of iodide to the surface, see Fig. 14(b)]. One sees that at this

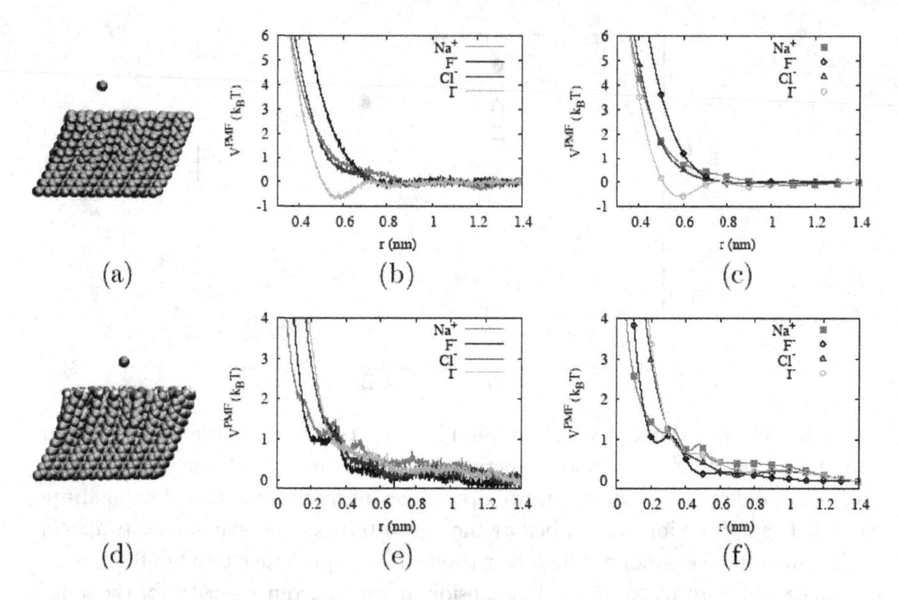

Fig. 14. (a) Simulation snapshots of one single ion at the CH_3-SAM/water interface. Potential of mean force for various ions at the CH_3-terminated SAM from (b) simulations and (c) heuristic fit function. (d) Simulation snapshots of one single ion at the OH-SAM/water interface. Potential of mean force for various ions at the OH-terminated SAM from (e) simulations and (f) heuristic fit function.

distance iodide in Fig. 15(d) has completely stripped off its hydration layer at the side facing the surface, while the other ions at that distance still have an almost intact hydration layer of water molecules. This suggests that the iodide adsorption to the surface is due to some type of hydrophobic attraction.

Figure 16 shows snapshots of the water hydration layer around a fluoride ion at varying distances from the hydrophilic surface. The local minimum in the fluoride PMF in Fig. 14(e) at small distances is shown to be caused by direct hydrogen bonds between the small fluoride ion and the hydroxyl surface groups [see snapshots in Figs. 16(a) and 16(b)]. At an intermediate distance in Fig. 16(c), water intrudes into the cleft between the ion and the surface, but the hydrogen-bond network is frustrated and pushes the ion away from the surface [corresponding to the distance range just to the right of the local maximum at $z = 0.3$ nm in Fig. 14(e)]. Finally, for the large distance shown in Fig. 16(d) the water is quite happy in the slab between ion and surface and the force acting on the ion is

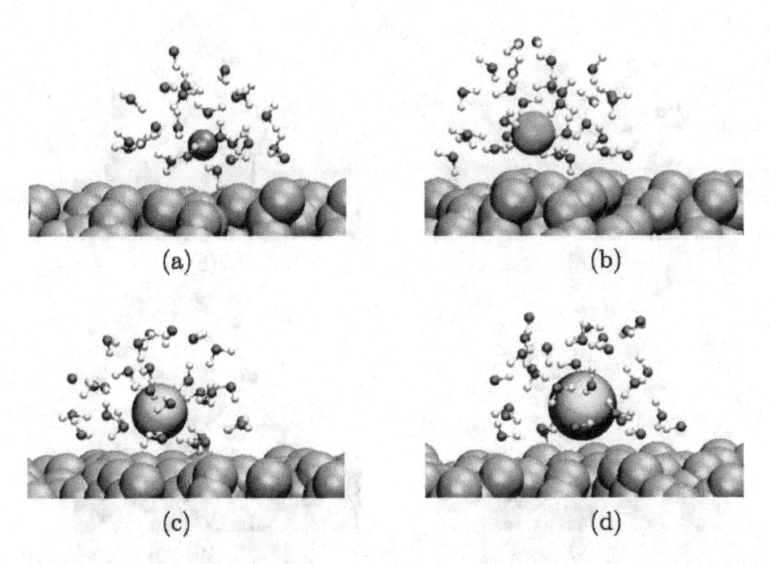

Fig. 15. Simulation snapshots of (a) Na^+, (b) F^-, (c) Cl^-, (d) I^- at the CH_3-SAM/water interface for a distance $z = 0.575$ nm, at which the PMF of I^- exhibits a minimum. All water molecules within 0.6 nm of the ion are shown. The sizes of the ions correspond to their Pauling radii from Table 1.

essentially zero [flat range of the PMF in Fig. 14(e) for distances larger than $z = 0.45$ nm]. In conclusion, the possibility of an ion to form hydrogen bonds with surface polar groups depends in the first place on the ion size and favours small ions (like fluoride) over large ions (bromide and iodide). This fact explains the inversion of the Hofmeister series displayed in Fig. 14 as one goes from hydrophobic to hydrophilic surfaces.

Finally, in Fig. 17 we show the surface potentials (upper panel) and surface tension (lower panel) for the various sodium halides at the hydrophobic (left graphs) and the hydrophilic (right graphs) surfaces. For all surfaces the surface tension is positive, meaning that the creation of the interface costs free energy such that there will be a short-ranged attractive contribution to the total interaction when two such surfaces come in contact (the range of that attraction will be of the order of the range of the PMF, i.e. a few Angstroms). The interfacial tension exhibits reversal of the Hofmeister series: while for the hydrophobic surfaces the large anions show the smallest tension [Fig. 17(c)], the situation is reversed on the hydrophilic surface [Fig. 17(d)]. The potential has more structure: on the hydrophobic surface, sodium and fluoride have roughly similar PMFs, meaning that

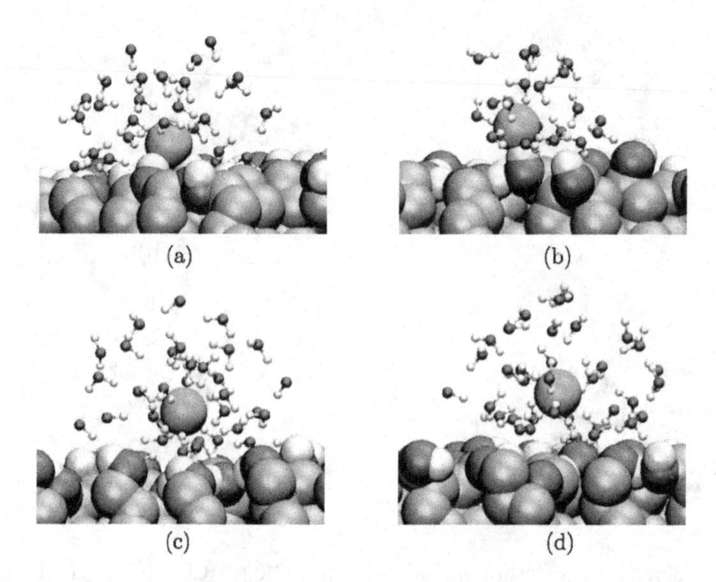

Fig. 16. Simulation snapshots of F^- at the OH-SAM/water interface for different distances z from the surface. At the distance (a) $z = 0.26$ nm and (b) $z = 0.29$ nm, roughly at the location where the PMF in Fig. 14(e) shows a minimum, H-bonds form between the ion and the surface hydroxyl groups. At the distance (c) $z = 0.38$ nm water molecules intrude between the ion and the surface and push the ion away from the interface; this is approximately the distance where the PMF shows a maximum. (d) At larger distance $z = 0.44$ nm from the surface, the PMF essentially is flat and the force acting on the ion is zero. Water molecules within 0.6 nm of the ion are shown. The size of the ion corresponds to its Pauling radius.

the surface potential for NaF is very small and the surface appears neutral, see Fig. 17(a). NaI on the other hand has a large negative potential, two hydrophobic surfaces in a NaI solution will therefore electrostatically repel for large distances. The situation is reversed on the hydrophilic surface [Fig. 17(b)] here sodium and iodide have roughly equal PMFs so that the net effect on the surface potential cancels. As one comes to fluoride, the potential is again negative due to the stronger adsorption of F^-. This should lead to repulsion and thus to electrostatic stabilisation of hydrophilic surfaces in NaF solutions. Some of the ion-binding effects on solid surfaces have been experimentally observed by different techniques such as electro-osmosis,[59] thin-film pressure-balance,[60] single-molecule desorption studies,[61] capacitance studies,[62] and surface-sensitive field-effect setups.[63] Others still await disclosure.

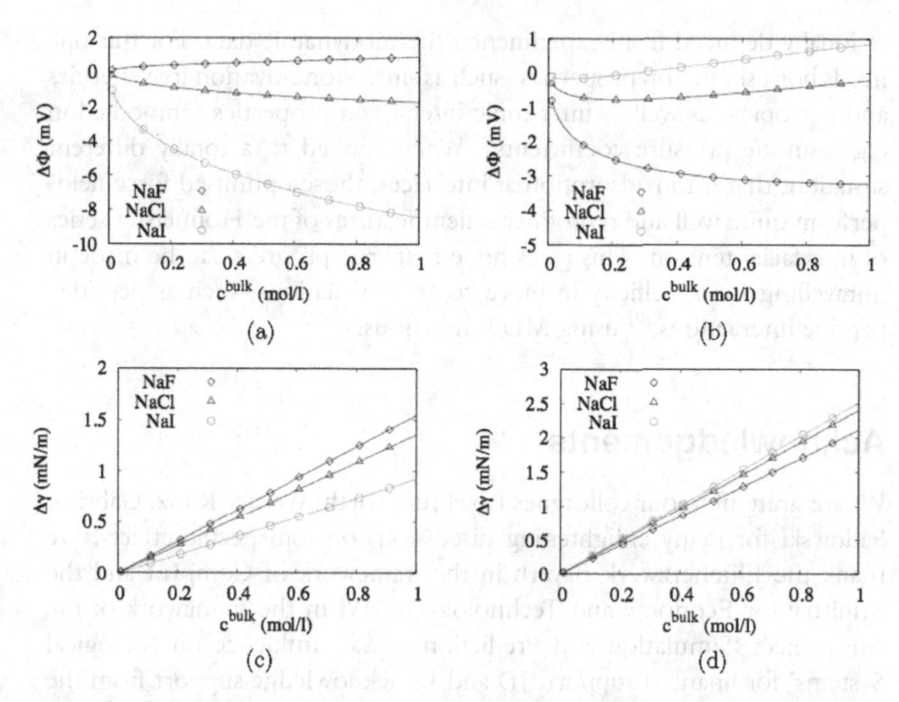

Fig. 17. Surface Potential $\Delta\phi$ and surface tension change $\Delta\gamma$ as a function of the bulk salt concentration c^{bulk} for various sodium halides at the [(a) & (c)] hydrophobic and [(b) & (d)] hydrophilic SAM-water interface.

7. Conclusion

This is not a finished story, but rather a preliminary report of work in progress. The problem of describing ion-specific effects within the framework of explicit-solvent, atomistic and classical molecular dynamics simulations is an ongoing endeavour that keeps evolving in close dialogue with new experimental results and systems. At the end of that journey, with faster computers, maybe optimal classical force fields will have emerged that closely reproduce all experimental ion-specific effects? Maybe one soon hits a limit where it must be realised that ion-specific effects cannot be cast into a classical description based on pair-wise Lennard–Jones potentials between all constituent atoms? Probably a mix of both scenarios will be realised.

What we tried to explain in this chapter is how the very fundamental of any MD simulation, the ionic force field, can at least in principle be rationally deduced from experimental thermodynamic data. For this one needs both single-ion properties (such as single-ion solvation free energies and entropies) as well as inter-ionic interaction properties (embodied in e.g. osmotic pressure coefficients). When applied to a totally different situation, that is ion adsorption at interfaces, those optimised force fields perform quite well and reproduce salient features of the Hofmeister series of interfacial tension. This gives hope that true progress can be made in unravelling ion specificity in more complex situations, such as peptide–peptide interactions,[30] using MD simulations.

Acknowledgements

We are grateful to our colleagues Pavel Jungwirth, Werner Kunz, Gabriele Sadowski for many enlightening discussions on ion-specific effects. We thank the Elitenetzwerk Bayern in the framework of CompInt and the Ministry for Economy and Technology BMWi in the framework of the AiF project 'Simulation and Prediction of Salt Influence on Biological Systems' for financial support. JD and IK acknowledge support from the Deutsche Forschungsgemeinschaft (DFG) within the Emmy-Noether-Program and computational resources from the cluster HLRB II in the Leibniz–Rechenzentrum (LRZ) Munich. MF acknowledges support by the Cluster of Excellence in Munich.

References

1. Gnanakaran S, Nymeyer H, Portman J, Sanbonmatsu KY, García AE. (2003) Peptide folding simulations. *Curr Opin Struct Biol* **13**: 168–174.
2. Chang TM, Dang LX. (2006) Recent advances in molecular simulations of ion solvation at liquid interfaces. *Chem Rev* **106**: 1305–1322.
3. Jungwirth P, Tobias DJ. (2006) Specific ion effects at the air/water interface. *Chem Rev* **106**: 1259–1281.
4. Hille B. (2001) *Ion Channels of Excitable Membranes*, 3rd ed., Sinauer Associates, Sunderland, MA.
5. Weissenborn PK, Pugh RJ. (1996) Surface tension of aqueous solutions of electrolytes: Relationship with ion hydration, oxygen solubility, and bubble coalescence. *J Coll Int Sci* **184**: 550–563.
6. Jungwirth P, Tobias DJ. (2001) Molecular structure of salt solutions: A new view of the interface with implications for heterogeneous atmospheric chemistry. *J Phys Chem B* **105**: 10468–10472.

7. Kunz W, Belloni L, Bernard O, Ninham BW. (2004) Osmotic coefficients and surface tensions of aqueous electrolyte solutions: Role of dispersion forces. *J Phys Chem B* **108**: 2398–2404.
8. Boström M, Williams DRM, Ninham BW. (2001) Surface tension of electrolytes: Specific ion effects explained by dispersion forces. *Langmuir* **17**: 4475–4478.
9. Horinek D, Netz RR. (2007) Specific ion adsorption at hydrophobic solid surfaces. *Phys Rev Lett* **99**: 2261041–2261044.
10. Hess B, Holm C. (2006) Modeling multibody effects in ionic solutions with a concentration dependent dielectric permittivity. *Phys Rev Lett* **96**: 1478011–1478014.
11. Kalcher I, Dzubiella J. (2009) Structure–thermodynamics relation of electrolyte solutions. *J Chem Phys* **130**: 134507.
12. Lyubartsev AP, Marcelja S. (2002) Evaluation of effective ion–ion potentials in aqueous electrolytes. *Phys Rev E* **65**: 0412021–0412026.
13. Dos Santos DJVA, Müller-Plathe F, Weiss VC. (2008) Consistency of ion adsorption and excess surface tension in molecular dynamics simulations of aqueous salt solutions. *J Phys Chem C* **112**: 19431–19442.
14. Harned HS, Owen BB. (1943) *The Physical Chemistry of Electrolytic Solutions.* Reinhold publishing corporation, New York.
15. Robinson RA, Stokes RH. (2003) *Electrolyte Solutions*, 2nd ed., Dover Publications Inc.
16. Soper AK, Weckström K. (2006) Ion solvation and water structure in potassium halide aqueous solutions. *Biophys Chem* **124**: 180–191.
17. Padmanabhan V, Daillant J, Belloni L, Mora S, Alba M, Konavalov O. (2007) Specific ion adsorption and short-range interactions at the air aqueous solution interface. *Phys Rev Lett* **99**: 0861051–0861054.
18. Petersen PB, Saykally RJ. (2006) On the nature of ions at the liquid water surface. *J Ann Rev Phys Chem* **57**: 333–364.
19. Petersen PB, Saykally RJ. (2005) Evidence for an enhanced hydronium concentration at the liquid water surface. *J Phys Chem B* **109**: 7976–7980.
20. Motschmann H, Koelsch P. (2010) Linear and non-linear optical techniques to probe ion profiles at the air–water interface. In: Kunz W (ed.), *Specific Ion Effects*, Chap. 4. World Scientific Publishing, Singapore.
21. Beattie JK. (2008) Comment on autoionization at the surface of neat water: Is the top layer pH neutral, basic, or acidic? *Phys Chem Chem Phys* **10**: 330–331.
22. Vacha R, Buch V, Milet A, Devlin JP, Jungwirth P. (2008) Response to comment on autoionization at the surface of neat water: Is the top layer pH neutral, basic, or acidic? *Phys Chem Chem Phys* **10**: 332–333.
23. Collins KD. (1997) Charge density-dependent strength of hydration and biological structure. *Biophys J* **72**: 65–76.
24. Lyklema J. (2003) Lyotropic sequences in colloid stability revisited. *Adv Coll Interf Sci* **100**: 1–12.
25. Kunz W, Neueder R. (2010) An attempt of a general overview. In: Kunz W (ed.), *Specific Ion Effects*, Chap. 1. World Scientific Publishing, Singapore.
26. Craig VSJ, Henry CL. (2010) Specific ion effects at the air/water interface: Experimental studies. In Kunz W (ed.), *Specific Ion Effects*, Chap. 7. World Scientific Publishing, Singapore.

27. Kunz W, Lo Nostro P, Ninham BW. (2004) The present state of affairs with Hofmeister effects. *Curr Opin Coll Interf Sci* **9**: 1–18.
28. Collins KD. (2006) Ion hydration: Implications for cellular function, polyelectrolytes, and protein crystallization. *Biophys Chem* **119**: 271–281.
29. Jungwirth P, Tobias DJ. (2002) Ions at the air/water interface. *J Phys Chem B* **106**: 6361–6373.
30. Lund M, Heyda J, Jungwirth P. (2010) Ion binding to biomolecules. In Kunz W (ed.), *Specific Ion Effects*, Chap. 8. World Scientific Publishing, Singapore.
31. Ashcroft NW, Mermin ND. (1976) *Solid State Physics.* Thomson Learning.
32. Bockris JOM. (1959) *Modern Aspects of Electrochemistry*, 2nd ed., Butterworth, London.
33. Parker VB. (1965) Thermal properties of aqueous uni-univalent electrolytes. *Natl Bur Stand (U.S.)* **2**: 1–8.
34. Marcus Y. (1997) *Ion Properties.* Marcel Dekker Inc., New York, Basel.
35. Ahlrichs R, Bär M, Häser M, Horn H, Kölmel K. (1989) Electronic-Structure calculations on workstation computers — The program system Turbomole. *Chem Phys Lett* **162**: 165–169.
36. Klamt A, Schuurmann G. (1993) Cosmo — A new approach to dielectric screening in solvents with explicit expressions for the screening energy and its gradient. *Journal of the Chemical Society — Perkin Transactions 2* **5**: 799–805.
37. Horinek D, Netz RR. unpublished.
38. Lamoureux G, Roux B. (2006) Absolute hydration free energy scale for alkali and halide ions established from simulations with a polarisable force field. *J Phys Chem B* **110**: 3308–3322.
39. Jensen KP, Jorgensen WL. (2006) Halide, ammonium, and alkali metal ion parameters for modeling aqueous solutions. *J Chem Theor Comp* **2**: 1499–1509.
40. Rajamani S, Ghosh T, Garde S. (2004) Size dependent ion hydration, its asymmetry, and convergence to macroscopic behaviour. *J Chem Phys* **120**: 4457–4466.
41. Joung IS, Cheatham TE. (2008) Determination of alkali and halide monovalent ion parameters for use in explicitly solvated biomolecular simulations. *J Phys Chem B* **112**: 9020–9041.
42. Horinek D, Mamatkulov SI, Netz RR. (2009) Rational design of ion force fields based on thermodynamic solvation properties. *J Chem Phys* **130**: 124507.
43. Hansen JP, McDonald IR. (1986) *Theory of Simple Liquids.* 2nd ed. Academic Press, London.
44. Dang LX. (1992) Development of nonadditive intermolecular potentials using molecular-dynamics — solvation of LI^+ and F^- ions in polarisable water. *J Chem Phys* **96**: 6970–6977.
45. Dang LX. (1995) Mechanism and thermodynamics of ion selectivity in aqueous-solutions of 18-Crown-6 ether — A molecular-dynamics study. *J Am Chem Soc* **117**: 6954–6960.
46. Dang LX. (1992) Fluoride–fluoride association in water from molecular-dynamics simulations. *Chem Phys Lett* **200**: 21–25.
47. Allahyarov E, D'Amico I, Löwen H. (1998) Attraction between like-charged macroions by Coulomb depletion. *Phys Rev Lett* **81**: 1334–1337.

48. Boström M, Lima ERA, Biscaia Jr. EC, Tavares FW, Kunz W. (2010) Modifying the Possion–Boltzmann approach to model specific ion effects. In: Kunz W (ed.), *Specific Ion Effects*, Chap. 11. World Scientific Publishing, Singapore.
49. Torrie GM, Valleau JP. (1977) Non-physical sampling distributions in monte-carlo free-energy estimation — Umbrella sampling. *J Comp Phys* **23**: 187–199.
50. Kumar S, Rosenberg JM, Bouzida D, Swendsen RH, Kollman PA. (1995) Multidimensional free-energy calculations using the weighted histogram analysis method. *J Comp Chem* **16**: 1339–1350.
51. Huang DM, Cottin-Bizonne C, Ybert C, Bocquet L. (2008) Aqueous electrolytes near hydrophobic surfaces: Dynamic effects of ion specificity and hydrodynamic slip. *Langmuir* **24**: 1442–1450.
52. Netz RR. (2004) Water and ions at interfaces. *Curr Opin Coll Interf Sci* **9**: 192–197.
53. Luo GM, Malkova S, Yoon J, Schultz DG, Lin BH, Meron M, Benjamin I, Vanysek P, Schlossman ML. (2006) Ion distributions near a liquid–liquid interface. *Science* **311**: 216–218.
54. Jarvis NL, Scheiman MA. (1968) Surface potentials of aqueous electrolyte solutions. *J Phys Chem* **72**: 74.
55. Marrink SJ, Marcelja S. (2001) Potential of mean force computations of ions approaching a surface. *Langmuir* **17**: 7929–7934.
56. Kreuzer HJ, Wang RLC, Grunze M. (2003) Hydroxide ion adsorption on self-assembled monolayers. *J Am Chem Soc* **125**: 8384–8389.
57. Zangi R, Engberts JBFN. (2005) Physisorption of hydroxide ions from aqueous solution to a hydrophobic surface. *J Am Chem Soc* **127**: 2272–2276.
58. Lyklema J, van Leeuwen P, van Vliet M, Cazabat AM. (2000) *Fundamentals of Interface and Colloid Science*. Academic Press, London.
59. Schweiss R, Welzel PB, Werner C, Knoll W. (2001) Dissociation of surface functional groups and preferential adsorption of ions on self-assembled monolayers assessed by streaming potential and streaming current measurements. *Langmuir* **17**: 4304–4311.
60. Ciunel C, Armelin M, Findenegg GH, von Klitzing R. (2005) Evidence of surface charge at the air–water interface from thin-film studies on polyelectrolyte-coated substrates. *Langmuir* **21**: 4790–4793.
61. Friedsam C, Gaub HE, Netz RR. (2005) Adsorption energies of single charged polymers. *Europhys Lett* **72**: 844–850.
62. Garcia-Celma JJ, Hatahet L, Kunz W, Fendler K. (2007) Specific anion and cation binding to lipid membranes investigated on a solid supported membrane. *Langmuir* **23**: 10074–10080.
63. Härtl A, Garrido JA, Nowy S, Zimmermann R, Werner C, Horinek D, Netz RR, Stutzmann M. (2007) The ion sensitivity of surface conductive single crystalline diamond. *J Am Chem Soc* **129**: 1287–1292.

Chapter 10

HNC Calculations of Specific Ion Effects

Luc Belloni* and Ioulia Chikina*

The theoretical approach based on the HNC integral equation is described in the context of ionic specificity. Two levels of description of the water medium are considered. Within the Primitive Model (continuous solvent), ionic specificity is introduced via effective, solvent-averaged, dispersion forces. The agreement with experimental data in bulk or at air–water interfaces is only partial and illustrates the limits of that approach. Within the Born–Oppenheimer model, the molecular HNC equation is solved with an explicit description of the solvent molecules (SPC water). Ionic and solvent profiles in bulk and at interfaces are enriched by short-range oscillated structures. The ionic polarisability is introduced via the self-consistent mean-field theory, the polarisable ions carrying an effective, fixed dipole moment. The study of the air–water interface reveals the limits of the conventional HNC approach and the needs for improved integral equations.

1. Introduction

The understanding of equilibrium, thermodynamical and structural properties in colloidal and interfacial systems, especially that of specific ionic effects in charged systems, can be attacked from the theoretical point of view using different levels of description and different, more or less approximated approaches.

Levels of description. Typical macromolecular systems contain big colloids or interfaces, small ions and polar solvent molecules. The first question which arises in a theoretical approach is: is it sufficient to consider *explicitly* the colloids and interfaces only? If yes, one usually treats the ionic fluid, ions + solvent, as a continuum which obeys the mean

*CEA/Saclay, Direction des Sciences de la Matière, SIS2M, LIONS, 91191 Gif-sur-Yvette Cedex, France.

field Poisson–Boltzmann (PB) equation and the boundary conditions imposed by the macromolecules. This simple description ignores the ion–ion correlations and leads to the famous *generic* DLVO analysis of colloidal interaction and stability.[1] Charged surfaces interact through screened coulombic repulsions which result from overlap of their double layers. Specificity can be introduced by modifying the external potential imposed by the surfaces on the ionic fluid. If the answer is no, is it then sufficient to treat *explicitly* the small ions as well, but still ignoring the molecular nature of the solvent? This intermediate level of description, called *Primitive Model* (PM), considers ions as hard or soft spheres immersed in a continuous solvent of given dielectric constant ε. The Coulombic potential of mean force between charged objects, namely the potential averaged over the solvent configurations, behaves as $1/\varepsilon r$. The correlations among the ions in bulk or near interfaces are naturally taken into account in that approach. In particular, when these correlations are strong as in the presence of multivalent ions, that level of description is able to predict non-DLVO, spectacular, *generic* phenomena like attraction between like-charged surfaces.[2] As recalled in the overview Chap. 1 of this book, ionic specificity can be introduced through the ionic radii and through more or less modern, extra, solvent-averaged potentials between ions and between surfaces and ions, which depend, for instance, on the ionic polarisability.[3] Since the solvent is not treated explicitly, these effective potentials are not precisely monitored and characterised and depend often on adjustable parameters like the hydrated ionic size, the Gurney coefficient, the *effective* polarisability, etc. Lastly, when one reaches the limits of the PM description, one must go to the next higher level that incorporates the discrete nature of the solvent (mainly water) molecules. This richer approach (called Born–Oppenheimer, BO) requires potentials between ions and molecules, this time in vacuum. The coulombic potential between charges and sites behaves now as $1/r$. The dielectric properties, the hydration phenomena are now *a posteriori* results of the BO model and ionic specificity can be introduced with more direct, less questionable, extra potentials. Note that this level of description, although very precise, is not the ultimate one. These potentials do not come from nowhere but have to be guessed (consider the number of water models in the literature!) or to be derived from the ultimate, *ab initio* level of description which solves the Schrödinger equation for the electrons! The three mentioned levels,

DLVO, PM and BO, are connected through rigorously stated, partial averaging (coarse graining as said in modern language). In principle, the BO level can feed the PM with solvent-averaged potentials and the PM can be illustrated in terms of colloid–colloid effective potentials. In practice, when one uses for instance the PM description, one hopes that simple solvent-averaged pair potentials with physically meaningful parameters will be sufficient to understand and rationalise the ionic specificity and that general rules and laws could be derived at that level. The PM limits are reached when the parameters become too numerous or less meaningful!

Technical treatment. As usual in statistical mechanics of liquids, once the level of description has been chosen with the (usually pair) potentials of interaction, it remains to solve the N-body problem! In practice, how to calculate the pair distribution functions $g_{ij}(r)$ (probability to find two particles of species i and j separated by the distance r) from the pair potentials of interaction $v_{ij}(r)$? The latter quantity is a two-body interaction while the former results from N-body correlations. Since the 1960s, one has the choice between 'exact' treatments based on numerical simulations (Monte Carlo, Molecular or Brownian Dynamics) and approximated ones, in particular those based on the Ornstein–Zernike and *integral equations*.[4] Choosing one or the other depends on a balance between validity, robustness, efficiency, and speed of execution. With the progresses of the computers, one could wonder why it is still interesting to use non-exact approaches. Even today, numerical simulations of systems with strong Coulombic ('infinite' range) interactions and high charge and number asymmetries are always difficult (read very long) to be performed. For instance, modern simulations of interfacial electrolytes at the BO level with tens of ions and thousands of H_2O require weeks of calculation on clusters of PC for a moderate precision (see Chaps. 8 and 9 devoted to simulation). So, as always in the history of statistical mechanics of liquids, there is room for an alternative approach, maybe 'less exact', but still precise enough, with a validity well documented and controlled by comparison with simulation in few test cases. This alternative approach can be numerically solved with high-level numerical procedures in, say, a few minutes on a simple PC, and could be routinely applied to explore a broad range of parameters and, ultimately, fit experimental data. Integral equations, in particular the bare HNC equation which is known to be well suited to charged systems, are certainly good candidates for that alternative. For 50 years, this approach has been improved and enriched, with approximated bridge functions (the functions missing in HNC), from homogeneous HNC to non-homogeneous versions, from 2-body to 3-body approaches, etc., allowing more and more complex systems (or levels of description) to be studied with higher validity, thanks to up-to-date numerical analysis. The present chapter will give a survey of what can be done (and what cannot) with that HNC-like approach in the context of ionic specificity, both at the Primitive Model and the Born–Oppenheimer levels.

2. Primitive Model Description

2.1. *Bulk Electrolytes*

At the PM level, ions are considered as charged hard *spheres*, of valency Z_i, diameter σ_i and number density ρ_i, immersed in a continuous, dielectric solvent. The bare PM pair potentials read:

$$\beta v_{ij}(r) = \begin{cases} +\infty & r < \sigma_{ij} = \dfrac{\sigma_i + \sigma_j}{2} \\ \dfrac{Z_i Z_j L_B}{r} & r > \sigma_{ij}, \end{cases} \tag{1}$$

where $\beta = 1/kT$ and L_B is the Bjerrum length (7.2 Å in water at room temperature). Modifying the hard sphere into soft sphere $1/r^n$, short range repulsion is straightforward. Note that, within this bare picture, ions of the same valency can be distinguished only through their sizes. Ionic specificity can be introduced with extra potentials, such as those proposed by Ninham and Yaminsky,[3] which depend on the ionic polarisability. Roughly speaking, these dispersion potentials vary as $-A_{ij}/r^6$ where the coefficient A_{ij} is proportional to the product of the effective polarisability of each ion, $\alpha_i^* \alpha_j^*$. The price to pay to use the PM picture is that the choice of the different potential parameter values is not obvious. Since the solvent is not taken explicitly into account, it must be introduced implicitly through these parameters. The sizes should represent *hydrated* values. The effective polarisability α_i^* should involve the contrast between ionic and solvent quantities....

The integral equation approach starts with the exact Ornstein–Zernike (OZ) equation which relates total $h_{ij} = g_{ij} - 1$ and direct c_{ij} correlation functions through convolution products[4]:

$$h_{ij}(r) = c_{ij}(r) + \sum_k \rho_k h_{ik} \otimes c_{kj}(r). \tag{2}$$

The second, formally exact, relation reads[4]:

$$g_{ij}(r) = \exp[-\beta v_{ij}(r) + \gamma_{ij}(r) + b_{ij}(r)], \tag{3}$$

where $\gamma_{ij} \equiv h_{ij} - c_{ij}$ and b_{ij} is the so-called bridge function, which has a complex diagrammatic expansion and cannot be expressed simply in terms of the previous functions. Approximation emerges when a choice is made for that function, defining a so-called integral equation or closure. The famous HNC equation simply neglects the bridge contribution, $b_{ij} = 0$. This approximation is able to account for the important phenomena which govern charged systems like screening, ion pairing, electrostatic condensation.... For the present bulk electrolytes, as long as one deals with *monovalent* ions, the HNC approach is *quantitatively* sufficient even in the molar range and does not necessitate improvements.

The numerical resolution is iterative. The OZ equation is expressed in Fourier space where the convolution product is replaced by an algebraic one. So, FFT Fourier transforms are intensively used to exchange r and q spaces. The functions are discretised on, say, $N = 256 - 2048$ r or q points. First, a guess is made for the (continuous) functions $\gamma_{ij}(r)$. The closure gives the functions $g_{ij}(r)$ and $c_{ij}(r) = g_{ij}(r) - 1 - \gamma_{ij}(r)$. A direct FT gives the functions $c_{ij}(q)$ [normalised by the density factor $(\rho_i\rho_j)^{1/2}$]. The OZ equation is inverted to give the (short range) $\gamma_{ij}(q)$ functions through a *matricial* relation:

$$\gamma = c^2(1 - c)^{-1}. \tag{4}$$

Lastly, an inverse FT leads back to the functions $\gamma_{ij}(r)$. In the presence of Coulombic, $1/r$ or $1/q^2$, 'diverging' potentials and correlation functions, reference functions of same divergence at large r or small q are subtracted from $c_{ij}(r)$ or $\gamma_{ij}(q)$ before Fourier transform, the numerical FT is performed for the non-diverging difference and the analytical FT of the reference is added afterwards.

The cycle is repeated until input and output $\gamma_{ij}(r)$ functions coincide. Powerful Newton–Raphson techniques are used to propose improved input functions from one cycle to the next one.[5,6] Convergence is reached with less than ten cycles in a few seconds.

An example of a systematic use of the PM/HNC approach in the context of ion specificity in bulk electrolytes can be found in Ref. 7. The first objective was to understand and quantitatively reproduce the equation of state (osmotic pressure, activity coefficients as a function of concentration) of numerous salts in the 0–4 M range. The osmotic pressure is derived from the virial or compressibility routes (both HNC routes are in good agreement for the present salts). The dilute regime, dominated by the screening effect, can be explained by the bare, *generic*, Debye–Hückel (DH) law, obtained in the context of integral equations by simply replacing the HNC equation with the DH one, $c_{ij}(r) = -Z_i Z_j L_B/r$ on the whole r-domain [so $h_{ij}(r) = -Z_i Z_j L_B \exp(-\kappa r)/r$ where κ is the screening constant], and exactly reproduced by the HNC treatment. The 0.1–1 M regime requires an account of ionic sizes. Many hydrated radii are proposed in the literature. Meanwhile, a good fit of the experimental equations of state needs to slightly play with these parameters and consider them as adjustable. That is the first illustration of the limits of the PM level of description. Finally, at higher concentration, the attractive dispersion interaction comes into play and tends to reduce the osmotic pressure. Also here, the effective ionic polarisabilities must be considered as adjustable parameters in order to get good fits of the different curves. Figure 1 presents such analysis for a 1-M $NaNO_3$ electrolyte. The addition of dispersion force leads to the formation of a narrow peak at contact in the NO_3–NO_3 pair distribution function.[7]

That semi-phenomenological approach could be considered as meaningless. Meanwhile, systematic fits of many salts on a large regime of concentration can be rationalised with a few parameters[7] and this representation of bulk electrolytes can now serve in a second step as a reference for the interfacial properties.

2.2. *Interfaces*

How to represent an interface between an electrolyte and a solid or fluid medium (of nanometric radius of curvature or of planar geometry) within integral equations? The simplest approach is to consider the solid–air object as a new species at *infinite* dilution, labelled '0' in the following, say a sphere of radius R which adds to the previous ion components, and to repeat the integral equation formalism for the new mixture.[8] Since the new object is at zero density, the ion–ion correlations are not perturbed

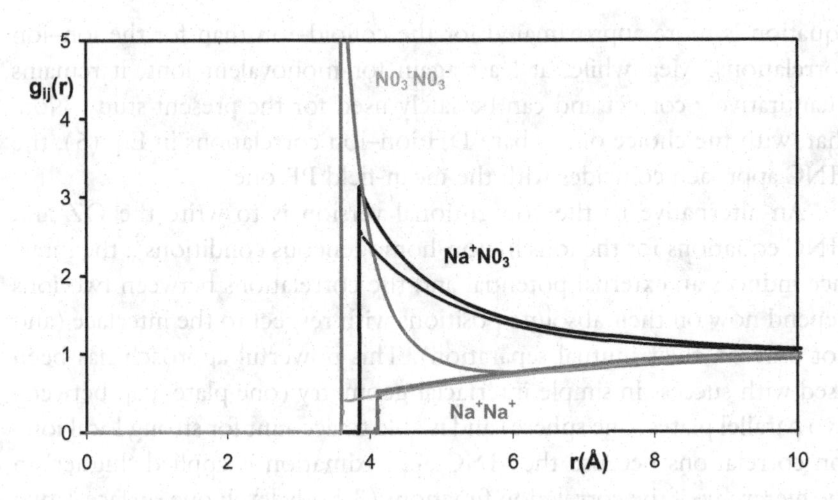

Fig. 1. HNC pair distribution functions for a 1-M $NaNO_3$ salt within the PM description. Thick lines: with dispersion potentials; thin lines: without. Note that, according to the electroneutrality condition, the three curves correspond to identical excess integrals. (From Ref. 7.)

and the OZ equation for the interface-ion $i0$ correlations becomes:

$$h_{i0}(r) = c_{i0}(r) + \sum_{k>0} \rho_k h_{ik} \otimes c_{k0}(r). \tag{5}$$

The sum involves the ionic components $k > 0$ only. Note that the last term of Eq. (5) could be alternatively expressed by exchanging h and c symbols. The bulk OZ equation guarantees the equivalence between both, technically and subtly different, versions. The OZ equation is then coupled with the $i0$ HNC closure [Eq. (3)] which depends on the imposed interface-ion potential. The resolution is iterative as before, simplified by the fact that the OZ Eq. (5) becomes linear in the unknown functions (the bulk ion–ion h_{ik}, calculated previously, are *given*). The final correlation functions h_{i0} leads to the desired local ionic densities at the interface $\rho_i(1 + h_{i0}(r))$. The planar geometry limit for the interface is obtained either by investigating big spheres, of radius R much larger than any characteristic distance (e.g. Debye length), or by explicitly writing the OZ equation (5) in 1D.

Equation (5) is formally exact. The fact that it involves the *bulk* ion–ion correlations *only*, non-perturbed at the approach of the interface, somewhere illustrates the importance of the neglected surface-ion bridge terms in this conventional, homogeneous version. It is expected that the HNC equation is more approximated for the colloid–ion than for the ion–ion correlations. Meanwhile, at least again for monovalent ions, it remains quantitatively correct and can be safely used for the present study. Note that, with the choice of the bare DH ion–ion correlations in Eq. (5), the HNC approach coincides with the mean-field PB one.

An alternative to the conventional version is to write the OZ and HNC equations for the ions in non-homogeneous conditions[9]: the interface induces an external potential and the correlations between two ions depend now on their absolute positions with respect to the interface (and not only on their mutual separation). This powerful approach has been used with success in simple interfacial geometry (one plate, gap between two parallel plates, one sphere) and is able to account for strong local ion–ion correlations because the HNC approximation is applied 'higher' in the hierarchy of the correlation functions (3-body level: one surface + two ions). The price to pay is the need of a much more difficult and complex numerical resolution. Again, this is not necessary for monovalent ions at the PM level and the homogeneous version is sufficient. In the same spirit, it is fair to say that the ion–ion correlations play a modest role in the ionic profiles, at least at moderate concentration, and that most of the HNC data could be reasonably reproduced with the PB treatment, described in Chap. 11.

In the context of air–water interface, the Ninham–Yaminsky dispersion potential felt by an ion localised at the distance z from the planar interface behaves as B_i/z^3 where B_i again, roughly speaking, is proportional to the effective polarisability α_i^*.[3] This specific potential is added to the short-range hard repulsion (hard wall at $z = \sigma_i/2$) and to the generic image force repulsion. Here are three remarks: firstly, the choice of *hard* repulsion, correct for solid–water interface, becomes questionable for the air–water soft interface. That illustrates again the limits of the PM picture. Playing with a soft repulsion is somewhat meaningless because the final local ionic profiles we are looking for are too much dependent on the choice for the *a priori* unknown softness. Secondly, the explicit description of the ions within the PM is in principle sufficient to account for the dielectric discontinuity and image forces. Meanwhile, the image force between two ions i, j depends on the *absolute* positions of both ions (ion i interacts with its own image, with j and with the image of j); that is not compatible with the conventional HNC version (but this can be perfectly taken into account within the inhomogeneous version evoked above[9]). This is why the DH, doubly screened $\exp(-2\kappa z)/z$, generic repulsion is explicitly added to the potential at the beginning. Thirdly, technically speaking, when the air–water interface is mimicked by the addition of a big (air bubble) sphere, it is not correct to simply consider the bubble–ion potential as $B_i/(r-R)^3$ (saying that $z \equiv r - R$). Indeed, this so constructed 3D potential becomes different at infinite range (its FT diverges at $q = 0$). So, one must return to the complete dispersion interaction between a sphere and an ion, $B_i 8R^3/(r^2 - R^2)^3$, which behaves as $B_i/(r-R)^3$ close to the interface as required, and as $B_i 8R^3/r^6$ far from the bubble.

Figure 2 presents the ionic interface profiles of the 1-M $NaNO_3$ salt of Fig. 1 ($R \approx 30$ Å is sufficient).[7] At that concentration, the ion–ion correlations have a non-negligible effect on the shape of the interface distributions. The addition of the air–ion dispersion potentials greatly modifies the bare profiles obtained with pure hard sphere + image force contributions.

Once the HNC equation is solved, the excess ionic adsorbed quantities Γ_i are derived by integrating the excess local profiles:

$$\Gamma_i = \rho_i \int_0^\infty (g_{i0}(z) - 1)dz. \tag{6}$$

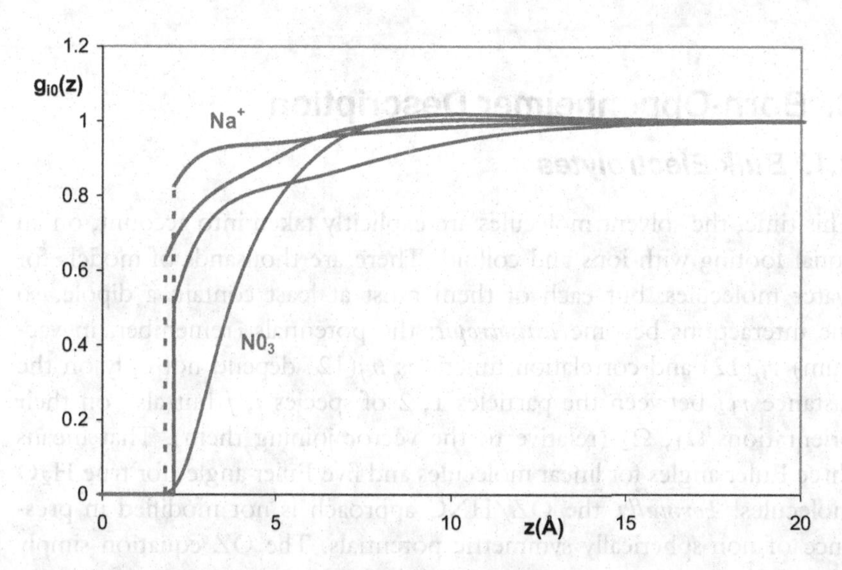

Fig. 2. HNC ionic profiles at the air–water interface within the PM description. 1-M $NaNO_3$ salt of Fig. 1. Solid lines: with air–ion dispersion potentials; dotted lines: without. Note that, according to the electroneutrality condition, the excess adsorbed quantity is identical for cation and anion. (From Ref. 7.)

In Ref. 7, the adsorbed quantities were expressed in terms of surface tension increments according to the Gibbs isotherm law and compared to the surface tension experimental data. With the size and polarisability values previously derived from the bulk properties (and without any further adjustment!), the agreement was found somewhat disappointing. More recently, a similar procedure was used to explain the ionic adsorbed quantities at air–water interface (in general $\Gamma_i < 0$, so it is a *desorbed* quantity) measured using grazing incidence X-ray fluorescence for different salts and salt mixtures[10] (see Chap. 5). Using the parameters of Ref. 7, it was found that the theoretical desorption is overestimated. Experimental values are recovered only after adding a short-range attraction whose strength depends on the nature of the anion, see Fig. 3 for KCl + KI mixtures. Although a few adjustable parameters are sufficient to reproduce many salt compositions and concentrations,[10] this semi-phenomenological treatment is the final illustration of the limits of the PM level of description: since the solvent molecules are absent, we cannot account for the proper hydration, for the perturbed water structure at the interface, we do not know or have to guess the proper effective potentials between ions in bulk and between interface and ions.

3. Born-Oppenheimer Description

3.1. *Bulk Electrolytes*

This time, the solvent molecules are explicitly taken into account, on an equal footing with ions and colloid. There are thousands of models for water molecules, but each of them must at least contain a dipole, so the interactions become *anisotropic*: the potentials (remember, in vacuum) $v_{ij}(12)$ and correlation functions $h_{ij}(12)$ depend not only on the distance r_{12} between the particles 1, 2 of species i, j but also on their orientations Ω_1, Ω_2 (relative to the vector joining them). That means three Euler angles for linear molecules and five Euler angles for true H_2O molecules! *Formally*, the OZ/HNC approach is not modified in presence of non-spherically symmetric potentials. The OZ equation simply becomes:

$$h_{ij}(12) = c_{ij}(12) + \sum_k \rho_k \int h_{ik}(13) c_{kj}(23) d3, \qquad (7)$$

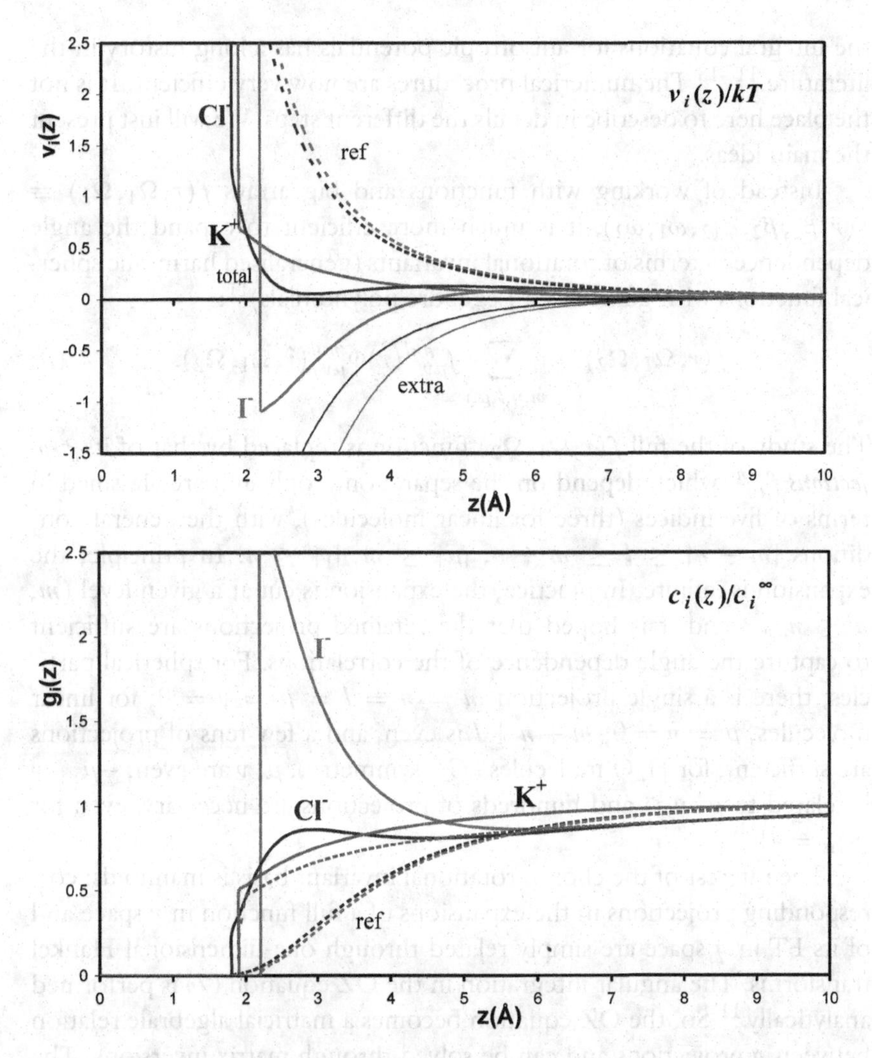

Fig. 3. KCl 0.1 M + KI 0.1 M mixture at air–water interface. Top: PM interface-ions potentials. Bottom: HNC ionic interface distributions. Dotted lines: image force + dispersion potentials of Ref. 7. Thick lines: an extra attraction must be added for the anions in order to reproduce the experimental adsorption. (From Ref. 10.)

where the convolution now involves integration over the position *and* orientation of the particle 3. The closure is written for each position and orientation. Of course, *technically* and *numerically*, the situation is completely different and becomes much more complex. The resolution of the integral equations for anisotropic potentials has a long history in the literature.[11–13] The numerical procedures are now very efficient. It is not the place here to describe in details the different steps. We will just present the main ideas.

Instead of working with functions and big arrays $f(r,\Omega_1,\Omega_2) \equiv f(r,\beta_1,\beta_2,\phi_{12},\omega_1,\omega_2)$, it is much more efficient to expand the angle dependence in terms of rotational invariants (generalised harmonic spherical functions with generalised Legendre polynomials)[11]:

$$f(r,\Omega_1,\Omega_2) = \sum_{m,n,l,\mu,\nu} f_{\mu\nu}^{mnl}(r)\Phi_{\mu\nu}^{mnl}(\hat{r},\Omega_1,\Omega_2). \tag{8}$$

The study of the full $f(r,\Omega_1,\Omega_2)$ function is replaced by that of its *projections* $f_{\mu\nu}^{mnl}$ which depend on the separation r only and are classified in terms of five indices (three for linear molecules), with the general conditions $|m-n| \leq l \leq m+n$, $|\mu| \leq m$, $|\nu| \leq n$. In principle, the expansion is infinite. In practice, the expansion is cut at a given level (m, $n \leq n_{\max}$) and it is hoped that the retained projections are sufficient to capture the angle dependence of the correlations. For spherical particles, there is a single projection $m = n = l = \mu = \nu = 0$; for linear molecules, $\mu = \nu = 0$, $m+n+l$ is even, and a few tens of projections are sufficient; for H_2O molecules (C_{2v} symmetry), μ, ν are even, $-\mu$, $-\nu$ is related to μ, ν,[14] and hundreds of projections are necessary, even for $n_{\max} = 4$!

The interest of the chosen rotational invariant basis is manifolds: corresponding projections in the expansions of a full function in r space and of its FT in q space are simply related through one-dimensional Hankel transforms. The angular integration in the OZ equation (7) is performed analytically.[11] So, the OZ equation becomes a matricial algebraic relation between q projections and can be solved through matrix inversions. The most critical and time-consuming part of the cycle concerns the closure where the projections of the exponential must be derived from those of the exponent (the exponentiation emphasises the anisotropy!). Different procedures have been proposed.[13,15] It seems that, for a perfect treatment of the angular dependence in that step, one must return to the bare definition: at each r separation where the coupling is highly anisotropic, the projections of the exponent are combined to reconstruct the angle dependence of the exponent according to Eq. (8), the exponential is calculated for each Euler triplet or quintet, and the angle-dependent result is projected onto the different rotational invariants of the basis using angular integration.[15]

The iterative resolution for our solvent molecules + ions mixtures is then similar to the PM case: one starts with $\gamma^{mnl}_{\mu\nu,ij}(r)$ projections for each pair, derive $\gamma^{mnl}_{\mu\nu,ij}(r)$ using the procedure described above and the imposed pair potentials, calculate the Hankel transforms of $c^{mnl}_{\mu\nu,ij}(r)$ using FFTs,[13] invert the OZ matrices to get the $\gamma^{mnl}_{\mu\nu,ij}(q)$,[11] and come back in r space using inverse Hankel transforms. Powerful Newton–Raphson techniques[6] improve the choice of the input projections for the next cycle and avoid numerical stability problems. For water-like systems, convergence can be reached in ten cycles within a few minutes on a simple PC. Although the numerical resolution requires a complex and robust code, it is reasonable to say that this numerical step is no more an obstacle to study anisotropic particles (as long as the particles keep a minimum symmetry like C_{2v} for H_2O; for arbitrary anisotropy, the number of projections rapidly explodes with $n_{\max}$). On the other hand, the validity of the HNC closure itself may become questionable before any BO calculation of electrolytes. Indeed, the pair interactions between charges in vacuum are much stronger than in the PM (in fact, $\varepsilon \approx 80$ as large!), the coupling between water molecules in bulk water and inside hydration layers involves many-body correlations and it is expected that neglecting bridge diagrams at that level should be more sensitive than at the PM level.

A starting example of the HNC treatment of bulk electrolytes at the BO level is discussed now. Water molecules are described using the SPC site model,[16] $d_{OH} = 1$ Å, $\theta_{HOH} = 109.47°$. A Lennard–Jones (LJ) potential is centred on the O atom (diameter = 3.1655 Å) and partial charges are put on the O,H,H sites (+0.41 on each H, −0.82 on O). Note that the resulting dipole is larger than that of an isolated molecule (in vapour) and implicitly accounts for polarisation effects in dense liquid phase. The monovalent anions and cations are represented by charged LJ spheres, with *equal* size for the moment, $\sigma_i = 4$ Å. The H_2O molecules are analysed with $n_{\max} = 4$, which has been proved to correspond to a rich enough angular basis.[17] That leads to one projection for each ion–ion function, nine for both ion–water functions and 250 for water–water function. Figure 4 presents the centre of mass–centre of mass projections $g^{000}_{000,ij}(r)$ of the distribution functions for 0.1-M salinity. The solvent–solvent peak and oscillations illustrate the local structure of bulk water while the ion–solvent

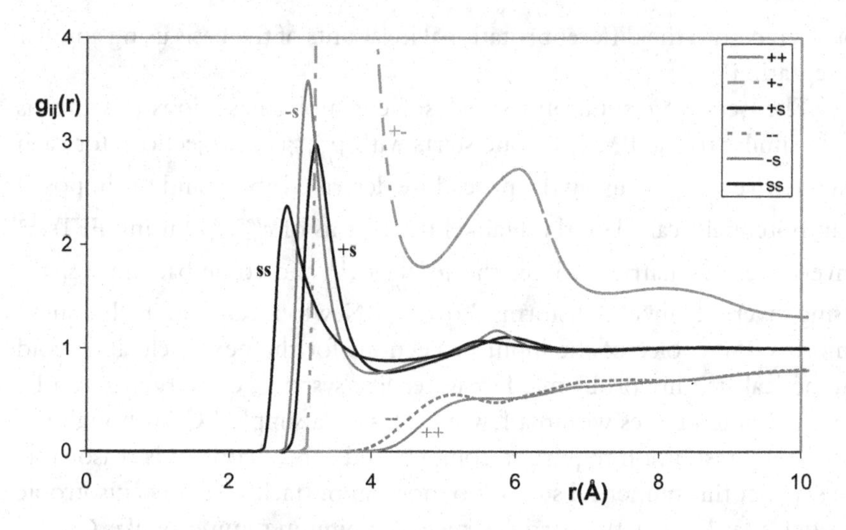

Fig. 4. Centre of mass–centre of mass pair distribution functions $g^{000}_{00,ij}(r)$ for a monovalent aqueous electrolyte at the BO level. LJ diameters = 4 Å (cation +), 4 Å (anion −), 3.1655 Å (H_2O/SPC solvent *s*). LJ energies = 0.2857 *kT*, 0.2857 *kT* and 0.26229 *kT*, respectively (T = 298.15 K). Densities = 0.1 M, 0.1 M, 55.3888 M. HNC results.

peak reveals the hydration layer around each ion.[18] The sign of the next projection $g^{011}_{00,\text{ion-solvent}}(r)$ confirms the expected preferred orientation of the water molecules around the ions (O against cations, H against anions). The differences observed between cation and anion illustrate the fact that SPC water molecules are more than simple dipoles. At equivalent ionic size, the hydration of anion is easier than that of cation (the hydration peak is higher and localised at shorter distance).

These BO results can be advantageously presented in terms of effective, solvent-averaged pair potentials $v^{eff}_{ij}(r)$. By definition, these potentials would lead to the same BO ion–ion pair correlations when used within the PM, at identical temperature and ionic densities.[19] They contain no more information than the ion–ion correlations but offer good illustration of what happens inside the electrolyte. Technically, the PM OZ/HNC procedure is inverted, the potentials being extracted from the imposed distribution functions. Figure 5 presents $v^{eff}_{ij}(r)$ for the starting electrolyte of Fig. 4. At long distances, the potentials exhibit the expected $1/\varepsilon r$ behaviour with ε being the dielectric constant of the BO electrolyte

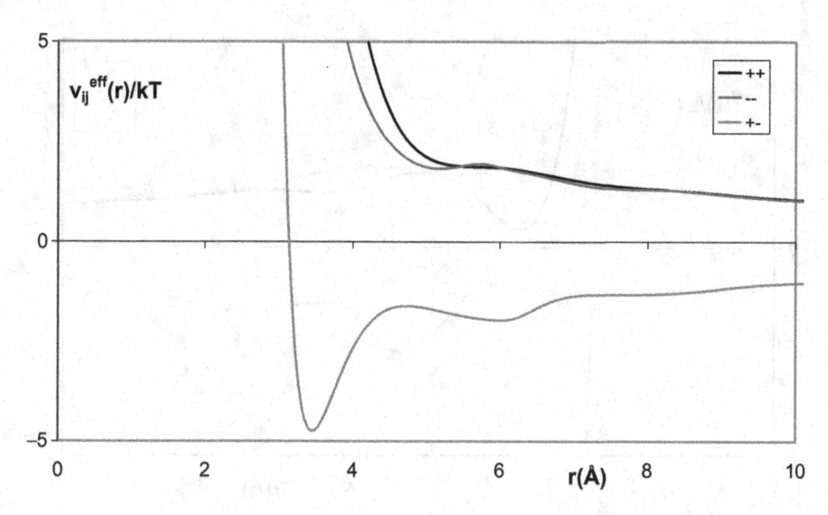

Fig. 5. Effective, solvent-averaged, PM ion–ion pair potentials $v_{ij}^{eff}(r)$ extracted from the BO ion–ion distribution functions of Fig. 4. Salinity 0.1 M. HNC results. The long distance behaviour is $1/\varepsilon r$, with $\varepsilon = 54$.

($\varepsilon = 54$ for the present SPC model with HNC,[17] instead of $\approx$ 60–75 obtained with simulation). At short distances, oscillations appear which result from the hidden water structure and which could not have been guessed *a priori* from bare PM picture. That illustrates the power and the richness of the BO description. It is also fair to add that this richness may appear somewhat confusing and that it is difficult to rationalise the BO results, the details of the oscillations being sensitive to the precise choice of the potential parameters (changing the diameters at the 0.1 Å level modifies the short-range structure). Figure 6 presents the similar ion–ion potentials of mean force for different ionic sizes $\sigma_+ = 2.4$ Å, $\sigma_- = 4.4$ Å characterising NaCl systems. Note the unexpected short range well (of positive value) between cations.

To what extent are the BO/HNC results valid? Few comparisons of HNC and simulation data in literature have shown the merits and limits of the HNC approximation. For less polar solvents, the HNC curves are in very good agreement with 'exact' ones.[14] For water, the situation is more critical. The number of first neighbours in the water structure is different between HNC and simulation,[17] illustrating the difficulty of HNC to account precisely for the tetrahedral local structure. One could conclude from these partial disagreements that (i) the bare

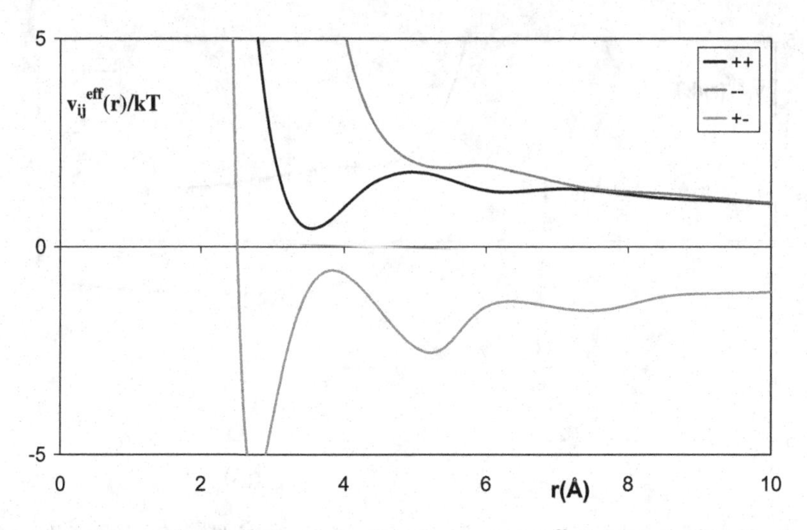

Fig. 6. Effective, solvent-averaged, PM pair potentials $v_{ij}^{eff}(r)$ for a 0.1-M NaCl electrolyte with ionic LJ diameters $\sigma_+ = 2.4$ Å, $\sigma_- = 4.4$ Å and LJ energies = 0.168 kT, 0.168 kT.

HNC data must be considered with caution when quantitative fitting is the objective, (ii) meanwhile, they capture most of the important phenomena, the BO/HNC approach is rich enough, and (iii) it will be fruitful to develop new approaches to propose improved integral equations and get quantitative precision.

3.2. *Ionic Polarisability — Self Consistent Mean-Field Theory*

The BO description is in principle well adapted to incorporate the 'zero-frequency' ionic polarisability and account for the coupling between charges, fixed and induced dipoles. A polarisable particle (solvent or ion) responds to the applied electric field exerted by its neighbours with an induced dipole, which will then exert a new field in the neighbourhood, and so on. The problem of the polarisable systems is that the interaction is no more pair-wise additive. The N-body problem can be explicitly treated in numerical simulation (with difficult and rather time-consuming iterative procedure at each configuration) but is not adapted to integral equations approaches. In order to circumvent the problem, the self-consistent mean-field theory[20] is used: the polarisability and all associated instantaneous dipole moments are replaced by a single effective dipole m^e, the value of which is calculated self-consistently from the ideal permanent dipole μ (if any), the polarisability α_i and the electrostatic coupling between the dipole and the neighbourhood. $(m^e)^2$ is identified with the mean square dipole moment $\langle m^2 \rangle$ and is given for the species i (solvent, ion) by the implicit, self-consistent relationship[20,21]:

$$(m_i^e)^2 = \frac{\mu_i^2}{(1 - \alpha_i C_i^e(m_i^e))^2} + \frac{3\alpha_i k_B T}{1 - \alpha_i C_i^e(m_i^e)}. \tag{9}$$

The first term accounts for the average total (permanent+induced) dipole moment and vanishes in the absence of ideal dipole μ, as for monoatomic ions; the second term accounts for fluctuations. For particles of C_{2v} or higher symmetry, the scalar $C(m_i^e)$ that contains the electrostatic coupling with the neighbourhood is defined by the expression

$$\langle E_l \rangle_i = C_i(m_i^e) m_i^e, \tag{10}$$

where $\langle E_l \rangle_i$ is the total average local electric field due to all other solvent particles and ions in solution acting on the particle i. Each contribution to the electric field and the constant C is calculated from the corresponding component of the electrostatic energy in the effective system. The constants C_i are thus given by[20,21]:

$$C_i = -(2\,U_{DiDi} + U_{Di-other})/(N_i m_i^{e^2}). \tag{11}$$

The first energy component U_{DiDi} results from the dipole–dipole interactions between molecules of the same kind i. The second one $U_{Di\text{-}other}$ stems from the interactions of the dipole i with all other sources of electrostatic coupling, the non-dipolar multipoles of the species i (charges, quadrupoles, etc.) as well as all multipoles (including dipoles) of the other species.

This theory, when applied to the SPC water geometry but with a fixed dipole equal to that of an isolated H_2O molecule, $\mu = 1.85D$, leads to the result $m^e = 2.62D$. This is in good agreement with the well-known extended simple point charged model SPC/E[22] which allows for the polarisation correction. This validates the self-consistent mean-field procedure. As an illustration of the anion polarisation effect, Fig. 7 presents the anion–anion and anion–water g_{--} and g_{-s} (centre of mass–centre of mass) distribution functions for the previous electrolyte model of Fig. 4 at two salinities, 0.1 M and 1 M, without any polarisability as before or with an *anionic* polarisability $\alpha_i = 3.416\,\text{Å}^3$ equal to that of a chloride ion. One notes that with such real polarisability, the induced effect remains modest.

3.3. Interfaces

As before for the PM case, a new object at infinite dilution is added to the BO electrolyte in order to mimic colloid–water, solid–water or air–water interfaces. For smooth surfaces, the added particle of radius R keeps its spherical symmetry. The previous electrolyte with $n_{\max} = 4$ for water requires one, one, nine projections for surface–cation, surface–anion and surface–water correlations, respectively. The local HNC profiles can be derived in a few seconds from the previously calculated bulk correlations $h^{mnl}_{\mu\nu,ij}$ and the imposed particle-surface potentials $v^{m0m}_{\mu 0,i0}(r)$.

For passive solids imposing an excluded volume, these potentials present a simple hard or soft repulsion, localised at $r = R+\sigma_i/2$. Figure 8 presents the local profiles for our symmetrical 0.1-M electrolyte of Figs. 4 and 5 near a hard sphere $R = 20\,\text{Å}$. The long distance behaviour can be interpreted in terms of image force repulsion due to the dielectric discontinuity between the water medium and the passive excluded volume. Peaks and oscillations near interface reveal the local depletion and structure effects of the water/ion fluid. As before, it is fruitful to express these data in terms of effective, solvent-averaged surface-ion potentials (those potentials which, within the PM picture, lead to the same BO ionic profiles), as shown in Fig. 8. We note that the differences in hydration between cations and anions (of the same size) previously noticed produce marked differences in local profiles and effective potentials, the anion being more expelled from the vicinity of the interface. Meanwhile, it is important to remember that the excess adsorbed quantities for cations and anions exactly coincide, thanks to the electroneutrality condition. So, the anion profile is above that of the cation at larger distance from the interface.

For air–water interfaces, the situation concerning the bubble-particle potentials to be imposed is less clear. Contrary to numerical simulations, the integral equation approach cannot treat true liquid–vapour interfaces and it is necessary to impose constraints of *a priori* unknown shape to prevent liquid water from penetrating the bubble medium. What type

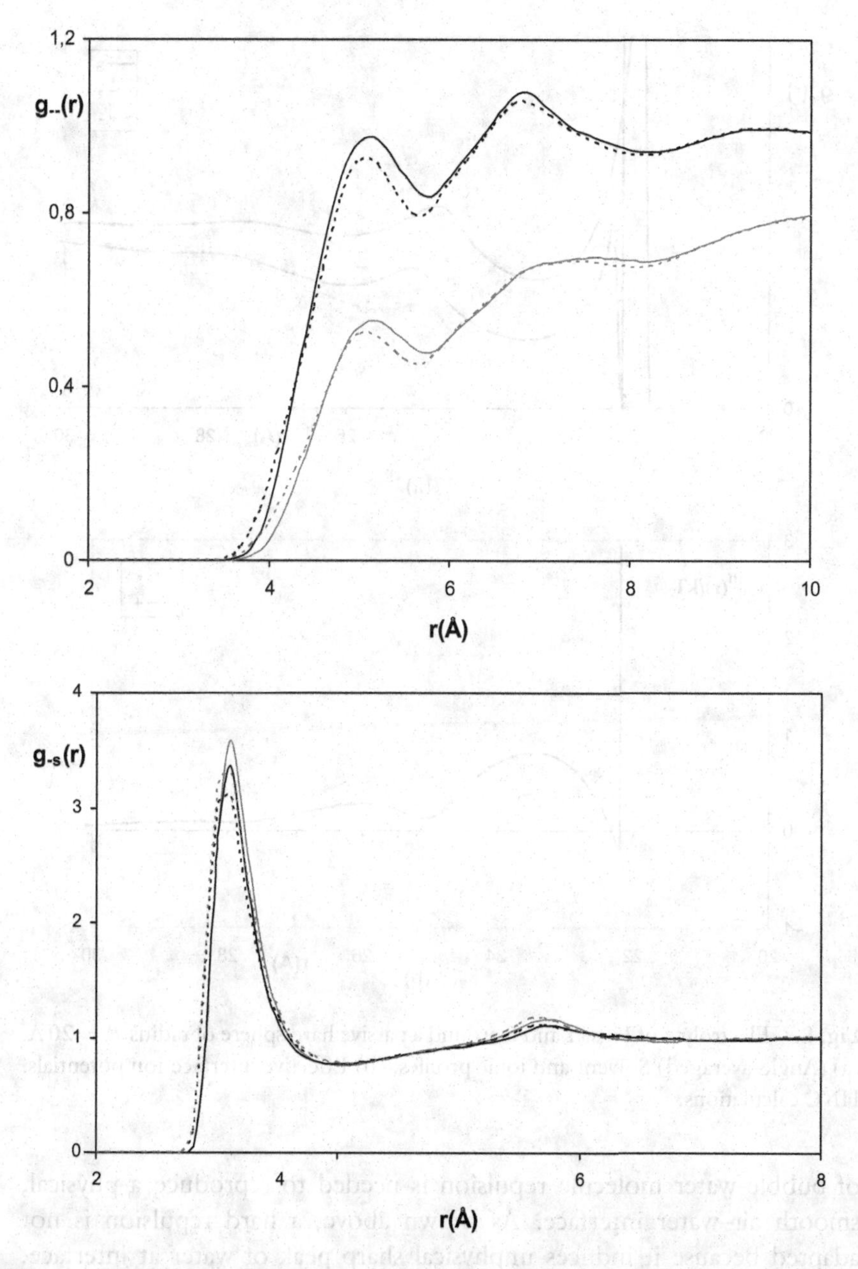

Fig. 7. HNC centre of mass–centre of mass pair distribution functions g_{--} and g_{-s} for the monovalent aqueous electrolyte of Fig. 4 at two salinities, 0.1 M (blue lines) and 1 M (green lines). Solid lines: non-polarisable ions; dashed lines: polarisable anions with $\alpha_{-} = 3.416\,\text{Å}^3$.

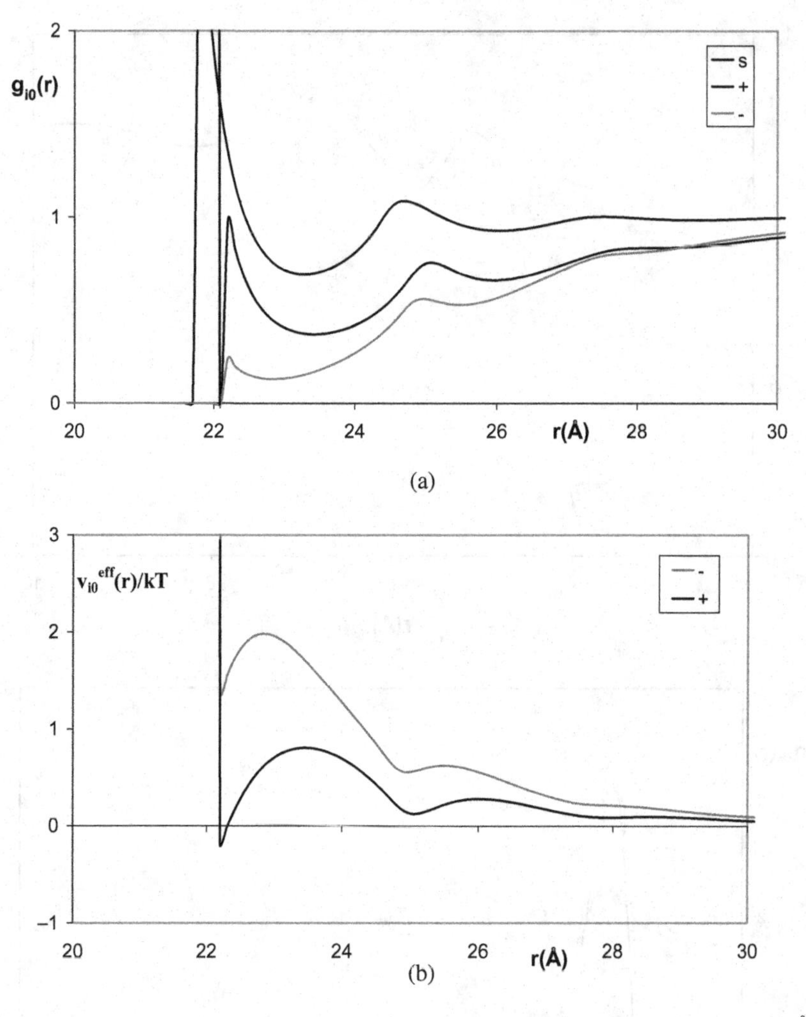

Fig. 8. Electrolyte of Figs. 4 and 5 around a passive hard sphere of radius $R = 20$ Å. (a) (Angle-averaged) Solvent and ionic profiles. (b) Effective interface-ion potentials. HNC calculations.

of bubble-water molecule repulsion is needed to reproduce a physical, smooth air–water interface? As shown above, a hard repulsion is not adapted because it induces unphysical sharp peak of water at interface. A much softer potential is required in order to get the reasonable 'hyperbolic tangent' shape. Figure 9 illustrates the role of the softness in the bubble-water potential in reproducing a standard air–water interface for

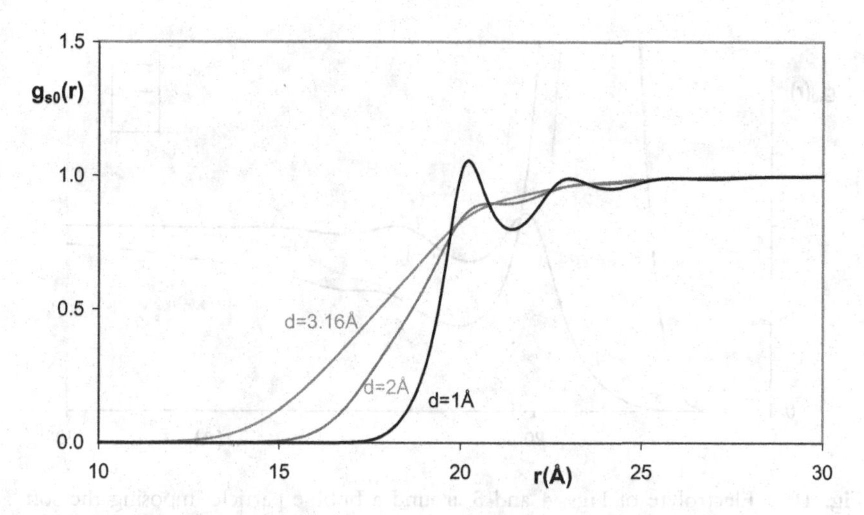

Fig. 9. Pure SPC water profiles around a sphere imposing the potential [Eq. (12)] with $R = 20$ Å and different ranges d. HNC calculations.

pure water. We have tried the following potential for water:

$$\exp(-\beta v^{000}_{00,s0}(r)) = \frac{1}{2}\left[1 + \tanh\left(\frac{r-R}{d/2}\right)\right], \tag{12}$$

where the parameter d monitors the extension of the liquid–vapour interface. It seems that the rough choice $d = 1$ Å gives a reasonable water profile.

Once the water profile has been constructed by adjusting this softness, it remains to choose the bubble-ion potentials. In fact, is it really necessary to impose such potentials? Is it not sufficient to let the ions respond to the imposed water profile? Indeed, in principle, since the ions are very well solvated in the liquid phase, they are not happy to go inside the bubble where there are no more water molecules. The chemical potential of the salt should be uniform; so, without imposing any excluded volume constraint for the ions, one expects that the concentration in the gas phase, equal to the (low) activity in the liquid phase, will spontaneously be very small and negligible. In practice, unfortunately, HNC ionic profiles present high density values inside the bubble without any ionic constraint, so high that, in fact, the HNC has no solution for the interface system! Figure 10 illustrates this HNC difficulty where the same soft potential [Eq. (12)] is offered to the ions but with a slightly lower bubble radius, $R = 19$ Å, than

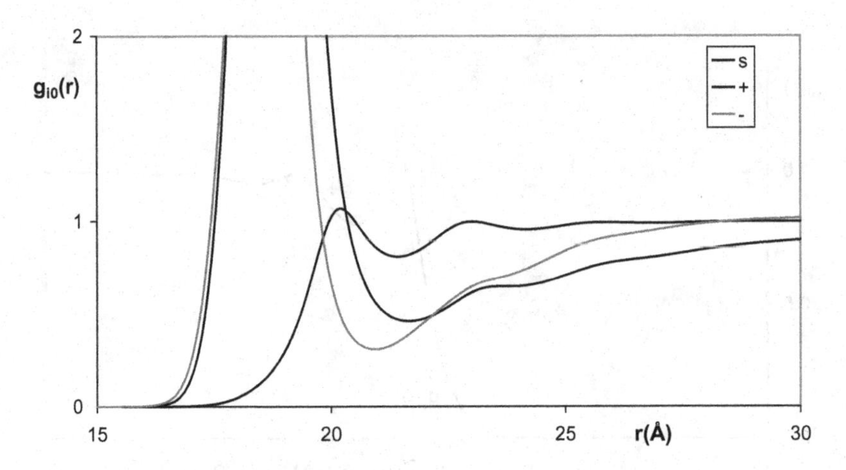

Fig. 10. Electrolyte of Figs. 4 and 5 around a bubble particle imposing the soft potentials [Eq. (12)] with $R = 20\,\text{Å}$, $d = 1\,\text{Å}$ for water and $R = 19\,\text{Å}$, $d = 1\,\text{Å}$ for ions. HNC calculations.

for water. The bubble is more penetrable to the ions and these accumulate in the extra layer despite the absence of solvent molecules. This illustrates the weakness of the bare HNC approximation for interfaces at the BO level and the importance of the neglected bridge interface-particles functions. More precisely, since the interfacial OZ equation involves explicitly the bulk ion–ion correlations only, it is difficult to account for the fact that the ions loose their hydration layers when they enter the empty bubble.

How to go beyond this intrinsic HNC problem? In the absence of the moment of improved integral equations (with some bridge functions in the conventional approach or with new, non-conventional ones), one is forced to use the standard procedure and impose reasonable constraints for the ions at will. Figure 11 illustrates what happens to the ionic distributions when the very same potential for water is offered to the ions as well (in a sense, this bubble-ion potential could be identified with the negative of the bridge function!). In this way, by construction, ions stay inside the liquid phase. The corresponding air–ion effective potentials are given in Fig. 11. With the same surface-particle potential, the more physical NaCl system of Fig. 6 exhibits quite a different interfacial property: despite the soft repulsion imposed on all ions, the chloride ions prefer now to stay at the interface, just 'on top' of water. To what extent does this

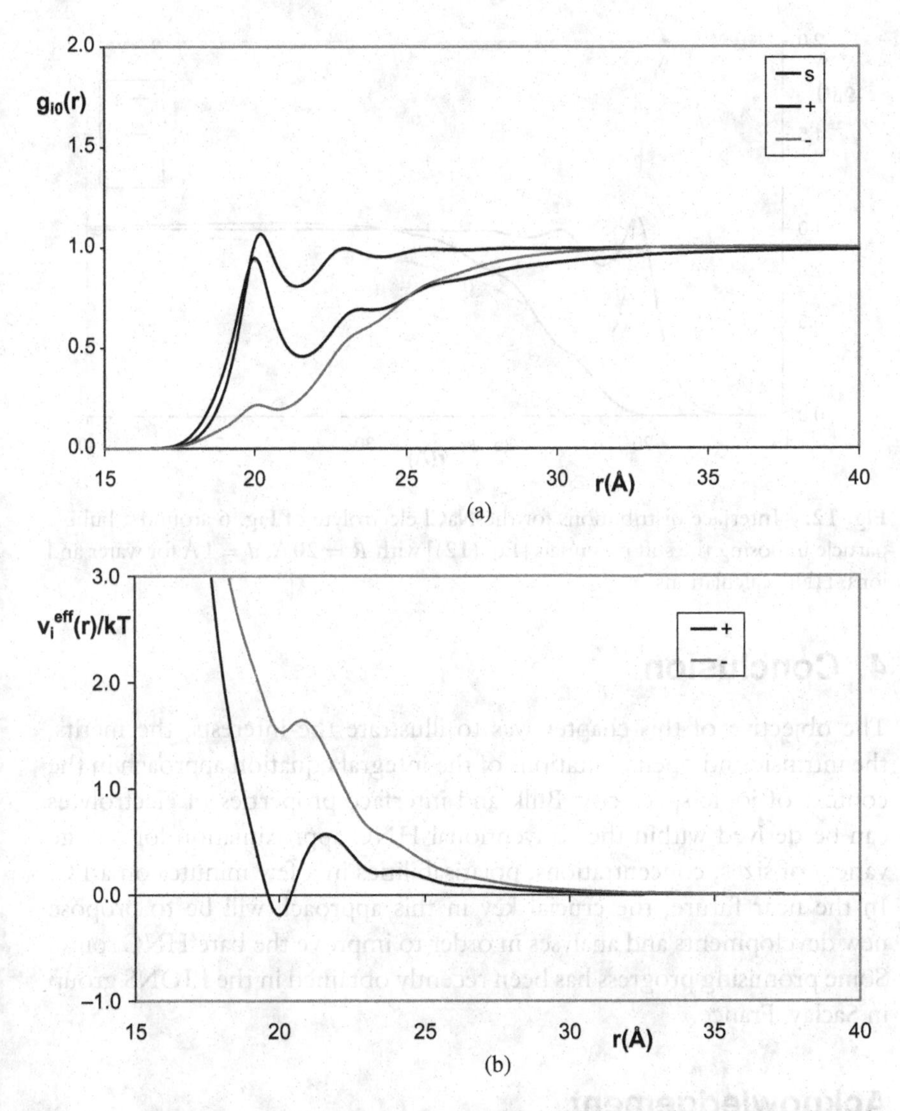

Fig. 11. Electrolyte of Figs. 4 and 5 around a bubble particle imposing the soft potentials [Eq. (12)] with $R = 20\,\text{Å}$, $d = 1\,\text{Å}$ for water and ions. (a) Local distributions. (b) Effective air-ion potentials. HNC calculations.

interesting behaviour, which looks like some simulation data described in other chapters, result from a correct HNC analysis or is it too dependent on the precise choice for the surface-ion potential? That must and will be analysed in further studies.

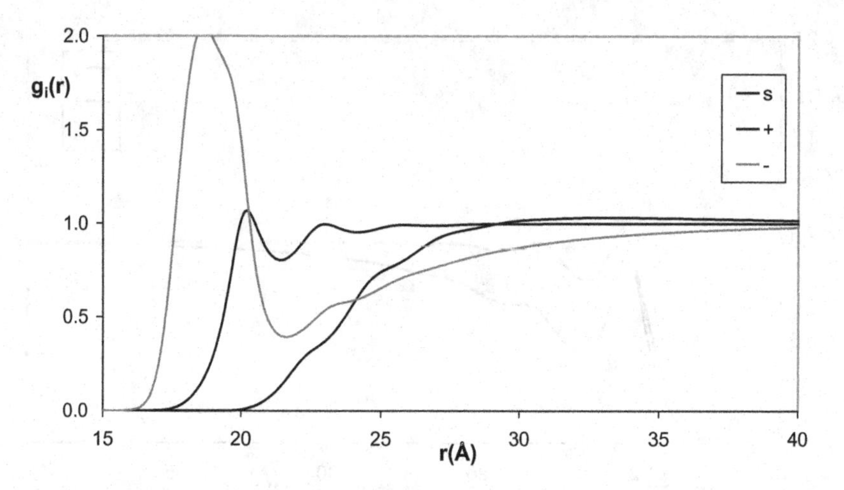

Fig. 12. Interface distributions for the NaCl electrolyte of Fig. 6 around a bubble particle imposing the soft potentials [Eq. (12)] with $R = 20\,\text{Å}$, $d = 1\,\text{Å}$ for water and ions. HNC calculations.

4. Conclusion

The objective of this chapter was to illustrate the interests, the merits, the intrinsic and open limitations of the integral equation approach in the context of ionic specificity. Bulk and interface properties of electrolytes can be derived within the conventional HNC approximation for a large variety of sizes, concentrations, polarisabilities in a few minutes on a PC. In the near future, the crucial key in this approach will be to propose new developments and analyses in order to improve the bare HNC route. Some promising progress has been recently obtained in the LIONS group in Saclay, France.

Acknowledgement

Support from ANR under Grant No. ANR-06-BLAN-0276 is gratefully acknowledged.

References

1. Verwey EJW, Overbeek JTG. (1948) *Theory of Stability of Lyotropic Colloids.* Elsevier, Amsterdam.
2. Belloni L. (2000) *J Phys Cond Matt* **12**: R549–R587.

3. Ninham BW, Yaminsky VV. (1997) *Langmuir* **13**: 2097.
4. Hansen JP, McDonald IR. (1986) *Theory of Simple Liquids*. Academic Press, London.
5. Zerah G. (1985) *J Comput Phys* **61**: 280.
6. Belloni L. (1988) *J Chem Phys* **88**: 5143.
7. Kunz W, Belloni L, Bernard O, Ninham BW. (2004) *J Phys Chem B* **108**: 2398.
8. Henderson D, Abraham FF, Barker JA. (1976) *Mol Phys* **31**: 1291.
9. Kjellander R, Marcelja S. (1985) *J Chem Phys* **82**: 2122.
10. Padmanabhan V, Daillant J, Belloni L, Mora S, Alba M, Konovalov O. (2007) *Phys Rev Lett* **99**: 086105.
11. Blum L, Torruella AJ. (1972) *J Chem Phys* **56**: 303; Blum L. (1972) *J Chem Phys* **57**: 1862.
12. Lado F. (1982) *Mol Phys* **47**: 283.
13. Fries PH, Patey GN. (1985) *J Chem Phys* **82**: 429.
14. Richardi J, Fries PH, Fisher R, Rast S, Krienke H. (1998) *Mol Phys* **93**: 925 and Ref. 29 therein.
15. Anta JA, Lomba E, Martin C, Lombardero M, Lado F. (1995) *Mol Phys* **84**: 743; Lado F, Lomba E, Lombardero M. (1995) *J Chem Phys* **103**: 481.
16. Berendsen HJC, Postma JPM, van Gunsteren WF, Hermans J. (1981) In: Pullmann B (ed), *Intermolecular Forces*, Reidel, Dordrecht.
17. Richardi J, Millot C, Fries PH. (1999) *J Chem Phys* **110**: 1138.
18. Kusalik PG, Patey GN (1988). *J Chem Phys* **88**: 7715.
19. Adelman SA. (1976) *J Chem Phys* **64**: 724.
20. Carnie SL, Patey GN. (1982) *Mol Phys* **47**: 1129.
21. Fries PH, Kunz W, Calmettes P, Turq P. (1994) *J Chem Phys* **101**: 554.
22. Berendsen HJ, Grigera JR, Straatsma TP. (1987) *J Chem Phys* **91**: 6269.

Chapter 11

Modifying the Poisson–Boltzmann Approach to Model Specific Ion Effects

Mathias Boström[*,†], Eduardo R. A. Lima[‡,§], Evaristo C. Biscaia Jr.[§], Frederico W. Tavares[‡] and Werner Kunz[*]

In this chapter, a short history is given about the introduction of non-electrostatic interactions in the Poisson–Boltzmann equation, starting from simple ionic dispersion forces up to very recent approaches, in which both water profile and ion-surface potentials inferred from molecular dynamics simulations are used.

1. Introduction

For many decades, DLVO (Derjaguin-Landau-Verwey-Overbeek) theory has been used to describe interactions between colloidal systems. Indeed it is a very successful theory which can reproduce many phenomena.[1,2] However, specific ion effects are not taken into account in this theory. This has different reasons: DLVO consists of a sum of electrostatic and van der Waals interactions. As Ninham pointed out,[3] this combination is not made in a rigorous way. We will discuss this in Sec. 2.

[*]Institute of Physical and Theoretical Chemistry, University of Regensburg, D-93040 Regensburg, Germany.
[†]Division of Theory and Modeling, Department of Physics, Chemistry and Biology, Linköping University, SE-581 83 Linköping, Sweden.
[‡]Escola de Química, Universidade Federal do Rio de Janeiro, Cidade Universitária, CEP 21949-900, Rio de Janeiro, RJ, Brazil.
[§]Programa de Engenharia Química, COPPE, Universidade Federal do Rio de Janeiro, 21941-914, Rio de Janeiro, RJ, Brazil.

Even when this basic theoretical problem is resolved, DLVO is still a so-called primitive model, in which the solvent structure is not explicitly taken into account. Only the dielectric constant describes it. However, the solvation of ions is a crucial factor that discriminates one ion from another. And the description of solvent structure, especially around ions, is not in the focus of DLVO.

Usually ion hydration is taken into account with different ion sizes. But such a simplified correction via 'decorated' ions is by no means sufficient to explain drastic differences in colloidal systems, when for example hydroxide ions are replaced by bromide ions.[4,5] Ternary phase diagrams of water, ionic surfactants and oil can be completely different, depending on the type of the counterion that is used with the surfactant.[6] Further, vesicular structures composed of surfactants may considerably swell or shrink, when counterions are exchanged, even if the charge of the ions is always the same.[7]

So what to do? Probably the simplest way of testing specific ion effects in colloid science is to consider the air–solution interface of simple aqueous salt solutions. In a pioneering work, Jungwirth and Tobias[8] could show that it is probably crucial to take into account the polarisability of the ions, such as Ninham and Yaminsky[9] proposed in their landmark paper several years earlier. But in contrast to Ninham and Yaminsky, Jungwirth and Tobias used molecular dynamics (MD) simulations in order to predict surface tensions. They found out that it is necessary to consider the influence of the ion polarisability on the solvent structure in order to get the right ordering of the surface tension. And they also could demonstrate that the surface tension of a salt solution can increase, despite the fact that in the very first surface layer ions are adsorbed. This is simply because, according to the Gibbs equation, the surface tension is an integral over the ion profiles and does not give information about local adsorption or desorption processes on the surface. This paper by Jungwirth and Tobias and several others initiated a renaissance in the research of specific ion effects and Hofmeister series.[10]

Nevertheless, some problems remained. Although MD simulations very roughly predicted the right Hofmeister series of surface tensions, such simulation techniques are heavy to carry out. Further, the simulation box is small, leading to possible artefacts in the calculation of the Gibbs integral. Due to the statistical noise, thermodynamic properties cannot be calculated with high precision. It is even very difficult to approximate the surface tension of pure water. In any case, MD simulations are not appropriate for rapid calculations of surface properties.

In this respect the approach by Ninham and Yaminsky is much easier to use. In principle the influence of solvent structure can be taken into account within the DLVO model by using a convenient Lifshitz-like ansatz.[9] There, all non-electrostatic interactions are taken into account via frequency summations over all electromagnetic interactions that take place in the solutions. If done rigorously, the result should be more or less exact. As a proof of principle, Boström and Ninham made a first attempt in this direction. The classical DLVO ansatz was replaced by a modified Poisson–Boltzmann (PB) equation, in which a simplified so-called dispersion term was added to the electrostatic interaction. In this way ion specificity came in quite naturally via the polarisability and the ionisation potential of the ions. However, it turned out that this first-order approximation of the non-electrostatic interactions was not sufficient to predict the Hofmeister series of surface tension.[11] Heavier ions such as iodide had to be supposed to have smaller polarisabilities compared to smaller ions such as chloride. Although the exact polarisabilities of ions in water are still under debate, this is not physical.

Further on, Kunz *et al.*[12] could show that this failure of the simplified dispersion model is not a consequence of the weakness of the Poisson–Boltzmann equation. More elaborate statistical mechanics, using the so-called hypernetted chain equation (HNC), yielded basically the same result. Obviously the problem comes from the neglect of ion–water interactions and their changes near the surface. To introduce such interactions in primitive model calculations, Boström *et al.*[13a] [see also Refs. 13(b)–13(d)] used Jungwirth's water profile perpendicular to the surface as a basis to model a distance-dependent electrostatic function, instead of a static dielectric constant. Such ideas were used several times over the years, for instance to model activity coefficients of electrolyte solutions.[14]

With such a modification, the description of ion profiles on surfaces became more physical within this modified DLVO approach. However, some of the profiles were still not satisfactory, probably because of the still too crude approximation of the so-called dispersion forces. In the following sections we will present the newest DLVO theory and discuss some exciting results.

2. Why DLVO Theory Is Incomplete and What to Do About It

Even at the level of the primitive (continuum solvent) model, it has been proved by Ninham and Yaminsky[9] that the extension of the Lifshitz theory that includes the contribution of ions is inconsistent. It does include ion specificity through its reliance on measured dielectric susceptibilities. But the effect is very weak. The theory is based on a linear response approximation. The ansatz that double layer forces treated by a non-linear theory and dispersion forces treated by a linear theory can be handled independently is inconsistent.[9] As discussed often by Ninham, this ansatz violates two thermodynamic requirements: the gauge condition on the electromagnetic field and the Gibbs adsorption equation. This is not just an esoteric point, but has major quantitative ramifications that are missing from classical theories. This is so also for interfacial effects. The dominating temperature-dependent, zero-frequency electrolyte-dependent many-body interaction in the extended Lifshitz theory must be precisely equivalent to Onsager's limiting law for the change of interfacial tension at a single air– or oil–water interface. And this, as remarked by Onsager himself, is invalid except perhaps at extremely low electrolyte concentrations. No ion specificity is captured. It is also so for electrolyte activities which involve ionic interactions. It is so for the standard Born free energy of transfer which again misses dispersion interactions due to the dynamic polarisability, contributions that are ion specific and range up to 25 to 50% of the whole free energy.[13(b)–13(d)]

Removal of the inconsistency in the manner Ninham, Boström and co-workers have described in a series of papers goes part of the way to a resolution, but only at the level of the primitive model. More is involved once the molecular nature of the solvent is taken into account. Ionic polarisability is combined at the same level with electrostatics to determine local induced water structure around ions. That is a key determinant of hydration, and of so-called ion specific Gurney potentials of interactions between ions due to the overlap of these solute-induced solvent profiles; so too for ionic adsorption at interfaces. The notions involved here embrace quantitatively the conventional ideas of cosmotropic, chaotropic, hard and soft ions. Some insights into this matter can be obtained via the alternative approach of computer simulation techniques of Jungwirth *et al.*[8] But the insights are hamstrung so far by a pragmatic restriction that limits simulation to too simple models of water. This approach works so far for ions like Br^-, Cl^-, K^+, Na^+, for which ionic hydration is not a problem. But for other ions some doubts remain. For example, present techniques and simulation models[8] predict that H_3O^+ is slightly more adsorbed to the air–water interface than OH^-, which is still a matter of debate and in apparent contrast to some experimental evidences.

Our point here is not to claim a complete description of the phenomena. It is that even at the primitive model level the inclusion of dispersion effects accounts for at least some specific ion effects. Inclusion of potential of mean force (PMF) from simulations improves the understanding of the Hofmeister effect significantly. Both ionic dispersion potentials and the PMF depend on the polarisability of the ions. It has been firmly established that the polarisability has an important role for the ion specificity observed in colloidal systems and in biology.

3. Modified Poisson–Boltzmann Equation (MPBE)

The distribution of ions at a charged surface is a fundamental problem of colloid and interface science. Gouy (1910)[15] and Chapman (1913)[16] were the first who considered this problem in a more quantitative fashion. The ions were considered as point charges embedded in a continuum of constant dielectric constants. The chemical potential, μ_i, of the ion 'i' with charge, ez_i, is given by[17,18]:

$$\mu_i(p, T) = \mu_{0i}(p, T) + k_B T \ln(c_i) + ez_i\phi, \tag{1}$$

where μ_{0i} represents the standard chemical potential of the ion of species 'i', k_B is the Boltzmann constant, T is the temperature, c_i stands for the ion concentration, e is the elementary charge, z_i is the ion valence and ϕ is the self consistent electric potential. In thermal equilibrium, μ_i remains constant throughout the system. Each sort of ion 'i' obeys the Boltzmann distribution.

$$c_i = c_{0i} \exp\left(-\frac{z_i e\phi}{k_B T}\right), \tag{2}$$

where c_{0i} is the ion concentration at the bulk reservoir.

The classical theory accounts only for electrostatics and thermal motion and neglects several important effects such as dispersion forces, fluctuation, hydration, ion size effects and the water structure at interface. Several extensions of the theory aim at including one of these contributions to the model in the framework of mean-field Poisson–Boltzmann theory.

The fundamental Poisson equation is used[18] to self-consistently relate the electric potential ϕ to the net excess charge density at a position $\mathbf{x}$:

$$\nabla(\varepsilon_0 \varepsilon_w \nabla \phi(\mathbf{x})) + 4\pi\rho(\mathbf{x}) = 0, \tag{3}$$

where $\varepsilon_0 = 8.854 \times 10^{-12} C^2/(Jm)$ is the dielectric permittivity of vacuum, ε_w is the dielectric constant of the solvent and $\rho(\bar{x})$ is the net charge density at a given position, defined as:

$$\rho(\bar{x}) = e \sum_i z_i c_i(\bar{x}). \tag{4}$$

The interaction between an ion and a surface is not only electrostatic: each ion experiences a further additive term that come partly from dispersion forces, U_i. Then, we have a modified Boltzmann distribution of ions in the solution.[11,13]

$$c_i = c_{0i} \exp\left(-\frac{[z_i e\phi + U_i]}{k_B T}\right). \tag{5}$$

The combination of the Boltzmann distribution to the Poisson equation for net excess charge density leads to a non-linear second-order differential equation for the electric potential ϕ. Therefore, the Poisson–Boltzmann equation is written as:

$$\nabla \cdot (\varepsilon_o \varepsilon_w(x) \nabla \phi(x)) + 4\pi e \sum_i z_i c_{oi} \exp\left(-\frac{z_i e\phi(x) + U_i(x)}{k_B T}\right) = 0. \tag{6}$$

4. Ion-Surface Non-electrostatic Potentials from Lifshitz Theory

Recently, Ninham and Parsons presented model calculations of polarisabilities of many different ions. This opens up for quantitative calculations of ionic dispersion potentials. However, for completeness we give also here some simple estimates of the non-electrostatic (NES) potential acting between an ion and a membrane surface:

$$U_\pm = \frac{B_\pm}{x^3}, \tag{7}$$

where the dispersion coefficient ($B_{\pm}$) will be different for different combinations of ion and membrane. The non-retarded dispersion coefficients can be calculated from Lifshitz theory as a sum over imaginary frequencies ($i\omega_n = i2\pi k_B T n/\hbar$, where $\hbar$ is Planck's constant),

$$B_{\pm} = \sum_{n=0}^{\infty} \frac{(2-\delta_{n,0})\alpha^{\pm}(i\omega_n)[\varepsilon_w(i\omega_n) - \varepsilon_{oil}(i\omega_n)]}{4\beta\varepsilon_w(i\omega_n)[\varepsilon_w(i\omega_n) + \varepsilon_{oil}(i\omega_n)]}. \tag{8}$$

The increment of the refractive index of water when a salt solution is added is different for different salt solutions.[19] The refractive index of pure water is $n_w = 1.3333$, for a 0.051-M KSCN solution the refractive index has increased to $n = 1.3339$. For a 0.067-M KCl solution the refractive index has increased to $n = 1.3337$. Similarly, to increase the refractive index of a salt solution up to 1.3404 one must add 0.763-M KCl, whereas if KSCN is used it suffices to add 0.419 M. The sum of static excess polarisabilities for the anion and the cation can then be estimated from the following approximation:

$$n^2 \approx n_w^2 + 4\pi c_{ion}(\alpha_+(0) + \alpha_-(0)). \tag{9}$$

We find that the sum of static excess polarisabilities is $\approx 3.68\,\text{Å}^3$ for KCl, and $\approx 6.22\,\text{Å}^3$ for KSCN. Anions with additional electrons are expected to be more polarisable than cations. If we assume that the static excess polarisability of Cl^- is at least as large as K^+, the static excess polarisability of SCN^- should be around 4.4 to $6.2\,\text{Å}^3$. We model the excess polarisability as

$$\alpha_{\pm}(i\omega_n) = \alpha_{\pm}(0)/(1 + \omega_n^2/\omega_0^2). \tag{10}$$

The effective resonance frequencies (ω_0) for different ions are not known, but should typically be in the range 1×10^{16} to 5×10^{16} rad/s.[12,20]

5. Comparison Between MPBE and Mean-Field Monte Carlo Simulations

The question that can be raised is whether ion size correlation and electrostatic correlation not accounted for in MPBE may influence the result. Solvent-averaged Monte Carlo (MC) simulations were performed by Boström *et al.*[21] for ion distributions outside a single globular macroion in different salt solutions. The model we used included both electrostatic and NES interactions between ions and between ions and macroions.

Simulation results were compared with the predictions of the Ornstein–Zernike (OZ) equation with the hypernetted chain (HNC) closure approximation and the non-linear Poisson–Boltzmann equation, both augmented by pertinent Lifshitz NES potentials. We show in Fig. 1 that there is very good agreement between modified Poisson–Boltzmann theory, MC simulations, and HNC calculations when the counterions and co-ions are monovalent. There is also good agreement between the different approaches with divalent co-ions (not shown here).[21] However, the results from MPBE cannot account for ion correlation effects that occur in Fig. 2 when the counterions are divalent. The reason is simply that the

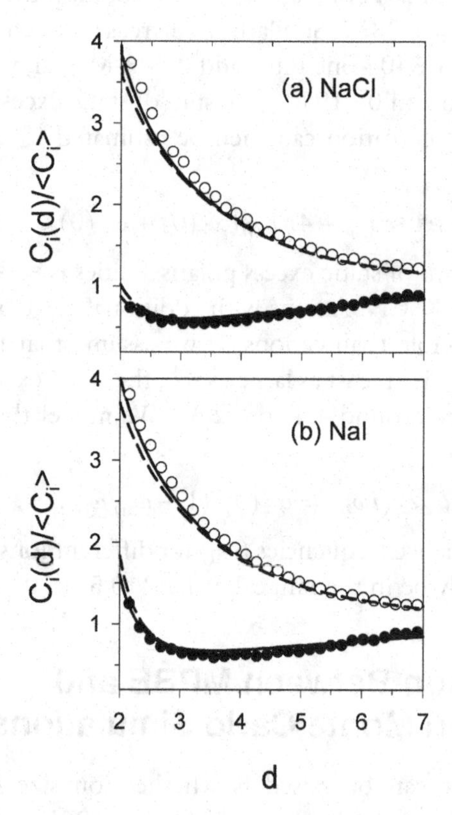

Fig. 1. Concentration profiles near a macroion ($\sigma_M = 30$ Å and $-20e_0$) in a monovalent electrolyte solution of ionic strength 1.0 M for (a) NaCl and (b) NaI. Open circles represent counterion concentrations and dark circles the co-ions. Solid lines are numerical solutions of the non-linear Poisson–Boltzmann equation and dashed lines are for the OZ–HNC integral equation.[21]

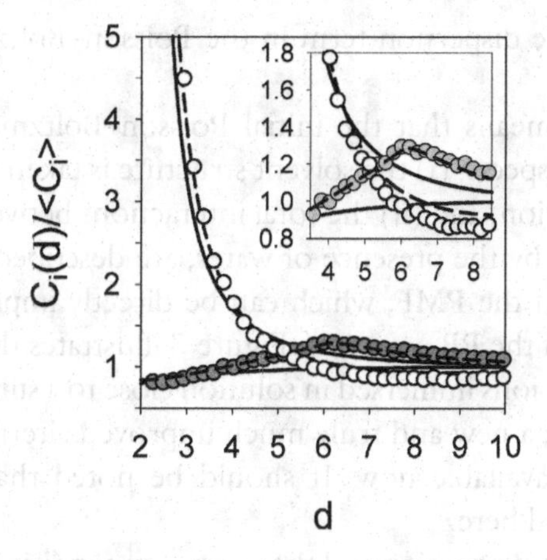

Fig. 2. Concentration profiles near a macroion ($\sigma_M = 30$ Å and $+20e_0$) in solution of ionic strength 1.0 M for Na_2SO_4.[21] Open circles represent counterion concentrations (sulfate) and gray circles the co-ions (Na^+). Solid lines are numerical solutions of the non-linear Poison–Boltzmann equation and dashed lines are for the OZ–HNC integral equation. The inset figure shows that the HNC calculations for co-ion distributions coincide with the simulation results.

divalent counterions accumulate close to the colloid surface, preventing by size correlation co-ions from coming close to the colloid surface. The co-ions are here seen to form an outer layer.

However, the general conclusion is that MPBE is in many cases a good approximation as compared to mean-field MC simulations or HNC calculations that account for ion size. This motivates us to use MPBE in improved calculations that account for monovalent ion potentials of mean force and water density profiles near surfaces.

6. NES Potentials from Simulations and Modified DLVO Theory

Recently, Horinek and Netz (Ref. 22 and Chap. 9 of this book) could successfully infer from MD simulations potentials of mean force (PMFs) of ions near the interface between a self-organised hydrophobic monolayer (SAM) and aqueous salt solutions. Interestingly it was by no means possible to interpret these PMFs by simple analytical physical models. They seem to be a complex balance of different interactions between ion and water and surface. Anyway, these PMFs can be used in numerical form to replace the dispersion term in the Poisson–Boltzmann equation previously used.

In total, this means that the initial Poisson–Boltzmann equation is modified in two respects: (i) the solvent structure is taken into account via the dielectric function, and (ii) the total interactions between ions and the surfaces, modified by the presence of water, are described by the electrostatic potential and the PMF, which can be directly implemented in the interaction term of the PB equation. Figure 3 illustrates these interactions experienced by the ions immersed in solution close to a surface. With these two modifications, a new and truly much improved alternative to the old DLVO theory is available now. It should be noted that no adjustable parameters are used here.

The modified PB equation and the boundary conditions are as follows:

$$\varepsilon_0 \frac{d}{dx}\left(\varepsilon(x)\frac{d\phi}{dx}\right) = -e\sum_i c_{0,i} z_i \exp[-(z_i e\phi + U_i(x))/k_B T], \qquad (11)$$

$$\left.\frac{d\phi}{dx}\right|_{x=L/2} = 0, \quad \left.\left(\varepsilon(x)\frac{d\phi}{dx}\right)\right|_{surface} = -\frac{\sigma}{\varepsilon_0}, \qquad (12)$$

here σ is the surface charge, U_i is the ionic non-electrostatic potential inferred from MD simulations (PMF) and acting between the surface and the ion. The local dielectric constant $\varepsilon(x)$ is defined as a function of the

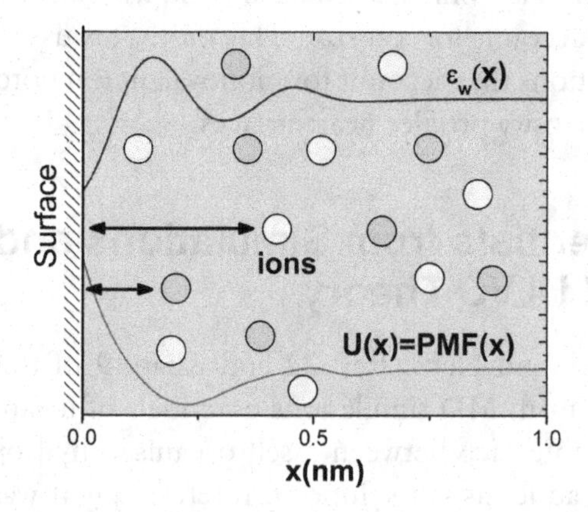

Fig. 3. Ions in solution close to a surface. Each ion experiences electrostatic and non-electrostatic (PMF) interactions with the surface. They also 'feel' the solvent structure, expressed in our model by the dielectric function (ε_w).

position x based on density simulation data, considering that these two parameters are highly correlated. The expression used here for SAM–water system is

$$\varepsilon(x) = \varepsilon_{SAM} + \rho_{wSAM}(x)(\varepsilon_w - \varepsilon_{SAM}), \tag{13}$$

where $\varepsilon_w = 78.5$ is the dielectric constant of water, $\varepsilon_{SAM} = 4$ is the dielectric constant of SAM, and $\rho_{wSAM}(x)$ is the density profile obtained from molecular simulation.[22] The density profile is expressed as

$$\rho_{wSAM}(x) \approx a_1 \tanh[b_1(x - c_1)] + a_2 \tanh[b_2(x - c_2)] + a_3 \tanh[b_3(x - c_3)] + a_4 \tanh[b_4(x - c_4)] + a_5, \tag{14}$$

where the parameters are $a_1 = -62.555$, $a_2 = 62.749$, $a_3 = -0.49722$, $a_4 = 0.80256$, $a_5 = 0.49923$, $b_1 = 9.5772$, $b_2 = 9.5454$, $b_3 = 14.061$, $b_4 = 14.029$, $c_1 = 0.46115$, $c_2 = 0.46067$, $c_3 = 0.21581$, $c_4 = 0.06589$.

Equation (11) can be solved using finite volume methods[23] to determine the ion distributions between two SAMs. In the following we show some results of this very new approach.

The obtained electrostatic potential profiles and ion distributions can in principle be used to calculate surface or interfacial tensions. However, up to now only few PMFs for ion–water surface interactions are available from MD simulations and there are no reliable experimental data of interfacial tensions for SAM–solution interfaces. Therefore it is not yet possible to check if the correct Hofmeister series can be obtained with this new approach.

But other properties can be inferred, such as the pressure between surfaces. The MD data from Horinek *et al.* allowed us to calculate the double layer pressure between two planar surfaces at a distance L simply by the differentiation of the free energy of the system[24–26]:

$$P = -\frac{\partial}{\partial L}\left(\frac{A}{area}\right), \tag{15}$$

and the free energy per unit of area is expressed by

$$\frac{A}{area} = \frac{e}{2}\int_{x_{\min}}^{L} \phi \sum_i c_i z_i dx + \int_{x_{\min}}^{L} \sum_i c_i U_i dx + k_B T \int_{x_{\min}}^{L} \sum_i c_i \left[\ln\left(\frac{c_i}{c_{0,i}}\right) - 1\right] dx. \tag{16}$$

The first two terms in the right-hand side of the Eq. (16) are the energy contributions (electrostatic and the ionic potential of mean force contribution, respectively) to the free energy of the system and the third term is the entropic contribution. The derivative of the free energy can be solved numerically or developed analytically to get[27]

$$P = k_B T \sum_i [c_i(L/2) - c_{0,i}] - 2 \sum_i \int_{x_{\min}}^{L/2} c_i \frac{dU_i}{dL} dx. \qquad (17)$$

7. Hofmeister Effect and Forces Between Hydrophobic Surfaces

Figure 4 shows the contribution of the ions to the double layer pressure of two self-assembled monolayers (SAM) as a function of their distance. Of course, this is just a calculation and no comparison to experiment is made at the moment. However, several interesting features can be seen. We observe that for distances greater than roughly 1 nm, the pressure between two SAM surfaces becomes more repulsive in the order NaCl < NaBr < NaI, following an inverse Hofmeister series, where the more polarisable the anion, the more repulsive the pressure. But there can also be more surprising effects such as a reversal of the Hofmeister series with plate

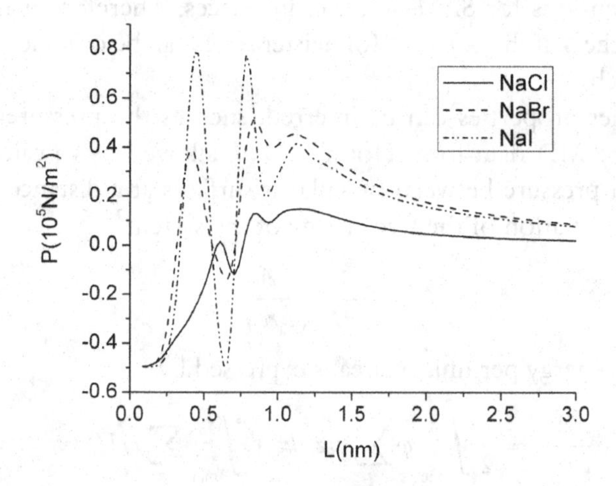

Fig. 4. Double layer pressure between two SAM surfaces interacting at different salt solutions: NaCl (solid line), NaBr (dashed line) and NaI (dash-double dotted line) at a temperature of 298.15 K and a concentration of 0.01 M.[24]

separation (at about 0.7 nm). Using surface force measurements between gold surfaces coated with self-assembled monolayers, one should be able to observe this transition from a direct to a reversed Hofmeister sequence. Even more, not only the series is reversed, but the pressure can go from a positive (repulsive) one to a negative (attractive) one. This model for Hofmeister effects exploiting simulations and mean-field theory, originally proposed by Luo *et al.*[28] and recently extended by Lima *et al.*,[26] truly predicts an interesting and complex picture.

The question is whether there is any experimental evidence in favour of these results. The answer is yes. To see this we consider the ion profiles near the surfaces, at a fixed distance of the layers, see Fig. 5. We observe that the ion specific PMF interactions give rise to a negatively charged adsorption layer near to each surface, so charging the surface and resulting in a repulsive double layer pressure. The concentration of bromide near the surfaces is around eight times the bulk concentration. Because of the higher concentration of anions near the surfaces, the pressure is more repulsive for bromide than for chloride. One can observe that the concentration of Na^+ between the plates increases when the anion is more polarisable. However, it is not sufficient to screen out the electrostatic repulsion between the two negative adsorption layers. This result is in reasonable agreement with very careful experiments that demonstrate preferential adsorption of bromide ions to neutral lipid membranes.[29]

The observation that bromide, and even iodide, can strongly adsorb on hydrophobic surfaces due to non-electrostatic forces, can also have a considerable impact in electrochemistry. There, for more than 80 years, Otto Stern's idea of an obscure non-electrostatically adsorbed layer is an *ad hoc* assumption that is not really explained.

8. Co-ion Effects and Ion Distributions near Hydrophobic Surfaces

In this section we present some unexpected and interesting results regarding ion competition and co-ion effects. At the uncharged SAM surface, the relative height of the adsorption peak of the anions decreases with increasing bulk concentration.[26] At the precharged surface this is not always the case. This is due to nontrivial competition of electrostatic forces and the attractive PMF that acts on the ions near the interface.

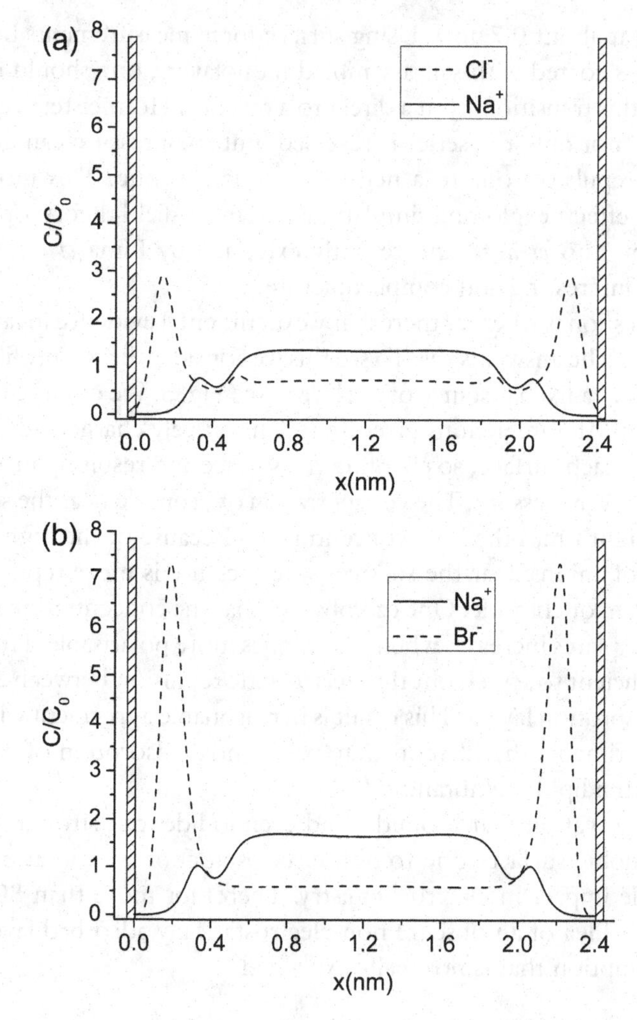

Fig. 5. Concentration profile of ions between two uncharged SAM surfaces at a bulk concentration $C_0 = 0.01$ mol/L and effective distance $L = 2.4$ nm. We consider two salt solutions: (a) NaCl and (b) NaBr. The hatched vertical bars represent the SAM surfaces.[24]

First of all, in Fig. 6 we show the influence of the surface charge density and bulk ion concentration on the distributions of Cl^- near a charged SAM surface immersed in a pure NaCl salt solution. We observe that for a bulk concentration of 10 mM [Fig. 6(a)], the first peak in the concentration of Cl^- is almost equal to the bulk concentration, while for a higher

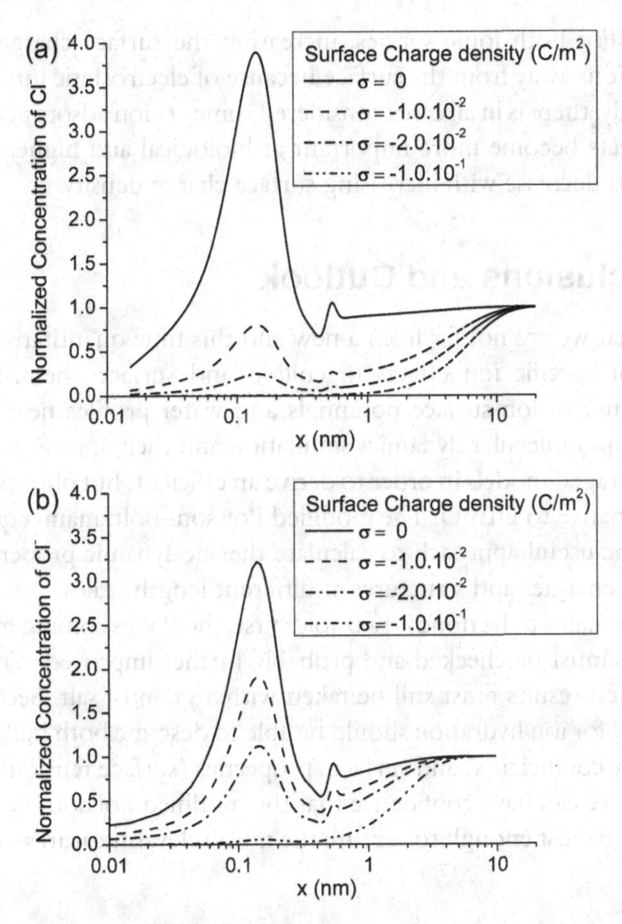

Fig. 6. Normalised concentration ($c_i(x)/c_{i,0}$) of chloride near charged SAM surfaces immersed in a NaCl salt solution with concentration (a) 10 mM and (b) 100 mM. Results show the influence of surface charge density.[26]

concentration of 100 mM [Fig. 6(b)], this peak reaches a value twice the bulk value. The relative enhancement is much higher at 100 mM than at 10 mM (1.9 versus 0.8). The reason for this apparently unintuitive effect is that there is a competition between the electrostatic repulsive force due to equal sign of the charge of the surface and the anions and the attractive PMF acting between the anions and the SAM. At low salt concentrations, the electrostatic repulsive force dominates. At biological and higher concentrations, the attractive ion specific PMF dominates. At even higher concentrations, the relative enhancement decreases due to self repulsion of anions. For both ionic species, increasing the surface charge density pushes co-ions away from the surface because of electrostatic interactions. Interestingly, there is in all cases considered some co-ion adsorption. These co-ion effects become more important at biological and higher concentrations and decrease with increasing surface charge density.

9. Conclusions and Outlook

It seems that we are not far from a new and this time quantitative understanding of specific ion effects in colloid and surface chemistry. The key is the use of ion-surface potentials and water profiles near surfaces inferred from molecular dynamics simulation and their appropriate use in solvent-averaged models in order to derive an efficient, but physically well-based alternative to DLVO. The modified Poisson–Boltzmann equation is shown to be useful approach to calculate thermodynamic properties that depend on energies and structures at different length scales.

What remains to be done is two-fold: first, the PMFs coming from MD simulations must be checked and probably further improved. Therefore, the presented results must still be taken with a grain of salt. Second, the same model for ion hydration should be able to describe both bulk properties (activity coefficients) and surfaces properties (surface tensions). If this is possible, we can have confidence that the modified Poisson–Boltzmann approach is robust enough to capture the most relevant features of specific ion effects.

Acknowledgements

We thank the Swedish Research Council, the German Arbeitsgemeinschaft industrieller Forschungvereinigungen Otto von Guericke e.V. (AiF), and FAPERJ, CNPq and CAPES (the last three Brazilian agencies) for financial support.

References

1. Marcelja S. (2004) *Curr Op Coll Interf Sci* **9**: 165.
2. Pashley RM, McGuiggan PM, Ninham BW, Brady J, Evans DF. (1986) *J Phys Chem* **90**: 1637.
3. Ninham BW. (1999) *Adv Coll Interf Sci* **83**: 1.

4. Ninham BW, Evans DF, Wie GJ. (1983) *J Phys Chem* **87**: 5020.
5. Ninham BW, Hashimoto S, Thomas JK. (1983) *J Coll Interf Sci* **95**: 594.
6. Nyden M, Söderman O, Hansson P. (2001) *Langmuir* **17**: 6794.
7. Karaman ME, Ninham BW, Pashley RM. (1994) *J Phys Chem* **98**: 11512.
8. Jungwirth P, Tobias DJ. (2001) *J Phys Chem B* **105**: 10468.
9. Ninham BW, Yaminsky V. (1997) *Langmuir* **13**: 2097.
10. Wilson EK. (2007) A Renaissance for Hofmeister, *Chem & Eng News* **85**: 47
11. (a) Boström M, Williams DRM, Ninham BW. (2001) *Langmuir* **17**: 4475; (b) Boström M, Williams DRM, Ninham BW. (2001) *Phys Rev Lett* **87**: 168103.
12. Kunz W, Belloni L, Bernard O, Ninham BW. (2004) *J Phys Chem B* **108**: 2398.
13. (a) Boström M, Kunz W, Ninham BW. (2005) *Langmuir* **21**: 2619; (b) Boström M, Ninham BW. (2004) *Langmuir* **20**: 7569–7574; (c) Boström M, Ninham BW. (2004) *J Phys Chem B* **108**: 12593–12595; (d) Boström M, Ninham BW. (2005) *Biophysical Chemistry* **114**: 95–101.
14. Kunz W, M'Halla J, Ferchiou S. (1991) *J Phys Cond Matt* **3**: 7907.
15. Gouy G. (1910) *J Phys Theor Appl* **9**: 455–468.
16. Chapman DL. (1913) *Philos Mag* **25**: 475–481.
17. Moreira LA, Boström M, Ninham BW, Biscaia Jr EC, Tavares FW (2006) *Colloids and Surfaces A: Physicochem Eng Aspects* **282**: 457–463.
18. Israelachvili JN. (1992) *Intermolecular and Surface Forces*, 2nd ed., Academic Press Inc., San Diego.
19. Weast RC. (ed.) (1971) *CRC Handbook of Chemistry and Physics*, 52nd ed., CRC Press, Cleveland, Ohio.
20. Tavares FW, Bratko D, Blanch H, Prausnitz JM. (2004) *J Phys Chem B* **108**: 9228.
21. Boström M, Tavares FW, Bratko D, Ninham BW. (2005) *J Phys Chem B* **109**: 24489–24494.
22. Horinek D, Netz RR. (2007) *Phys Rev Lett* **99**: 226104/1.
23. Lima ERA, Biscaia Jr EC, Tavares FW. (2007) *Phys Chem Chem Phys* **9**: 3174–3180.
24. Lima ERA, Horinek D, Netz RR, Biscaia EC, Tavares FW, Kunz W, Boström M. (2008) *J Phys Chem B* **112**: 1580–1585.
25. Horinek D, Serr A, Bonthuis D, Boström M, Kunz W, Netz RR. (2008) *Langmuir* **24**: 1271–1283.
26. Lima ERA, Boström M, Horinek D, Biscaia Jr EC, Kunz W, Tavares FW. (2008) *Langmuir* **24**: 3944.
27. Edwards S, Williams DRM. (2004) *Phys Rev Lett* **92**: 248303.
28. Luo GM, Malkova S, Yoon J, Schultz DG, Lin BH, Meron M, Benjamin I, Vanysek P, Schlossman ML. (2006) *Science* **311**: 216.
29. Petrache HI, Zemb T, Belloni L, Parsegian VA. (2006) *Proc Natl Acad Sci USA* **103**: 7982.

Part D

SUMMARY AND CONCLUSIONS

Chapter 12

An Attempt of a Summary

Werner Kunz* and Gordon J. T. Tiddy†

In its origins, the Hofmeister series is simply an empirical ranking of the extent to which added electrolytes modify protein solubility. The very wide range of systems and phenomena to which this concept has now been applied is amply demonstrated by the chapters in this volume. Almost all of the examples involve the adsorption of ions to surfaces. For those that do not, there is the adsorption of ions to each other ('complex' formation). Most of the applications also involve some reorganisation of the system, whether it is a phase change, a change in the molecular configuration (usually within a surface layer) or in the composition of the species present (e.g. a change in the number of water molecules bound to a solute). Occasionally there is a change in the chemical nature of the species involved. The obvious example is that of proteins in aqueous solution, where each different ionic structure is a different chemical. The concentrations of all of these change as the pH is altered. Hence the chemical species above and below pI are different. A less obvious example is the lipid lecithin. Here the alkyl chains can adopt hundreds of thousands of different conformations — each one a different compound. Fortunately, as far as ion-head group interactions are concerned, they are all broadly similar. However, the head group too can adopt different conformations, in this case the number being about ten. Some of these are likely to have very different interactions with ions. Indeed, ion adsorption may well change their relative populations. Important contributions to the understanding of ion-specific effects on phospholipids can be found in Chap. 2.

*Institute of Physical and Theoretical Chemistry, University of Regensburg, 93040 Regensburg, Germany.

†School of Chemical Engineering & Analytical Science, University of Manchester, PO Box 88, Manchester, M60 1QD, UK.

Faced with this complexity, the question arises, can we draw any general conclusions from the present state of the art?

Doubtlessly, the concept of matching water affinities as introduced by Kim D. Collins turns out to be most valuable as a first rule of thumb. That big ions preferentially associate with big counterions whereas small ions associate with small counterions is a useful pictorial qualitative interpretation of many experiments. It is also fairly predictive and can be used when further information is not available.

Of course, it cannot be more than a rule of thumb. As shown in Chap. 9, the actual situation of ion hydration is much more complicated. It is a trivial statement that the detailed geometry of the ions is of utmost importance. We only cite two examples for this evidence: as discussed in Chap. 6, the behaviour of guanidinium ions may be due to very special self-association, which strongly depends on the unique electron charge distribution of this ion. Therefore, it is difficult to put it into a universal Hofmeister series that would hold for many different experimental manifestations. As a second example, we may consider choline ions. This organic cation has the positively charged nitrogen atom surrounded by three 'hydrophobic' methyl groups and one ethyl group having an OH-group at its end. It can therefore be expected that its orientation depends on the given system, e.g. with either the charged part or the hydroxyl group pointing towards a surface.

The last example leads us to a very important statement: we believe that the specific behaviour of an ion cannot be described by only one parameter. In fact, the specificity of a certain type of ion is always dependent on the environment, be it the counterion, the interface or the interacting part of a macromolecule. A chloride ion, for example, can be either hard or soft, depending on the hardness or softness of the counterion. Consequently, it is utopian to describe the ions only with a single set of parameters. This is also the reason why correlations of specific ion effects with ion radii, polarisabilities, viscosity B coefficients, lyoptropic numbers, etc. never give satisfactory results for a broad range of experimental findings.

One of these ion properties is their polarisability. Ninham and coworkers[1–3] proposed that polarisability was the key missing parameter further to ion sizes and charges. Also Jungwirth and Tobias based their remarkably predictive simulation results for ions near surfaces on potentials that took ion polarisabilities into account.[4,5] Therefore, for several years after these papers, people were optimistic that the mystery of the

Hofmeister series could finally be unravelled with a simple parameter. Now, we are much less optimistic. In 2004, Kunz *et al.*[6] showed that a simplistic use of the ion polarisabilities is misleading for the interpretation of bulk effects. It turned out that it is not possible to reproduce surface tensions and osmotic coefficients with the same parameters.

The ion interaction parameters generally employed in the simulations were not the optimal ones. Recently, Horinek *et al.*[7] proposed new improved force fields based on thermodynamic solvation properties, and it is likely that these new parameter sets will challenge the current importance ascribed to ion polarisability. Of course, Jungwirth's idea that the ion polarisabilities influence the surrounding water molecules is still of significant importance, but it seems that many of the calculated effects can also be described by simply changing the surface charge density of the ions or by recalculating the basic Lennard–Jones parameters for the ions, like it was done by Horinek *et al.*[a]

Indeed, nature is much more complicated. Even systems that show the same Hofmeister series may have different interactions that underpin the behaviour. As ion behaviour depends on the environment, it is also difficult to interpret widely different experiments by the same type of interactions. That the same series is found in two experiments is not a proof that the same interactions govern both experiments. This is a frequent mistake. How subtle these interactions can be is illustrated in Chap. 11, where the potentials of mean force from molecular dynamics simulations are taken and inserted into Poisson–Boltzmann calculations. Figure 4 of this chapter demonstrates that the double layer pressure increases in the non-Hofmeister sequence NaCl < NaI < NaBr at large separations (> 1.0 nm). Frequent changes in this sequence occur as the surfaces approach closer to each other, including NaCl < NaBr < NaI and NaCl > NaBr > NaI. Even if the underlying approximations do not fully reflect the real world, this result is far too complicated to be interpreted by simple rules. Note that integral equations such as the hypernetted chain theory are a more powerful alternative to the Poisson–Boltzmann equation, see Chap. 10.

[a]It is often argued that the computer simulation results are arbitrary, because they depend on the potentials chosen for the interaction potentials. This is only in part true. It seems that the molecular dynamics results for specific ion effects are quite robust and at least qualitatively independent of the water model, as long as this is more or less reasonable. In particular, the various polarisability models for water all lead to more or less similar conclusions.

In Fig. 3 of Chap. 11, a model for water distribution is given in terms of the dielectric function close to interfaces. We should recall that three types of interfaces can be distinguished: a soft surface like the water–air one, a hard surface (solid–water) and molecular surfaces (e.g. macromolecules). For a soft surface the water has much more freedom to adapt to the interface than in the case of hard surfaces, where the water exhibits a significant layering. And as far as proteins or macromolecules are concerned, there is no defined water 'surface' at all; there is rather an individual hydration of each segment. As a result, the specific ion effects will reflect also these different water arrangements.

So, is there nothing to generalise and to learn further to Collins' simplistic rules? Fortunately, the situation is not so desperate. First of all, it is good news that properties like osmotic and activity coefficients as well as solubilities and densities at least can be modelled with few parameters, although care should be taken not to over-interpret the physical meaning of these, see Chap. 3. And second, some new rules emerge: Collins' concept of matching water affinities is a special case of the general rule in chemistry 'like seeks like'. Lund and Jungwirth find also some very interesting new rules by their molecular dynamics simulations that fit well with experimental findings (and also Collins' rules), see Chap. 8: it seems that ions of low charge density are preferably attracted to hydrophobic surfaces or particles and that this attraction can be even stronger than the attraction to oppositely charged entities. Ions of low charge density have only loosely bound water of relatively high energy, while hydrophobic groups have no directly bound water. The pairing of the low-charge density ions with the hydrophobic parts releases the contact water, which then comes into a favourable state of much lower energy. This process can be energetically more favourable than a direct ion-pairing, which would require the dehydration of both ions. Similar conclusions can be drawn from the results shown in Chap. 8 where iodide binding to an idealised cationic protein occurs mainly at hydrophobic patches while chloride binding is at the positive charges.

In Chap. 11 another rule appears: in contrast to what is commonly accepted in physical chemistry, mostly inspired by the Debye–Hückel theory, electrostatic interactions may not always dominate at small interparticle distances. At small distances, the structure of water (or in general of the solvent) cannot be neglected, nor can all types of non-electrostatic interactions. The latter may even dominate over electrostatics. As a result it may occur that anions are attracted by slightly negatively charged surfaces (the same for the case of positive charges close to positively charged surfaces). When air is the surrounding medium, this is widely accepted (charged dust can aggregate), but within water this was rigorously neglected. Of course, the dielectric properties of water are very different from those of air, but such a categorical neglect is not valid and is often misleading.

Of course, ion effects are concentration dependent. At very low concentrations (< 0.1 M) no significant specificities are usually measured for bulk properties. Here electrostatics dominates and the Debye–Hückel screening parameter is often sufficient. This is not the case for surface phenomena, where ion specificities may occur even at concentrations far below 0.1 M. In protein solutions at very high salt concentrations (> 1–2 M), all salts become salting-out, simply because very little free water is present in the solution. All the water molecules are more or less involved in ion hydration and are no longer available for protein solvation. The stronger the ion charge density, the higher the precipitation tendency at such high concentrations. Consequently, the most pronounced specific ion effects mostly occur at intermediate concentrations, i.e. roughly between 0.1 and 2 M.

In this context, there is a long-standing debate about the mechanism of 'structure-making' and 'structure-breaking' by ions that are commonly called 'cosmotropic' and 'chaotropic'. According to recent work by Bakker,[8] Cremer,[9] and Pielak,[10] it seems that small 'simple' ions do not affect water structure beyond the first hydration shell.[11] However, it should be added that the water bound in this first hydration shell can be a significant amount of all water, when ion concentration exceeds 1 M. And this hydration water is very dependent on ion concentration and on the location of the hydrate, either close to a surface, to another molecule or headgroup or in the bulk. If it is close to macromolecules, the resulting salting-in or salting-out process is a very subtle balance between several different effects (different for chaotropes and cosmotropes as well as for cations and anions).[11] In any case, specific ion effects are very probably the result of a direct interaction of ions with other species mediated by water. But the bulk water structure is not significantly altered.

What happens if other solvents are used? Of course, every solvent is a special case, although some rules seem to be roughly general, see Chap. 7. In this book, the main focus is on water, not only because it is the most important solvent, but also one of the most complicated ones, as far as the interpretation of specific ion effects are considered. The behaviour of electrolytes in non-aqueous media is described in part in other works, see e.g. Ref. 17, 22, and 34 in Chap. 1.

Let us come back to water. At interfaces, Collins' concept as well as what was written before on hydrophobic surfaces seem to be good approaches. In total, it turns out that specific ion effects are more pronounced at interfaces compared to bulk properties. Whereas powerful experimental techniques exist for several years for the characterisation of the molecular distribution of ions in water, see Chap. 6, experiments of comparable precision for interfacial ion specific properties only emerged in the last years, especially for the solution–air interfaces. The present state of the art of the most relevant of these techniques is summarised in Chaps. 4 and 5. Second harmonics and sum-frequency generation, and grazing incidence X-ray diffraction will more and more provide us with quantitative details of specific ion adsorption (and their concentrations) at interfaces and their profiles perpendicular to the surface. They help us to check the validity of molecular dynamics simulation results and will soon be the new experimental reference standards for all kinds of models and simulations. However, as always with newly emerging techniques, there are still some ambiguities in the interpretation of the measured signals.

Finally, it should be mentioned that there is still at least one ion-specific phenomenon that today cannot be explained at all, despite some encouraging and exciting ideas: it is the phenomenon of gas bubble–bubble coalescence in water in the presence of different salts, see Chap. 7. This is not only a theoretical problem and a challenge for scientists; it is probably related to the evolution of life on earth, since the oxygen content of sea water dramatically changed, whenever the salt concentration was higher or lower than a certain threshold level during the lifetime of the earth in the last billion years.

At the end we have good and bad news: the good news is that several promising models came up within the last ten years and these models significantly help us not only to interpret known experimental results, but also to qualitatively predict ion specificities in many cases. The bad news is that the seemingly universal Hofmeister series made us believe that there must be a simple and universal rule at its origin. This is truly not the case. An accurate description of any Hofmeister phenomenon must include *all* the different molecular contributions. Many of these are extremely difficult to measure independently of other effects, while it is likely that some are not yet recognised. It is a real challenge to produce an overview that accurately reflects the current understanding of the science. To present a ‘road map’ of the way forward is as challenging as the production of any other ‘road map’! Here we made such an attempt.

References

1. Ninham BW, Yaminsky V. (1997) *Langmuir* **13**: 2097–2108.
2. Boström M, Williams DRM, Ninham BW. (2001) *Langmuir* **17**: 4475–4478.
3. Boström M, Williams DRM, Ninham BW. (2004) *Prog Coll Poly Sci* **123**: 110–113.
4. Jungwirth P, Tobias DJ. (2001) *J Phys Chem B* **105**: 10468–10472.
5. Jungwirth P, Tobias DJ. (2006) *Chem Rev* **106**: 1259–1281.
6. Kunz W, Belloni L, Bernard O, Ninham, BW. (2004) *J Phys Chem B* **108**: 2398–2404.
7. Horinek D, Mamatkulov SI, Netz RR. (2009) *J Chem Phys* **130**: 124507/1–124507/21.
8. Omta AW, Kropman MF, Woutersen S, Bakker HJ. (2003) *Science* **301**: 347–349.
9. Zhang YJ, Furyk S, Bergbreiter DE, Cremer PS. (2005) *J Am Chem Soc* **127**: 14505–14510.
10. Batchelor JD, Olteanu A, Tripathy A, Pielak GJ. (2004) *J Am Chem Soc* **126**: 1958–1961.
11. Zhang Y, Cremer PS. (2006) *Curr Opin Chem Biol* **10**: 658–663.

Index